Springer Tracts in Civil Engineering

Springer Tracts in Civil Engineering (STCE) publishes the latest developments in Civil Engineering - quickly, informally and in top quality. The series scope includes monographs, professional books, graduate textbooks and edited volumes, as well as outstanding PhD theses. Its goal is to cover all the main branches of civil engineering, both theoretical and applied, including:

- Construction and Structural Mechanics
- Building Materials
- Concrete, Steel and Timber Structures
- Geotechnical Engineering
- Earthquake Engineering
- Coastal Engineering; Ocean and Offshore Engineering
- Hydraulics, Hydrology and Water Resources Engineering
- Environmental Engineering and Sustainability
- Structural Health and Monitoring
- Surveying and Geographical Information Systems
- Heating, Ventilation and Air Conditioning (HVAC)
- Transportation and Traffic
- Risk Analysis
- Safety and Security

Indexed by Scopus

To submit a proposal or request further information, please contact:
Pierpaolo Riva at Pierpaolo.Riva@springer.com (Europe and Americas)
Wayne Hu at wayne.hu@springer.com (China)

Luís Bragança · Meri Cvetkovska · Rand Askar ·
Viorel Ungureanu
Editors

Creating a Roadmap Towards Circularity in the Built Environment

COST
EUROPEAN COOPERATION
IN SCIENCE AND TECHNOLOGY

circular B

Springer

Editors
Luís Bragança
Department of Civil Engineering
Institute for Sustainability and Innovation in Structural Engineering
University of Minho
Guimaraes, Portugal

Rand Askar
Department of Civil Engineering
Institute for Sustainability and Innovation in Structural Engineering
University of Minho
Guimares, Portugal

Meri Cvetkovska
Faculty of Civil Engineering
Ss. Cyril and Methodius University in Skopje
Skopje, North Macedonia

Viorel Ungureanu
Department of Steel Structures and Structural Mechanics
Civil Engineering Faculty
Politehnica University of Timisoara
Timişoara, Romania

ISSN 2366-259X ISSN 2366-2603 (electronic)
Springer Tracts in Civil Engineering
ISBN 978-3-031-45979-5 ISBN 978-3-031-45980-1 (eBook)
https://doi.org/10.1007/978-3-031-45980-1

This Springer imprint is published by the registered company Springer Nature Switzerland AG
The registered company address is: Gewerbestrasse 11, 6330 Cham, Switzerland

Contents

Part I
Circular Economy Best Practices in the Built Environment

Chapter 1
Circular Construction Principles: From Theoretical Perspective to Practical Application in Public Procurement

Ana Andabaka

Abstract This paper aims to highlight the role of government and other public authorities in promoting the implementation of circular economy principles in the built environment. Circular construction ecosystem is based on circular building design and the coordinated action of stakeholders along the value chain. Government and local authorities have a special role to play in the built environment as they manage infrastructure and act as regulators and enablers as well. Moreover, circular building design strategies should be encouraged through the government policy tools like public procurement allowing public authorities to formulate tenders that incorporate circular economy principles when purchasing goods, services, and works related to the built environment. The case studies presented show that public procurement can be a valuable tool for promoting circular use of materials and reducing waste, encouraging the construction of circularly designed zero-emission buildings, and stimulating market innovation that reduces the environmental impact of construction while providing benefits to society. Thus, public authorities can be a powerful force driving change and innovation in the construction industry towards more circularity in the built environment.

Keywords Circular construction · Circular economy · Public procurement

1.1 Introduction

According to the European Commission [1], the full life cycle of buildings in the European Union (EU) including extraction, manufacture, transport, construction and end of life accounts approximately for 50% of the total energy use, 50% of the raw material extraction, 40% of the total greenhouse gas emissions, and a third of all water consumption and waste generated. There is a need to break this vicious cycle

---

A. Andabaka (✉)
Faculty of Economics and Business, University of Zagreb, Zagreb, Croatia
e-mail: aandabaka@efzg.hr

L. Bragança et al. (eds.), *Creating a Roadmap Towards Circularity in the Built Environment*, Springer Tracts in Civil Engineering,
https://doi.org/10.1007/978-3-031-45980-1_1

and create a new one by applying the circular economy principles in construction. Key elements of circular construction are circular design and the circular use of materials that stimulates the application of reused and bio-based materials [2, 3].

The way a building is designed and constructed has a crucial role regarding its environmental and social impact over its life cycle and the potential for new life cycles [4]. Circular design enables creating a sustainable built environment by making buildings more adaptable and facilitates the high-value reuse of a structure's products and materials at the end of their life [3]. Reversible building design is design of buildings which can be easily deconstructed, or where parts can be removed and added easily without damaging the building or the products, components or materials, thus focusing on their future use [5, 6]. Different layers like windows, floors, inner walls, ventilation can be accessed without damaging other parts of the building enabling resource efficient repair, replacement, reuse, and recovery of products, building materials, and components [6]. Durmisevic [7] presents a new philosophy where reversible building design means designing for circular value chains, while construction and demolition waste is considered a design error. She identifies three dimensions of reversibility within a building—spatial, structural, and material which should be integrated into the building's design to ensure high transformation capacity and reuse potential of building and its structure.

However, the implementation of the circular economy in the built environment is possible only through coordinated action of all stakeholders involved in the circular construction ecosystem. The aim of this paper is to present the role of the government as a key stakeholder that can use its purchasing power to promote circularity in the built environment.

The remainder of this paper is structured as follows. Section 1.2 presents different circular design strategies as part of the circular construction ecosystem. Benefits and challenges of the circular construction ecosystem are described in Sect. 1.3. In the fourth Section two case studies will be presented to demonstrate the role of public procurement in promoting circular construction. The final section contains concluding remarks.

1.2 Circular Design as a Foundation of Circular Construction Ecosystem

Circular design is the centrepiece of the circular construction ecosystem. Not only that it influences the carbon footprint of the built environment and its users, but it can also allow adaptation to changing use patterns [8]. Ellen MacArthur Foundation and Arup have developed the circular design framework [9] encompassing design for longevity, design for adaptability, and design for disassembly as different approaches to circular design of buildings (Table 1.1).

Design for longevity seeks to achieve timeless architecture while using durable products and materials that can be adapted and reused in the future [10]. Longevity

Table 1.1 Circular design strategies [10]

Circular design strategy	Description
Design for longevity [10]	Aims at maximizing the value of the building and its components over time, optimising value retention and value recovery potential. It ensures a long-life cycle of components by setting the baseline for an element's quality, maintenance need, necessity for repair, adaptability and residual value when removed
Design for adaptability [10]	Aims at enabling buildings to adapt to new functions during the use stage to retain their value, which is especially important considering the short functional life span of buildings
Design for disassembly [10]	Allows for the practical disassembly of components in order to recover residual value at the end of their service life

allows the resources used in building construction to last a long time, thus contributing to the circular economy by slowing down the loop [11].

Design for adaptability enables building to redefine its purpose without major interventions and to use considerably less material compared to major renovation, it reduces the need to demolish and avoids a considerable amount of construction and demolition waste [8]. Adaptable buildings can undergo spatial change and functional changes during its lifespan, thus allowing for multiple life cycles [3]. Increased adaptability is provided by using of modular concepts, easy-to-change façades allowing for changes in building appearance and functionality, and plug-and-play technical installations [8].

Design for deconstruction and disassembly facilitates deconstruction of a building at the end of its useful life, in such a way that components and parts that outlast their service life as part of a system (building) can be recycled, reused, or recovered for further economic use [10, 12]. Bertino et al. [13] suggest that the building deconstruction can be facilitated by reducing building complexity through favouring the modularity and lightness of the components, prefabrication and the simplification of the connections between the structural and non-structural elements, and minimizing the number and types of components; choosing reusable and eco-compatible materials whilst minimizing the use of hazardous and composite materials; and providing the information on the building construction and deconstruction. Deconstruction also includes securing the current construction, the analysis of building's contents, the decontamination and removal of any hazardous waste, the demolition activities and the recycling operations to recover the value of the existing materials [12].

During construction 10–15% of built material is wasted, 54% of demolition materials are landfilled, while most materials are not suitable for reuse due to toxic elements [14]. Circular material use in construction is based on principles of maximizing the use of virgin materials and bio-based materials, maximizing the potential for high-value reuse and the amount of recycled materials used [3]. In 2021, the estimated circular material rate in the European Union was 11.7%, which leaves plenty of room for improvement, especially considering that the circularity rate recorded in Netherlands was 33.8% [15].

Sustainable circular construction adds to the circular construction ecosystem by including the social component and presents an integrated approach to "design, construction, operation and occupancy, maintenance, renovation, and demolition processes of structures, all of which are environmentally responsible and resource efficient throughout a building's life cycle, limiting the environmental impacts and ensuring optimal efficiency whilst creating a high level of quality of life for its occupants." [16].

1.3 Benefits and Challenges of the Circular Construction Ecosystem

Circularly designed buildings offer several benefits, one of them being the decreased environmental impact, which is especially important in the context of limited resources [3]. In the long term such buildings provide reduced material extraction, reduced embodied carbon emissions, enable future reuse of building components and recovery of an asset's value at the end of its service life [10]. Circularly designed buildings cost less over their life span than traditional buildings because they generate value during their maintenance and replacement phase, entail modular products that can be replaced in parts, interior walls that can be moved (enabling flexibility and adaptability), and avoid demolition costs due to detachability [3].

Two main challenges for the first use of circularly designed buildings are that they may require higher material quantities and may result in higher environmental impact [10]. However, over the full life-cycle these materials would provide benefits because they could be reused, replaced or recycled, thus reducing negative impact on the environment. Real challenge lies in persuading the stakeholders (Fig. 1.1) to embrace circular business models and the full life cycle approach that goes beyond economic cost of the building construction and considers environmental aspects as well. European Commission [12] identifies seven target groups that can contribute to meeting durability, adaptability, and waste reduction objectives in the construction industry: 1. building users, facility managers and owners; 2. design teams (engineering & architecture of buildings); 3. contractors and builders; 4. manufacturers of construction products; 5. deconstruction and demolition teams; 6. investors, developers and insurance providers; and 7. government/regulators/local authorities. In order to facilitate the future circular use of building materials, stakeholders should be familiar with the concept of circular construction and an integrated approach along the value chain is needed [12].

Carra and Magdani [17] stress the importance of developing integrated value chains in providing following competitive advantages for the companies in the future. Building users, facility managers and owners will benefit from an increased value of their assets due to more efficient, productive, and adaptive spaces [17]. Investors will

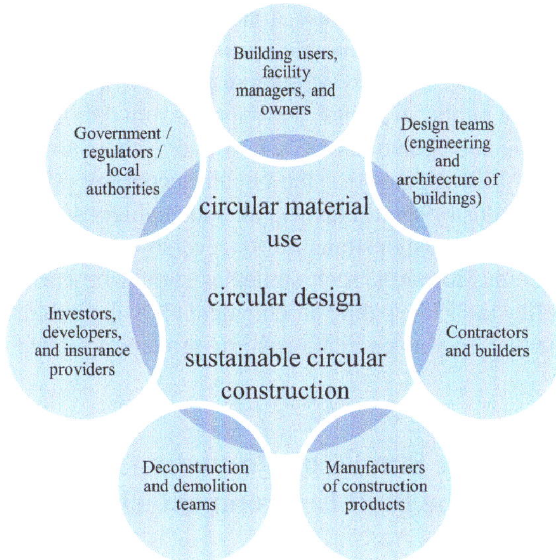

Fig. 1.1 Circular construction ecosystem (based on [3, 12])

benefit from the increased value of the space which can be quickly adapted to the demands of new tenants, thereby minimising the period of vacancy [17]. Designers will need to ensure building design that allows for durability, disassembly, and adaptability by working closely with product manufacturers and suppliers [17]. The use of materials and products that can be recycled and reused should be foreseen in the planning stage of a renovation or a new building in order to be designed and built as a resource bank [2]. Designers need to consider the whole life cycle of the building from the decision to build new or refurbish to the eventual demolition or deconstruction of an obsolete building [16]. The possibility of on-site reuse of reclaimed building materials should also be considered by architects during the renovation design process [18].

Contractors and builders will need to ensure users, facility managers and developers implement the circular solutions during the building lifecycle [17]. Manufacturers and suppliers have an opportunity to earn additional income by recovering materials at the end of a product's life and reselling or repurposing them [17]. Deconstruction and demolition teams will have an opportunity to become material reuse providers and potentially join forces with material extractors or producers to ensure a continuity of supply. [17]. Prior to demolition, material inventory should be made through audits to properly identify quantity, dimensions, technical function, maintenance, and the presence of harmful substances in the materials [2].

Public authorities manage large building stock and urban infrastructure, and act as regulators and enablers through zoning and permitting [2]. Raph [8] suggests that future regulatory European Union initiatives addressing buildings should encourage the construction of adaptive buildings that are aligned with circular economy principles and ensure that major renovations take adaptability into account. Design teams should discuss the future strategy of the building with local authorities to ensure reconfiguration is possible by using a modular approach allowing for easy disassembly and assembly of components [17]. According to Bilal et al. [19], there is a lack of public awareness and support from public institutions, while the lack of environmental regulations and laws triggers other barriers to the circular economy. Giorgi et al. [20] suggest that the circular economy policy development in the construction industry should involve all stakeholders and emphasise the need to define and harmonize design strategies for reversible buildings, preferably within a regulatory system. Moreover, such strategies should be encouraged through the government policy tools like public procurement allowing public authorities to formulate tenders that incorporate circular economy principles related to the built environment.

1.4 Case Studies on Promoting the Circular Economy in the Built Environment Through Public Procurement

These two case studies present good practice and provide an overview of how public authorities can promote the reuse of building materials and design for circularity. Although selected by the European Commission as examples of green public procurement [18, 21], they also include a social component, demonstrating that public procurement can serve as an important tool for promoting social inclusion and achieving other benefits that improve the well-being of society.

In France, circular construction principles were implemented in public procurement by setting reclamation targets in a redevelopment tender to ensure the reuse of materials. The public developer SERS (Société d'Aménagement et d'Equipement de la Région de Strasbourg) was supported by the non-profit organization Rotor to find the market for reclaimed materials. As a partner in the Interreg project "Facilitating the circulation of reclaimed building elements in Northwestern Europe", Rotor had access to hundreds of companies dealing with reclaimed building materials. The various steps taken with the help of an expert included creating an inventory of reusable materials, conducting dismantling tests, and performing a market study to identify companies interested in reclaiming the materials from an old building dating from the 1930s. As a result, the contracting authority integrated ambitious reclamation targets in its call for tenders, providing general guidance on reclamation and reuse (dismantling and conditioning tips, useful tools, references, etc.) in annex. The reclamation audit contained three different categories [18]: (A) most promising materials (present in large quantities in the building and with high reuse potential confirmed by the dismantling tests and/or stable market demand for these materials); (B) materials with an estimated reuse potential (for which the potential was not further investigated and/or present in smaller quantities in the building); and (C) other materials (with low reuse potential depending on specific opportunities rather than stable market demand). The reclamation targets for three major batches of materials with very high reuse potential (Category A) were formulated as technical specifications: 1. cast-iron radiators, for which at least 80% of the pieces had to be reclaimed for reuse; 2. structural timber in the roof, for which at least 50% of the total volume had to be reclaimed for reuse; 3. enamelled wall tiles, where at least 50% of the surface area had to be reclaimed for reuse. An award criterion related to the reclamation efforts encouraged bidders to commit to achieving better reclamation rates than the minimum specifications for the three main batches of reusable materials, but they could also commit to reclaiming other batches from Categories B and C, which included metal stairs, handrail, guardrail, antique windows, lighting, doors, and sanitary equipment. A social clause was also included to help people who have difficulty finding work to access employment or return to work. The minimum number of hours for an integration activity was set at 300, but the bidders knew that proposing more hours would improve their score. Before submitting their bids, candidates had the opportunity to visit a site. Ultimately, four bidders submitted an offer and complied with the minimum requirements. The sale of the materials was implicitly factored in the offers as the bidders had to set their prices taking into account the possibility of selling the materials afterwards. Demolition companies with less experience in reclamation found it more difficult to estimate the potential retail value of the batches of reclaimed materials, so the price of some operations, such as dismantling of the structural timber varied considerably. The winning bidder submitted an offer that was the best value for money. Although they had committed to reaching the minimum target, they managed to outperform the initial plan and more batches of dismantled materials could successfully be reclaimed for reuse. They subcontracted the integration structure in charge of dismantling the wall tiles for 328 hours. Finally, 51 tons of materials could be reclaimed for reuse and most of these materials were purchased or recovered by professional dealers who ensured their subsequent resale. The environmental benefits go beyond the waste reduction as the reuse of materials avoids the raw resources extraction and the associated greenhouse gas emissions [18].

In the Czech Republic, the Supreme Audit Office (SAO) decided to build its first permanent headquarters reusing a brownfield land in the centre of Prague. The SAO will share the building with the Parliamentary Library and the Archives of the Chamber of Deputies and will offer their services to the public [22]. The project was driven by the SAO's desire to lead by example and promote the construction of a public building with net-zero energy consumption, the lowest life cycle costs (LCC), and the longest service life. The implementation of circular building design principles is reflected in the flexibility and adaptability of spaces, which are also accessible. The building should also provide modularity in terms of control and replacement of utilities, and the floor designed to allow easy access to utilities. Prior to launching the tender, the project team consulted with academic experts at the Czech Technical University regarding the evaluation of the building's service life and maintenance costs, and conducted two preliminary market consultations [21]. The innovative approach to public procurement was reflected in the first major public project in the Czech Republic to be realised in accordance with the Fédération Internationale des Ingénieurs-Conseils (FIDIC) Yellow Book for Design & Build, and to use Building Information Modelling (BIM) methods for the evaluation of LCC of the building [22]. According to PORR a.s. [23], the company that was awarded the contract, it has relied on BIM technology for many years to give users a complete overview of all relevant aspects of a project and add value to the entire construction value chain. It highlighted that when all project stakeholders use the same model, design aspects and dimensions can be viewed in the context of the entire project life cycle [23].

The SAO published three separate tenders for the principal architect, the civil engineering company, and the construction company. Because it was not easy to meet the innovative requirement of prior experience with BIM projects, SAO required evidence of experience with six separate segments of BIM. The traditional approach to procurement of construction projects, with separate tenders for a designer and a contractor, was replaced with the Design-Build tender providing a single point of responsibility to reduce risk and overall cost. The SAO received four bids and the construction company PORR a.s. won the tender due to its low bid price, although it was the second best in terms of LCC criteria. The winner stated that it was able to offer a price 13% lower than the estimated value of the public contract due to its long experience with the Design-Build method and the ability to manage the implementation of the construction using the Design-Build method supported by BIM, the professional team of in-house designers or close cooperation with design offices whose professional experience is related to practical implementation, and the favourable prices of suppliers [21]. During construction, all stages of the project documentation are processed in BIM and a detailed digital model is created, which enables the use of a drilling robot that transfers the data from the model directly to the building structures. The use of a drilling robot on the construction site significantly speeds up the process of drilling holes in a monolithic structure and marking the holes for each type of pipe. PORR a.s. uses augmented reality to easily check the accuracy of the work performed [24]. In addition, BIM facilitates the use of material passports, which contain valuable digital information about all building materials used. Finally, environmental and social benefits can be achieved (Table 1.2) through good contract management and communication between the contracting authority, the architect, and construction company [21].

This project showed the importance of collaboration with academic experts. The experts from the Faculty of Civil Engineering developed the methodology for calculating the LCC which in this case included operating costs, renovation costs, and maintenance costs for 30 years. Finally, they verified the costs calculated by all four construction companies that submitted their bids.

Preliminary market consultations proved very useful as they allowed potential suppliers to prepare bids or prepare for recycling, sale and reuse of materials, while public authorities could inform the market of their intentions, receive feedback from the market, and prepare the procurement procedure [18, 21]. In the latter case, it became clear that the market was not ready to meet the intended requirement of submitting bids in the BIM form due to a lack of standards and problems with handing over the BIM original format [21]. Innovative and smart solutions need time to overcome problems related to new technologies and market developments. However, public procurement should not be merely a financial and administrative task, but a means of finding sustainable solutions that meet the needs of communities

Table 1.2 Benefits for the environment and society [21]

Environmental benefits	
Resulting from the construction process:	The use of recycled concrete from the previous structures on site; Using a concrete facility on the nearby river bank and transport by water; The use of sand from the construction pit for the construction itself; Saving 22.5 tons of CO2 through sustainable construction; Wireless lighting control saved wiring material
Resulting from the design phase using BIM:	The building will consume almost net-zero amount of energy; Smart heating and cooling system with thermohydraulic solutions, supplemented by heat recovery, stored in the summer; By determining the ratio between the glazed and the fixed part of the façade, complemented by external, centrally controlled blinds, it is possible to optimise the light and heat quality of the interior, saving energy and money and reducing pollution; Green roofs will benefit the top floors, provide additional thermal insulation, and will also be useful for neighbouring buildings
Social benefits	
Benefits for the employees:	Openable windows in offices; The Child Care Centre will serve as a benefit for employees with kids facilitating their return to work after parental leave; Charging stations for electric cars and e-bikes in the parking lot; A fitness centre shall improve health and well-being of employees; Accessible facilities will allow institutions to easily employ people with disabilities (with light signals for emergencies complementing sirens)
Benefits for the neighbourhood:	New SAO headquarters will contribute to the revitalisation of the area; The area will be walkable with a small park open to the public; The Parliamentary Library, the Archive of the Chamber of Deputies, and the cafeteria within the building will be available to the public

and create a better quality of life for all. Given that this paper focused on the good practice of selected public institutions, future research could explore the potential of local government procurement practices to demonstrate how cities can contribute to the implementation of the circular economy in the built environment.

1.5 Conclusion

Circular building design lies at the heart of the circular construction ecosystem. Translating circular economy principles into design enables the construction of buildings that last longer, produce less waste, facilitate management and maintenance, and allow for the reuse or high-quality recycling of major building components.

Implementing the circular economy in the built environment requires an integrated approach by all stakeholders along the value chain. Governments and local authorities have an important role to play in circular construction as they manage a large building stock and urban infrastructure, and act as regulators and enablers as well. Public authorities can also encourage demand for circular design of buildings and secondary materials through public procurement and benefit from allies such as reclamation audit experts and academic experts during the tendering process.

The case studies presented show that public procurement can be a valuable tool for promoting the circular use of materials and reducing waste, encouraging the construction of circularly designed zero-emission buildings, and stimulating market innovation that reduces the environmental impact of construction while providing benefits to society. In this way, public authorities can be a powerful force driving change and innovation in the construction industry towards more circularity in the built environment.

References

1. European Commission (2019) LEVEL(S)—taking action on the TOTAL impact of the construction sector. https://ec.europa.eu/environment/eussd/pdf/LEVEL(S)%20CONFERENCE%20REPORT.pdf
2. ICLEI, Eurocities (2020) Public procurement of circular construction materials—key takeaways from the Big Buyers Initiative working group. https://circulars.iclei.org/wp-content/uploads/2020/09/BBI-CCM-lessons-learned.pdf
3. Metabolic (2022) The circular design of buildings—the circular tool box. https://www.metabolic.nl/publications/the-circular-design-of-buildings/
4. Heinrich M, Lang W (2019) Materials Passports—Best Practice. https://www.bamb2020.eu/wp-content/uploads/2019/02/BAMB_MaterialsPassports_BestPractice.pdf
5. Durmisevic E (2018) Reversible building design guidelines and protocol. https://www.bamb2020.eu/wp-content/uploads/2018/12/Reversible-Building-Design-guidelines-and-protocol.pdf
6. BAMB, Reversible building design. https://www.bamb2020.eu/topics/reversible-building-design/. Accessed 15 Feb 2023
7. Durmisevic E (2019) Circular economy in construction—design strategies for reversible buildings. https://www.bamb2020.eu/wp-content/uploads/2019/05/Reversible-Building-Design-Strateges.pdf
8. Raph O (2019) The zero carbon and circular economy challenge in the built environment—policy options for the European Union and its Member States. https://www.bpie.eu/wp-content/uploads/2019/06/Policy-Options-for-the-EU-and-its-Member-States-on-circularity.pdf
9. d.Hub Circular Buildings Toolkit. https://ce-toolkit.dhub.arup.com/framework. Accessed 7 Feb 2023
10. d.Hub Circular Buildings Toolkit. https://ce-toolkit.dhub.arup.com/strategies. Accessed 7 Feb 2023
11. Figge F, Thorpe AS, Givry P, Canning L, Franklin-Johnson E (2018) Longevity and circularity as indicators of eco-efficient resource use in the circular economy. Ecol Econ 150:297–306
12. European Commission (2020) Circular Economy—Principles for Buildings Design. https://ec.europa.eu/docsroom/documents/39984
13. Bertino G, Kisser J, Zeilinger J, Langergraber G, Fischer T, Österreicher D (2021) Fundamentals of building deconstruction as a circular economy strategy for the reuse of construction materials. Appl Sci 11(3):939

14. Arup, Ellen MacArthur Foundation (2018) From principles to practices: first steps towards a circular built environment. https://emf.thirdlight.com/link/ufe6ol7qbkm-a9mzju/@/preview/1?o
15. Eurostat, Circular material use rate. https://ec.europa.eu/eurostat/databrowser/view/ENV_AC_CUR/default/table?lang=en. Accessed 25 Feb 2023
16. ACR+ (2019) Sustainable construction guidelines for public authorities—a circular economy perspective. ACR+, Brussels
17. Cara G, Magdani N (2019) Circular business models for the built environment. https://www.arup.com/-/media/arup/files/publications/a/8436_business-models-low-res.pdf
18. European Commission (2022a) Adopting circular economy principles: setting reclamation targets in a redevelopment tender. https://ec.europa.eu/environment/gpp/pdf/news_alert/Issue114_NewsAlert_CaseStudyStrasbourg.pdf
19. Bilal M, Khan KIA, Thaheem MJ, Nasir AR (2020) Current state and barriers to the circular economy in the building sector: towards a mitigation framework. J Clean Prod 276:123250
20. Giorgi S, Lavagna M, Wang K, Osmani M, Liu G, Campioli A (2022) Drivers and barriers towards circular economy in the building sector: stakeholder interviews and analysis of five European countries policies and practices. J Clean Prod 336:130395
21. European Commission (2022b) Building net-zero energy innovation through procurement in the construction of the headquarters of the Supreme Audit Office. https://ec.europa.eu/environment/gpp/pdf/news_alert/Issue115_NewsAlert_CaseStudyPrague.pdf
22. Czech Republic Supreme Audit Office, The Supreme Audit Office has launched the construction of its new head office building which will also house the Parliamentary Library and the Archive of the Chamber of Deputies. https://www.nku.cz/en/for-media/the-supreme-audit-office-has-launched-the-construction-of-its-new-head-office-building-which-will-also-house-the-parliamentary-library-and-the-archive-id11468/. Accessed 20 Feb 2023
23. PORR AG, Press release—PORR to build the new site of the Supreme Audit Office. https://porr-group.com/en/press/press-release/news/czech-pilot-project-in-accordance-with-fidic-yellow-book/#:~:text=2020%20%E2%80%93%20PORR%20has%20been%20entrusted,Book%20(design%20%26%20build). Accessed 20 Feb 2023
24. PORR Group, PORR Czech Republic: BIM Technology @ Supreme Audit Office in Prague. https://www.youtube.com/watch?v=uIRA13Xr3q4. Accessed 20 Feb 2023

Chapter 2
Procedure for Waste Prevention and Management to Implement Best Practices in the Design and Construction of a Building

Paola Villoria Sáez and Mercedes del Río Merino

Abstract This research develops a Waste Management System (WMS), which can be used during the design and construction stages of a building, and provides best practices for CDW prevention and management. For this, a literature review of studies dealing with best practices for CDW prevention and recycling was performed in order to develop the WMS, which is composed by eight procedures. Furthermore, the WMS was implemented in a construction company and individual interviews were conducted in order to know the main benefits and disadvantages of its implementation. The results of this study show that implementing a WMS improves CDW quantification, because the company obtains its own CDW generation ratio. The ratios will help construction agents to develop and write the CDW Management Plans and Reports and also achieve more accurate waste management costs. Finally, the integration of this CDW system with the company's environmental management system supports the cohesion of the construction process organisation at all stages, establishes responsibilities in the field of CDW, and providing greater control of the entire process.

Keywords Best practices · Circular economy · Execution phase · Management systems

P. Villoria Sáez
Universidad Politécnica de Madrid, Escuela Técnica Superior de Edificación, TEMA Research Group, Avenida Juan de Herrera 6 28040, Madrid, Spain
e-mail: paola.villoria@upm.es

M. del Río Merino (✉)
Escuela Técnica Superior de Edificación. TEMA Research Group, Universidad Politécnica de Madrid, Avenida Juan de Herrera 6, 28040 Madrid, Spain
e-mail: mercedes.delrio@upm.es

L. Bragança et al. (eds.), *Creating a Roadmap Towards Circularity in the Built Environment*, Springer Tracts in Civil Engineering,
https://doi.org/10.1007/978-3-031-45980-1_2

2.1 Introduction

The intense activity of the construction sector in the European Union, in recent years and specifically in Spain, is giving rise to the generation of huge volumes of Construction and Demolition Waste (CDW). Specifically, in 2020, 4.8 tons of construction and demolition waste per EU inhabitant, from out of which 39.2% of the waste was recycled and 32.2% were deposited in landfills [1]. Therefore, new initiatives are emerging in Europe to transition to a more efficient circular model that shifts toward the minimization, recycling, and reusing of CDW. In this regard, developing new regulations and policies, decreasing the exploitation of natural resources, optimising production processes, establishing commercial competitiveness for recovered and recycled building materials, and diminishing the generation of CDW throughout the construction process are actions aligned with circular economy that must be considered by government agencies and private companies active in the construction sector to implement and accomplish best practices. Furthermore, the EU member states have been drafting specific regulations for several years to establish a legal framework for the production and handling of CDW and boost the circular economy in the construction industry [2–5]. Among all the measures required in Spain, it should be highlighted that two written reports are needed. One is presented during the design phase of the project (Waste management report (WMR)) and the second during the planning of the construction works (Waste management plan (WMP)).

In addition, it is increasingly common that current projects are designed considering the criteria of environmental certificates which are currently on the market such a BREEAM, LEED, VERDE, etc. However, other environmental certificates which are more especial and consider particular issues for waste prevention and circular economy are still scarce [6]. For this reason, large construction companies are implementing Environmental Management Systems (EMS), seeking the implementation of best environmental practices in their work [7]. However, the vast majority of construction companies (small and medium companies) are still very far from adopting measures to promote CDW minimization and implement circular economy criteria. In this sense, the development of tools for CDW estimation which help to plan an adequate CDW management and of a specific section in the EMS of the construction companies, incorporating best practices for CDW management–, are essential to achieve construction works with zero waste generation.

This situation has been of great interest to researchers in the field that have paid special attention to CDW management and minimisation in recent years, specially during the design stage [7]. Best practices during the design can be clustered into five strategies: design for deconstruction; design for adaptability and flexibility; design to standardise; design recycled materials; and design for waste prevention.

- Designing for deconstruction includes strategies [8], such as: Develop and design a deconstruction plan already in the design process, allowing, when possible, parallel disassembly; Use mechanical joints instead of other types of joints; and Avoid the use of binders, adhesive, resin, and secondary finishes.
- Design for adaptability and flexibility; by using open spaces and regular shapes.

- Design to standardize, using simple, modular designs, avoiding complex shapes and also the use of prefabricated materials [9].
- Desing recycled materials: Use of materials with high recycled content [10] or reuse materials within construction work. Detecting activities that can incorporate reusable materials from the site itself.
- Desing for waste prevention, including a wide range of specific strategies such as:
- The use of waste management plan templates [11];
- Waste prediction tools [12–14];
- Promote construction systems that generate less waste [9, 15];
- Development of detailed projects to minimize design changes and errors.
- Increase coordination and communication → using BIM tools [16, 17].
- Planning waste management—by implementing onsite best practices.

One of the key strategies for preventing CDW generation is planning waste management during the execution of the work, incorporating best practices such as:

- CDW estimation and follow-up (type and quantity) [18, 19].
- Purchase of bulk materials to reduce packaging [20].
- Source separation [21]
- Set the time of the materials' reception, according to the construction site.
- Follow the manufacturer's instructions on the transportation and storage of materials.
- Using small containers in work areas [22].
- CDW training [23]
- Control and follow-up [24].

Finally, although there are numerous guides and manuals of best practices for CDW management, no references have been found implementing these measures in the Environmental Management System of construction companies. Furthermore, even though best practices regarding CDW management have been suggested and studied, specifying these practices according to the type of construction and the different waste categories generated remains a research to be performed. Therefore, this work aims to improve current Integrated Quality Management Systems (IQMS) by determining a specific CDW management system (CDWMS) for construction sites, providing consistent measures to prevent CDW generation in order to achieve circular construction projects.

2.2 Methods

Despite the CDWMS should include the entire building lifecycle—from the design stage until deconstruction—the scope of the CDWMS developed here has been limited to the building construction stage. However, a procedure to develop the CDW

Management Report, written in the design phase—is also considered, as it has a direct connection with the document CDW Management Plan. Moreover, CDWMS should be a useful and operational tool for the construction agents of the company. To do so, the proposed CDWMS follows the same format and structure as commonly used IQMS [18]. The content and structure of each procedure were established after analyzing the IQMS procedures of several Spanish construction companies.

Finally, a procedure was chosen to be implemented in a real construction company. First, several meetings with the construction company were established in order to choose the procedure to be implemented and the reasons that led to his selection. After two years of implementation, an interview was carried out with the Head of Quality and Environment of the Company and the technician responsible for carrying out the procedure, in order to know the advantages and drawbacks encountered after its implementation and to identify potential areas for improvement. Throughout the interview, the agents answered the following questions: In how many construction sites has the procedure been implemented? What advantages/drawbacks were found after implementing the procedure? These advantages/drawbacks were found in all the construction sites? Can you suggest areas for improvement? and Based on your opinion, which procedure will be the next to be implemented? Why?

2.3 Results and Discussion

The CDWMS proposed here follows one of the eight main principles of quality management, i.e., management based on Procedures. The procedures define the activities needed to achieve a proper CDW management and are normally structured one or more procedures in order to describe and facilitate the proper management of waste generated on site. Each procedure contains several forms, which are Word or Excel documents. The CDWMS proposed unfolds in eight procedures (Fig. 2.1).

Procedure 1: Development of the CDW Management Report: This procedure is divided into eleven forms. The CDW management report will be defined once all the forms are completed:

Form PR1.1/0: Introduction. This form describes general issues of the construction site, the Company, and the different regulations.

Form PR1.1/1: Description of work. This form describes: type of construction (new/refurbish), construction site address, number of dwellings, total built area, etc.

Forms PR1.1/2 and PR1.1/3: CDW quantification. This form identifies and quantifies the amount of each waste category expected to be generated both in weight and volume, as described by RD 105/2008.

Form PR1-1/4: CDW management. This form describes the percentages of each waste category that will be allocated to each management option (reuse, recycling, recovery, or disposal).

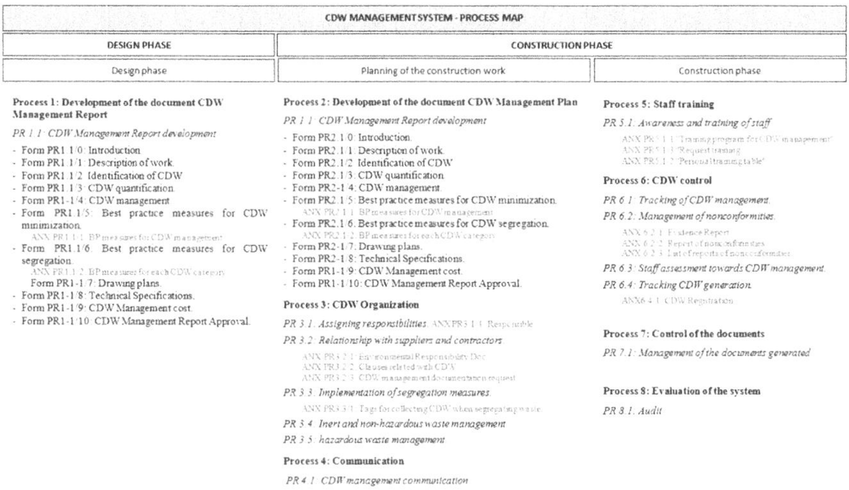

Fig. 2.1 General structure of the Construction and Demolition Waste Management System (CDWMS)

Form PR1.1/5: Best-practise measures for CDW minimization. This form must define all the measures to be used for reducing the amount of CDW generated on site. This should be done to consider the possible waste to be generated in each activity of the work and prioritizing the waste to be generated in larger quantities.
Form PR1.1/6: Best-practise measures for CDW segregation. This section should define all the measures that will be used to separate different waste categories.
Form PR1-1/7: Drawing plans. This form allocates in the working site the installations and storage sites needed for a proper segregation of CDW.
Form PR1-1/8: Technical specifications. This form describes the Technical Specifications for Waste Management.
Form PR1-1/9: CDW management cost.
Form PR1-1/10: Approval of the CDW Management Report. This form must be approved by the Company's Director of Environment and the project developer.

Procedure 2: Development of the CDW Management Plan: This document specifies the CDW Management Report document, which is more general, to a specific construction site. This procedure is also divided into eleven forms as in Procedure 1 numbered as Form PR1.2/0.

Procedure 3: CDW organization: This Procedure describes how to organise CDW management generated on site.

PR 3.1: Assigning responsibility. The people responsible for each action on waste management are designated in PR 3.1 by completing Annex ANX PR3.1/1.
PR 3.2: Relationship with suppliers and contractors. When contacting suppliers and contractors, procedure PR 3.2 will be followed, establishing the minimum requirements and best practices for suppliers and subcontractors by filling three

annexes: ANX PR3.2/1: Environmental Responsibility Doc, ANX PR3.2/2: Clauses related with CDW and ANX PR3.2/3: Request for CDW management documentation.
PR 3.3: Implementation of segregation measures. By filling in and using labels in Annex ANX PR3.3/1: Tags to collect CDW when segregating waste.
PR 3.4: Inert and nonhazardous waste management and PR 3.5: hazardous waste management. In order to manage the waste generated, one of the two procedures will be followed, depending on the type of waste.

Procedure 4: Communication: consists of a single procedure which describes the methodology to communicate CDW Management issues between the different construction stakeholders.

Procedure 5: Staff training: consists of a single procedure which describes CDW training for construction workers and staff. The training activities planned are recorded in Annex ANX PR5.1/1 "Training programme for CDW management". Subsequently, those responsible for the training activities will make a request, following Annex ANX PR5.1/3 "Request training", to the Head of Human Resources stating the reasons for training, the proposed training activities, and the staff concerned. Finally, the Head of Human Resources will have a personal record of all training activities of the Company by completing Annex ANX PR5.1/2 "personal training table".

Procedure 6: CDW control: Controlling CDWMS documents is essential to achieve correct CDW management [19]. Therefore, the CDWMS has a specific tracking procedure based on the development of traceable records by filling out the forms. The procedure contains four procedures in order to describe how to monitor and control the CDW management carried out on site.

PR 6.1: Tracking of CDW management. This procedure provides the methodology for monitoring and controlling various activities related to waste management. For this, the form 6.1/1 needs to be completed. The best practices anticipated in the CDW management plan will be scored and determined if they are being implemented or not.
PR 6.2: Management of nonconformities. This procedure describes how to identify non-conformities and implement corrective and preventive actions. To this end, the following three annexes need to be completed: ANX 6.2/1: Evidence Report; ANX 6.2/2: Report of non-conformities and ANX 6.2/3: List of reports of non-conformities.
PR 6.3: Assessment of staff towards CDW management. This procedure establishes the attitude of the process for assessing the staff and evaluates their commitment to the CDW environmental management policies.
PR 6.4: Tracking CDW generation. This procedure provides the methodology for recording the quantities of waste generated on site, allowing one to identify, through the actual on-site data, the volume and the weight of the CDW generated daily at each construction site. For this, each time a waste container leaves the site Excel file of Annex ANX6.4/1: The generated CDW registration is completed, using the delivery notes corresponding to when the work containers left the site

(service delivery), as well as the delivery notes issued by the waste manager once the heavy containers arrived at the recycling plant (receipt of admission).

Procedure 7: Document control. This procedure describes the process to ensure the identification, maintenance and storage of the different documents and files generated in the CDWMS.

Procedure 8: Evaluation of the system. Consists of a single procedure "PR 8.1: Audit" which describes how to evaluate the CDWMS developed in order to test if the system is effective and appropriate to achieve the goals and targets of the company.

2.3.1 Structure of the Procedures

It is important that the CDWMS developed here uses a standard structure for the different processes to facilitate the identification and understanding of the information contained in the CDWMS. To determine the structure of the procedures, various internal quality management systems from four construction companies were taken as examples. Based on the findings, the procedures developed in the CDWMS follow a standardised structure: Aim; Scope; References; Responsibilities; Description/Phases; Flowchart; Changes to the procedure; and Annexes.

2.3.2 Implementation of the Procedure

Procedure 6.4 "Tracking CDW generation" was chosen to be implemented in the company, because:

Particular CDW generation ratios are obtained after its implementation, which serve as the basis for calculations of Procedures 1 and 2 in order to develop the CDW Management Plans and Reports.

It can be immediately incorporated into the company organisation.

The company has implemented this procedure in all their construction works (a total of fifty), not only in residential building construction works but also in nonresidential projects. According to the benefits found during its implementation, the company highlighted that this procedure allows determining accurate CDW quantification ratios (amount CDW per m^2 of built surface), in order to set particular and more reliable indicators. The establishment of these indicators helps to develop CDW Management Plans and Reports in order to achieve a higher adjustment in CDW management costs. Another advantage is that no construction agents had a high resistance or opposition to implement the method. However, the company highlighted that some construction teams were more involved than others. In fact, the company organises an annual award for the best construction team of the year. This award considers and evaluates, among other factors, the CDW management carried out on site.

The main drawbacks found during its implementation were focused on reorganization, as CDW management documents control—i.e., the delivery notes issued by the waste manager once the heavy containers arrived at the recycling plant (receipt of admission)—was relocated from the site to the central office. Initially, the delivery notes were received at each construction site and each construction team was responsible for completing ANX6.4/1 "CDW Registration". After a trial period, the Company observed that there was a lack of coordination to complete and return to the central office the annexes. Therefore, to solve this, the company decided to centralise the procedure at the central office. To this end, admission receipts from all the works were received at the central office in order to improve the control and coordination of CDW data collection. However, this change in the company's organization led to an increase in time and resource consumption, as the company had to assign an employee to fill out the ANX6.4/1 "CDW Registration" (ie the monthly CDW data generation) for each site. In this regard, the company considers it essential that waste managers provide CDW information in digital format (Excel file or similar) rather than sending them on paper.

Furthermore, the company considered that the completion of the Excel file containing Annex ANX6.4/1 should be simpler, so they developed a software application that automates the process. The computer tool shows a summary of the CDW containers generated to date (marked with a red arrow) and specific information about each CDW container. Finally, the information is automatically incorporated into the Excel file ANX6.4/1 "CDW registration".

Today, the tool does not allow entering the construction work activities running at that time: structure, masonry, and finishings. This information is manually included, as planned in the original procedure. However, the company considers automating this information in the future. In short, the area of improvement proposed and developed by the company is geared towards automation of the process. Finally, the company considered that the next procedure to be implemented would be 5.1 "Staff training", which they consider fundamental to achieving a proper CDW management on site.

2.4 Conclusions

A CDW management system for newly built residential building constructions has been developed, based on eight processes with their own procedures, formats, and annexes. This CDW management system can be implemented in the construction companies in order to help CDW management. In general, this CDW management system establishes control, preventive, and corrective measures on aspects that have previously been identified as the most significant for reducing potential problems.

The procedure 'PR6.4 CDW generation" was chosen as the first procedure of the CDW Management System to be implemented in the company. This procedure allows to obtain rigorous CDW quantification ratios and also helps to achieve greater adjustments for their CDW management costs. Stablishing personal CDW ratios helps them with the development of CDW management plans and Reports. The area of improvement proposed and developed by the company focuses on automating the process, in order to obtain a complete tool that allows easier filling. Finally, procedure PR5.1 'Staff training' can be the next procedure to be implemented, because the training on CDW is essential in order to achieve proper on-site CDW management.

Finally, the results obtained revealed that the CDW Management System developed allows to improve and optimise construction processes, favoring the control and savings of raw materials through reuse and minimization of waste, and giving CDW responsibilities to their employees and managers. Therefore, determining and implementing an on-site CDW Management System, integrated with the EMS and in turn with the QMS of the companies, will definitely improve existing management systems.

Acknowledgements The authors thank ARPADA Company, particularly the Department of Environment and Quality, for implementing the CDW Management developed in this work, which was basic for this study.

References

1. Eurostat (2022) Eurostat waste statistics Available: https://ec.europa.eu/eurostat/statistics-explained/index.php?title=Waste_statistics/es#Generaci.C3.B3n_total_de_residuos
2. European Parliament (2008) Directive 2008/98/CE of the European Parliament and of the council of 19 November 2008 on waste and repealing certain directives, ed. Off J Eur Union, pp. 3–30
3. European Commission (2015) Closing the loop-An EU action plan for the Circular Economy. Communication from the Commission to the European Parliament, the Council, the European Economic and Social Committee and the Committee of the Regions COM, vol. 614, no. 2, p. 2015
4. Spanish Government () Real Decreto 105/2008, de 1 de Febrero, por el que se Regula la Producción y Gestión de los Residuos de Construcción y Demolición. (Royal Decree 105/2008, of February 1, on the Production and Management of Construction and Demolition Waste)," ed. Official state bulletin (BOE), 2008, pp. 7724–7730.
5. Ley 7/2022, de 8 de abril, de residuos y suelos contaminados para una economía circular (Law 7/2022, of April 8, on waste and contaminated soil for a circular economy), 85, 2022.
6. Haselsteiner E, Rizvanolli BV, Villoria Sá P, Kontovourkis O (2021) Drivers and Barriers leading to a successful paradigm shift toward regenerative neighborhoods. Sustainability 13(9):5179
7. Osmani M, Glass J, Price ADF (2008) Architects' perspectives on construction waste reduction by design. Waste Manage 28(7):1147–1158
8. Akinade OO et al (2017) Design for deconstruction (DfD): critical success factors for diverting end-of-life waste from landfills. Waste Manage 60:3–13
9. Tam, Tam, Zeng, Ng (2007) Towards adoption of prefabrication in construction. Build Environ 42(10):3642–3654
10. del Río-Merino M, Vidales-Barriguete A, Piña-Ramírez C, Vitiello V, Santa Cruz-Astorqui J, Castelluccio R (2022) A review of the research about gypsum mortars with waste aggregates. J Build Eng 45:103338

11. Masudi AF, Che Hassan CR, Mahmood NZ, Mokhtar SN, Sulaiman NM (2012) Waste quantification models for estimation of construction and demolition waste generation: a review. Int J Global Environ Issues 12(2–4):269–281
12. Yuan L, Lu W, Xue F (2021) Estimation of construction waste composition based on bulk density: a big data-probability (BD-P) model. J Environ Manage 292:112822
13. Villoria Sá P, Porras Amores C, del Rio Merino M (2020) Chapter 3: Estimation of construction and demolition waste. In: Pacheco-Torgal F (ed) Advances in construction and demolition waste recycling: management, processing and environmental assessment. Elsevier, England, pp 13–27
14. Coskuner G, Jassim MS, Zontul M, Karateke S (2021) Application of artificial intelligence neural network modeling to predict the generation of domestic, commercial and construction wastes. Waste Manag Res 39(3):499–507
15. Li Z, Shen GQ, Alshawi M (2014) Measuring the impact of prefabrication on construction waste reduction: An empirical study in China. Resources Conserv Recyc 91:27–39
16. Shi Y, Xu J (2021) BIM-based information system for econo-enviro-friendly end-of-life disposal of construction and demolition waste. Autom Constr 125:103611
17. Göçer, Ö.; Y. Hua, and K. Göçer, "A BIM-GIS integrated pre-retrofit model for building data mapping," in Building Simulation, 2016, vol. 9, pp. 513–527:.
18. Cheng JC, Ma LY (2013) A BIM-based system for demolition and renovation waste estimation and planning. Waste Manag 33(6):1539–1551
19. Coronado M, Dosal E, Coz A, Viguri J, Andrés A (2011) Estimation of construction and demolition waste (C&DW) generation and multicriteria analysis of C&DW management alternatives: a case study in Spain. Waste Biomass Valor 2(2):209–225
20. González Pericot N, del Río Merino M (2011) Management of waste from packaging of construction materials. Open Constr Build Technol J 5(Suplemento 2):156–161
21. Begum RA, Siwar C, Pereira JJ, Jaafar AH (2007) Implementation of waste management and minimisation in the construction industry of Malaysia. Resour Conserv Recycl 51(1):190–202%@ 0921–3449
22. Villoria Saez P, del Río Merino M, San-Antonio González A, Porras-Amores C (2013) Best practice measures assessment for construction and demolition waste management in building constructions. Resour Conserv Recycl 75:52–62
23. Begum RA, Siwar C, Pereira J, Jaafar AH (2009) Attitude and behavioral factors in waste management in the construction industry of Malaysia. Resour Conserv Recycl 53(6):321–328
24. Al-Saggaf A, Jrade A (2015) Benefits of integrating BIM and GIS in construction management and control. In: The CSCE International Construction Specialty Conference, 2015, vol. 2015
25. ARPADA S.A. Available: https://www.arpada.net/

Chapter 3
Reuse of Steel Sheet Piles—Best Practice

François Fohl and Oliver Hechler

Abstract Steel sheet piles are used for retaining walls. Due to their modularity, they can be easily installed and extracted after their service life. After their first use, they can be either directly recycled or reused several times and then recycled. The reuse of steel sheet piles allows to avoid new production and thus CO_2-emissions for their production, reducing the environmental impacts per use. In temporary works, like construction pits, the reuse of sheet piles is common practice. Also, for certain permanent projects, there is no disadvantage in using second-hand sheet piles. This paper shows, on the basis of two case studies the best practice for reuse of sheet piles. The environmental impacts for a temporary project in Germany are discussed based on a Life Cycle Assessment for the sheet piling tonnage. Over the life cycle of the steel, 1,535 t of CO_2-eq are emitted. Reuse of the sections saved 79% of greenhouse gases. For a dyke reinforcement in the Netherlands, the project owner partly chose second-hand sheet piles to reduce the environmental impacts of the infrastructure project.

Keywords Steel · Sheet piles · Reuse · Environmental impacts

3.1 Steel Sheet Piles—a Short Introduction

Steel sheet piles are widely used in infrastructure applications. Functioning mainly as a retaining wall, they are found for example in quay walls, cofferdams, bridge abutments, underground car parks or temporary pits. Steel sheet piles are sections that are driven into the ground and are connected over interlocks (or clutches), to form a continuous wall. There is a wide variety of sections, but U- and Z-piles are the most common (see Fig. 3.1). Common widths for one sheet pile (one module) are ranging

F. Fohl (✉) · O. Hechler
ArcelorMittal Commercial RPS, Esch-Sur-Alzette, Luxembourg
e-mail: francois.fohl@arcelormittal.com

O. Hechler
e-mail: oliver.hechler@arcelormittal.com

L. Bragança et al. (eds.), *Creating a Roadmap Towards Circularity in the Built Environment*, Springer Tracts in Civil Engineering,
https://doi.org/10.1007/978-3-031-45980-1_3

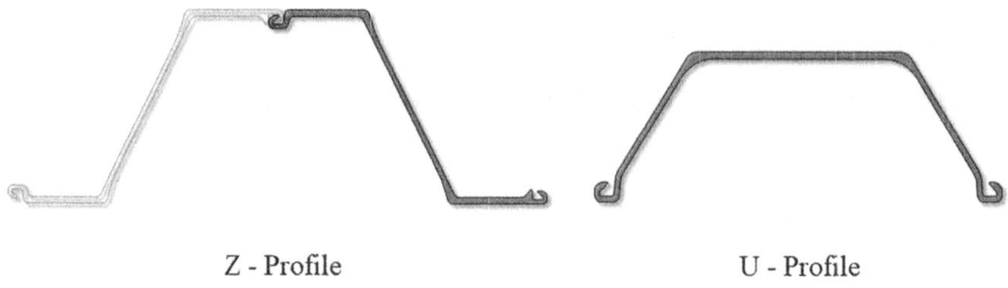

Fig. 3.1 Common types of steel sheet piles—Cross section

from 600 to 800 mm. Steel sheet piles are completely impervious, an infiltration of water through a sheet pile wall is only possible along the interlocks. Due to the tight shape of the interlocks of type 'Larssen', a high seepage resistance is already given for ArcelorMittal sheet piles. For applications with high requirements to watertightness, like cut-off walls for contaminated sites, several sealant systems are available. Welding of the interlocks can provide complete watertightness. The durability of steel sheet piles is well documented. Unprotected steel in the atmosphere, water or soil is subject to corrosion. However, based on the life-time requirements for the steel structure, the integrity can be assured by several methods. 'Sacrifical steel' to build a structural design reserve is common, but the corrosion process can also be limited by coatings, low-corrosion steel grades, cathodic protection and others. Corrosion rates for steel sheet piles for all relevant environments are well documented in Eurocode 5—Part 5 (EN 1993–5) [1]. Sheet pile structures can be easily dimensioned for 50–100 years lifetime depending on the application. Steel sheet piles are prefabricated products, hence very efficient in installation due to their modular system [2].

3.2 Steel Sheet Piles—a Product in the Sense of Circular Economy

Circular economy is a concept that aims to minimize the use of resources, energy and waste flows. Unlike the linear economy, which follows a take—make—use—throw-away approach, the circular economy is looking to extend the lifecycle of products. The 9 R framework is describing the way from a linear to a circular economy, hence closing material loops, following several key words, such as Reduce, Reuse, Refurbish, Recycle [3]. Especially for construction products, the end-of-life credentials are of high importance as they are consuming important material resources. These resources are lost unless they are maintained in use through reuse or genuine recycling. The Construction Product Regulation (CPR) [4] or initiatives from European Commission such as:

- The European Resource Efficiency Platform [5],
- Towards a circular economy: A zero waste programme for Europe [6],
- Resource efficiency opportunities in the building sector [7],

are an opportunity for steel construction methods in general, due to the inherent properties of steel and the possibility to deliver sustainable structures, with components which are demountable and reusable [8].

Steel sheet piles perform especially well in the following categories, relevant for a transition to circular economy:

- Reduce: Resource efficient design;
- Reuse: Steel sheet piles can be reused multiple times;
- Recycle: Steel can be 100% recycled.

Through optimization of sheet pile solutions, the profiles have become lighter and lighter over the past decades, while still meeting the same requirements. This is mainly due to optimized and efficient profiles and high strength steel grades. Steel sheet piles can be used and reused up to 10 times for temporary applications, thus reducing the environmental impact each time the sheets are reused. ArcelorMittal also offers rental and sale of second-hand sheet piles. In addition, steel is a permanent material. It can be 100% recycled without any loss of quality. Steel sheet piles from ArcelorMittal, that are produced in Luxembourg are out of 100% recycled steel [2]. In this paper, the focus is on the advantages on reusing sheet piles.

3.3 Reuse of Steel Sheet Piles

3.3.1 How-To

Firstly, the reuse of steel sheet piles is possible due to their modularity forming the system. The sections can be extracted after their specific service life. Contrary to temporary applications, sheet piles in permanent applications are generally not reused after their service life, but directly recycled and in consequence, reintroduced into the steel loop. However, for temporary applications, sheet piles have much shorter life cycles; normally less than 2 years. It is common practice for contractors to use sheet piles for temporary excavation pits and reuse them again in following projects. Depending on the soil, the sections can be reused up to 10 times, however a small portion ($\leq$50 cm) of the length of the section is generally cut-away after each use, due to deformations respectively damages at the head. The symmetric U-sections are well suited for reuse, as they offer high stiffness due to their compactness. The extraction of the sheet piles can be realized with vibratory hammers, which are usually also used also for the installation (see Fig. 3.2).

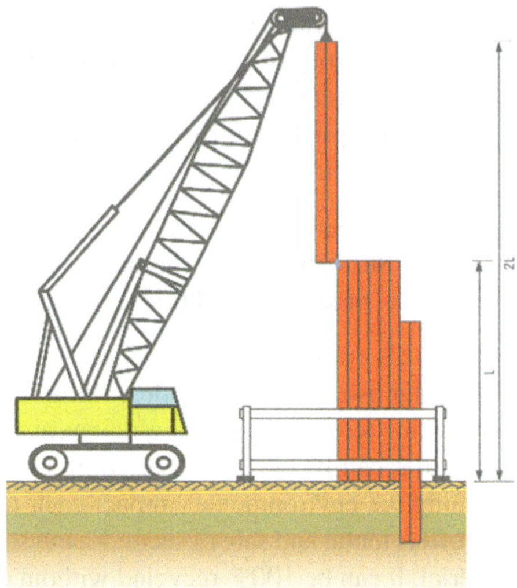

Fig. 3.2 Procedure of sheet piles installation (panel driving) [9]

3.3.2 Global Warming Potential Over the Life Cycle of Steel Sheet Piles

Production of steel represents over 7% of global greenhouse gas emissions [10]. One way to reduce these, is the reuse of used steel products such as sections and sheet piles. Their reuse avoids the production of new products, hence the consumption of raw materials, energy and the average emission of 1.89 tonnes of CO_2 per tonne of crude steel [11]. The environmental impacts of building materials can be documented by type III Environmental Product Declarations (EPDs), following EN 15804 [12] and ISO 14025 [13]. An EPD is a document that describes the environmental impacts of a product based on a life cycle assessment with quantitative data. It enables manufacturers to transparently communicate the environmental impacts of their products. EPDs form the basis for the ecological assessment of buildings or other construction projects. ArcelorMittal's EPDs are verified and published by independent experts, such as the *Institut Bauen und Umwelt e.V.* (IBU).

In case new sheet piles are reused several times on the same project and recycled afterwards, then the analysis of the environmental impacts is straightforward. The total quantity of new material purchased for the project can be used. However, in most cases, the sheet piles will return to a storage yard, will be cleaned and/or repaired (damaged portions or piles will be scrapped), and later reused on a different project. Based on feedback from customers from ArcelorMittal, around 25% of the sheets produced are reused several times. Hence for temporary applications, the LCA practitioner can assume that steel sheet piles pertaining to the "rental business" will be used 5 times with some small losses during the total life cycle due to damages during installation. This approach reduces the environmental impact for each use phase to around 1/5 of the overall impact. Note that generally speaking, the lifecycle of a used

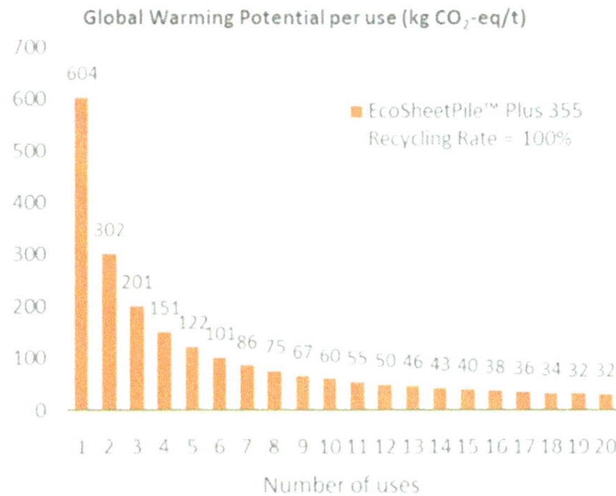

Fig. 3.3 Global Warming Potential over the entire life cycle—Influence of reuse

sheet pile is quite short (rentals can be as short as a few weeks to two or three years) and rarely above five years. Figure 3.3 shows the decreasing environmental impact of sheet piles, when used multiple times. The calculation is based on ArcelorMittal's EcoSheetPile Plus EPD and shows the Global Warming Potential (GWP) for one tonne of sheet piles considering modules A1-A3, C3, C4 and D according to EN 15804. Module A4 (transport gate to site), A5 (installation), C1 (deconstruction) and C2 (transport) are not considered.

Based on the EPD, one single use of one tonne of sheet piles emits as much as 604 kg CO_2-eq over the lifecycle. Reusing the sheet pile once, results in a GWP of 302 kg CO_2-eq for each use, whereas using them 5 times leads to a GWP of 122 kg CO_2-eq per use. When doing life cycle assessments for temporary works, the business practice of cutting of the head of the sheet pile due to damages may also be considered.

According to the EPD EcoSheetPile Plus, the production (modules A1-A3) of one tonne of sheet piles emits 370 kg CO_2-eq. Around 96% of all the relevant greenhouse gas emissions for a sheet pile structure, are emitted during the production stage of the steel. Energy emissions for installation as well as transport shouldn't be neglected but have a rather small influence on the total emissions [14].

3.3.3 End-of-Life Assumptions for Steel Sheet Piles

The calculation shown in Fig. 3.3 considers a final recycling rate of 100%. However, the basic end-of-life assumptions for the EPD EcoSheetPile Plus are the following, representing an average use of ArcelorMittal's sheet piles [15]:

- Reuse: 25%;
- Recycling: 60%;

- Landfill: 15%.

The reuse and recycling rate contribute together to the final recycling rate. The assumptions are based on a preliminary standard elaborated by CEN TC 135 WG 17 [16] and a detailed analysis of ArcelorMittal, considering the:

- Application type of the sheet pile (quay wall, bridge abutment, ...);
- Sheet pile type (Heavy or light sections);
- Corrosion depending on the environment;
- Use for permanent or temporary applications;
- Possibility to extract the sheet piles;
- Design life (50 years, 100 years).

Note that the corrosion rate, which depends greatly on the environment and the service life of the sheet pile structure, has a large impact on the landfill rate.

3.4 Case Study: Reuse of Sheet Piles for Temporary Works

Today's *Schwarze Pumpe* Industrial Park is located about 120 km southeast of Berlin on the territory of the German states of Brandenburg and Saxony in Spremberg. The site was founded as the *VEB Gaskombinat Schwarze Pumpe* in 1955 with the task to utilize the nearby mined brown coal for energy supply or to refine it as briquettes for domestic heating. There were gas production plants, coking plants, combined heat and power plants and briquette factories with the necessary ancillary facilities on the site. By 1989, this had created the world's largest brown coal processing plant with 15,200 employees. By 1992, this figure had already been reduced to just 6,600 employees, together with the start of dismantling obsolete plants and the simultaneous construction of new power plants in line with current environmental protection standards. The remediation of the contaminated sites that had accumulated over decades required careful planning and preparation. Starting in 2017, over a period of several years, 430,000 tonnes of contaminated soil will be processed in total. Part of the material, which is mainly contaminated with benzene and phenols, is cleaned using soil treatment processes. The heavily contaminated areas are excavated, thermally treated and reinstalled. In order to reduce the necessary transports, a soil cleaning plant is built on site.

The main argument to use a sheet pile solution for this project, is their modularity and the possibility to reuse them on the jobsite several times. A PU 22^{-1} profile was chosen for this purpose, as it's a robust section suitable for reuse. It has a weight of 136.5 kg/m^2 and an elastic section modulus of 2,060 cm^3 per m of wall (Fig. 3.4).

The lengths of the sheet piles are optimized for the areas of application and range from 14.0 m to 23.0 m. Length reduction due to multiple uses is already included in the length determination. Figure 3.5 shows the plan view of the project.

Inside the surrounding primary pit, secondary pits are constructed. In these, the contaminated soil is excavated, decontaminated, and then placed again. Once the

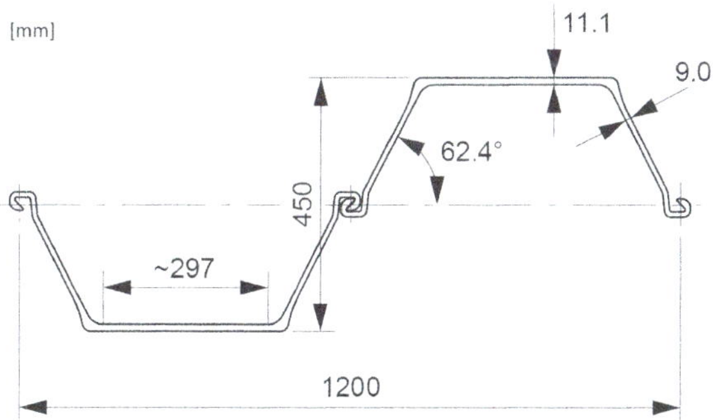

Fig. 3.4 Steel sheet pile PU 22^{-1} [17]

Fig. 3.5 Left: Plan view—Excavation pit; Right: Installation of sheet piles for the secondary excavation pit [18]

process is finished for one of these areas, the sheet piles are extracted. After inspection, washing, and cutting in case of deformations, the sections are used for another secondary pit. One sheet pile is used on average 5 times on the site, before it gets recycled.

The environmental impact of the sheet piles tonnage for this project is assessed with an internal Life Cycle Assessment (LCA) tool. Based on the actually delivered sheet piles, the GWP and other relevant indicators are assessed following EN 15804 and ISO 14025, on the basis of ArcelorMittal's EPD EcoSheetPiles [19]. In the analysis, modules A1-A3, A4, C3 and D are considered. Table 3.1 shows the parameters and assumptions, which are the basis for the Life Cycle Assessment.

Figure 3.6 compares the Global Warming Potential for both scenarios, the first being an approximation of the as-built condition with sheet piles that are used 5 times in total, and the second one representing the case with no reuse of the sections. In this case, the non-reuse of the sections is approximated by multiplying the required tonnage by a factor of 5. Based on the considered modules, the as-built scenario emits 1,535 t of CO_2-eq. This represents a reduction of 79% in greenhouse gases over the life cycle, due to the multiple reuse of sections.

Table 3.1 Parameters for the life cycle assessment

	Reuse of sheet piles as-built	No reuse of sheet piles
Tonnage	2,000 t	10,000 t
Length of sheet piles	18.5 m (avg. length)	
Minimum length before recycling	16.5 m (cut 0.5 m after each use)	*Not applicable*
Number of uses	5×	1×
EPD	EcoSheetPiles	
End-of-Life Assumptions	Reuse: 78.9%, Recycling: 20.9%, Landfill: 0.2% Final recycling rate: 99%	Reuse: 0%, Recycling: 99%, Landfill: 1% Final recycling rate: 99%
Transport by truck	Belval (LU)—Spremberg (DE)—793 km	

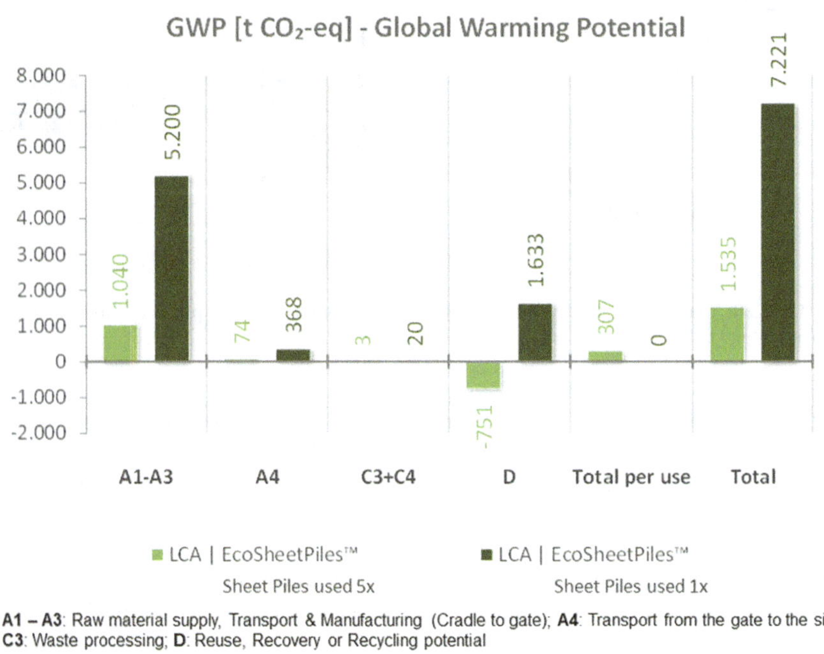

Fig. 3.6 GWP based on EPD EcoSheetPiles—Influence of reuse

3.5 Case Study: Reuse of Sheet Piles for Permanent Application

For the reinforcement of the Gorinchem-Waardenburg (GoWa) dyke in the Netherlands, steel sheet piles are used on a large scale. Many efforts are being made to reduce the environmental impact of the project by the project alliance *Graaf Reinald Alliantie*, including the use of used sheet piles, made from recycled steel.

Water regulation in the Netherlands has always been important to the survival of the country: about two-thirds of the Netherlands is vulnerable to flooding and marine submersion, while one-third of the country's surface is below sea level. The first major hydrological works to gain territory over the water date back to the thirteenth century, and today a network of more than 17,500 km of dykes protect the 6,000 km of canals that run through the country. In particular, 3,750 km of dykes form the primary flood defence, protecting the Netherlands against flooding from the North Sea, the Wadden Sea and the major rivers. Following the dramatic floods of 1953, the major 'Delta' plan was launched to improve the safety of the reclaimed land. It is estimated that the risk of flooding has been reduced to less than once every 4,000 years in Zeeland and even once every 10,000 years in the Rotterdam area. Dyke strengthening and raising work continues across the country, especially on the 1,850 km of dykes that do not yet meet the safety standards required by 2050.

For the reinforcement of the GoWa dyke, steel sheet piles are used on more than 6 km of the dyke, 23 km in total length. They are installed with the aim to protect the dyke against the risk of failure due to general stability and/or piping. In this case, sheet piling is an ideal solution to support and reinforce the dyke against these two types of failure, especially where the surrounding space is limited and does not allow the extension of the dyke footprint. In this case, the construction of a sheet pile wall can provide both reinforcement tasks (see Fig. 3.7).

For the Gorinchem-Waardenburg project, the project owner deliberately chose to install used sheet piles at locations where there is a risk of hydraulic failure by piping. In order to protect the dyke against piping, the length of the seepage path

Fig. 3.7 Dyke reinforcement with sheet piles in Wolferen-Sprok [20]

must be extended so that the phenomenon can no longer occur. A sheet pile wall can perfectly fulfil this function. In fact, in this application, a used sheet pile section can perform this function just as well as a new sheet pile. The main difference is that the total greenhouse gas emissions of the steel sheet pile production phase do not have to be included in the environmental balance of the project, as they have already been partially included in the previous temporary applications. The scope of the project, and the recent inclusion of environmental criteria in the Dutch tendering process, means that the environmental impact of the selected solutions must be reduced as much as possible. The project owner and ArcelorMittal worked together to reduce the carbon footprint of the project by applying the three important principles of the circular economy: Reduce, Reuse, Recycle. If technically possible, reuse of steel products should always be considered, even for permanent applications.

3.6 Conclusion

Steel sheet piles, especially produced from recycled steel, offer the potential to reduce the carbon footprint of a project. Due to the modularity, steel sheet piles can further be reused and may contribute significantly to a circular economy concept in foundation solutions. The projects 'Schwarze Pumpe' and 'GoWa' show perfect examples how the concept may be applied. Considering 'Schwarze Pumpe', the sheet piling tonnage emits in total 1,535 t of CO_2-eq, based on the assumption that the sheet piles are used 5 times on average and taking into account a shortening of the sections after each use, due to damages and deformations. The choice of the right sheet pile section is important in terms of reduction of the environmental impacts over the entire life cycle. If the sections are reused several times before recycling, robust sections should be chosen, like the PU's with reinforced shoulders from ArcelorMittal. These sections have higher environmental impacts than AZ sections with similar mechanical properties in the production stage, however due to their stiffness they can be reused more often, hence the impact over the entire life cycle is reduced. For permanent applications, efficient AZ sections are often the right choice. Due to their efficient shape, lighter sections can assure the same mechanical properties. Furthermore, the current method from the Worldsteel Association, which allows the repartition of the environmental impacts on the number of usages (see Fig. 3.3) can be discussed. Another approach would be, that the environmental impacts are taken into account for the first use, and for consecutive usages, there is no significant impact.

References

1. EN 1993-5:2007, Eurocode 3—Design of steel structures—Part 5: Piling
2. ArcelorMittal Commercial RPS, Piling Handbook, Esch-sur-Alzette, 9th edition (2016), revision (2022)

3. Potting J, Hekkert M, Worrell E, Hanemaaijer A (2017) Circular economy: measuring innovation in the product chain. PBL Netherlands Environmental Assessment Agency
4. Regulation (EU) No 305/2011 of the European Parliament and of the Council of 9 March 2011 laying down harmonised conditions for the marketing of construction products and repealing Council Directive 89/106/EEC (2011)
5. European Commission (2014)European resource efficiency platform, Manifesto and policy recommendations
6. COM (2014) 398 final towards a circular economy: a zero waste programme for Europe
7. COM (2014) 445 final on resource efficiency opportunities in the building sector
8. European Commission (2019) Research Programme of the Research Fund for Coal and Steel, TGA4, Reuse and demountability using steel structures and the circular economy, REDUCE
9. ArcelorMittal Commercial RPS (2022) Rammfibel für Stahlspundbohlen, Esch-sur-Alzette
10. McKinsey & Company (2018) Decarbonization of industrial sectors: the next frontier
11. Worldsteel Association (2021) Sustainable steel
12. EN 15804 + A2:2019, Sustainability of construction works—Environmental product declarations—Core rules for the product category of construction products
13. ISO 14025:2006, Environmental labels and declarations—Type III environmental declarations—Principles and procedures
14. EFFC DFI (2014) Carbon Calculator Methodological Guide—V2.2
15. ArcelorMittal Commercial RPS (2021) Environmental Product Declaration—EcoSheetPile Plus, Institut Bauen und Umwelt e.V
16. prEN17662 (2020) Execution of steel structures and aluminium structures—Environmental Product Declarations—Product category rules complementary to EN 15804 for Steel, Iron and Aluminium structural products for use in construction works. (CEN-TC135 N1000 TC135 WG17)
17. ArcelorMittal Commercial RPS (2008) Info sheet PU 22^{-1}
18. ArcelorMittal Commercial RPS (2020) Case Study, Schwarze Pumpe
19. ArcelorMittal Commercial RPS (2018) Environmental Product Declaration—EcoSheetPiles, Institut Bauen und Umwelt e.V.
20. Dike reinforcement project Wolferen—Sprok, Waterschap Rivierenland (2021)

Part II
Design Strategies and Tools for Circular Buildings

Chapter 4
Design Strategies and Technical Practices for Reusable Steel–Concrete Composite Structural Systems

Florentia Kavoura and Milan Veljkovic

Abstract The technical solutions and design strategies for deconstruction are reviewed and investigated in the building sector in order to achieve the sustainability goals set in the EU Commission´s "Green Deal" towards net zero greenhouse gas emissions by 2050. The demountable and reusable steel–concrete composite structures could contribute immensely to the achievement of these sustainability goals. The main technical practice which allows for demountability and reusability is the use of bolted connections on the skeleton of the steel structure with the parallel use of demountable shear connectors in their floor systems with the steel load-bearing beams. However, there is a very limited number of studies and methods into specific demountable and steel–concrete composite structural systems, and even fewer focus on their practicability and feasibility. Since these systems have the potential to reduce embodied carbon impacts, encourage resource efficiency and, reduce construction waste, their feasibility and technical practices should be investigated. This paper focuses on technical solutions and design strategies that currently have been developed for steel and steel–concrete composite reusable structural systems.

Keywords Demountable shear connectors · Reusability of composite floors · Sustainable structural design

4.1 Introduction

The structural design for deconstruction is estimated that it is going to contribute considerably to the sustainable development of the built environment [1]. In response to the sustainability requirements set in the EU Commission´s "Green Deal" [2] towards reduction of the greenhouse gas emissions the preferred strategy in order to achieve a sustainably built environment is to design for demounting, and reuse of

F. Kavoura (✉) · M. Veljkovic
Delft University of Technology, 2628 CD Delft, Netherlands
e-mail: f.kavoura@tudelft.nl

L. Bragança et al. (eds.), *Creating a Roadmap Towards Circularity in the Built Environment*, Springer Tracts in Civil Engineering,
https://doi.org/10.1007/978-3-031-45980-1_4

the components before thinking about repair, remanufacturing, or recycling. Consequently, to follow Cramer's [3] structural scheme on how a circular economy can be incorporated for sustainable construction.

Therefore, the aim for a sustainable design needs to focus on material-efficient use and the production of layouts that allow a whole or partial structural component to be reused and achieve longer structural life spans [4]. With respect to efficient material use, the research is directed toward the use of multi-material structural components. With these components, we could achieve optimum material use by combining the advantages of mechanical properties of different construction materials (e.g. concrete in compression and steel in tension). The main focus areas in the construction sector are the demountability and reuse since they allow the potential for cost reduction in extended life cycles, for value retention of the structural members, a smaller environmental impact and consequently enhancing sustainable construction. Current developments of the steel–concrete composite structures underline their advantages with respect to their speed of construction, the decrease of the self-weight of the structure, and achieve larger floor span when compared with other types of structural systems (e.g. reinforced concrete) [1]. The demountability of steel–concrete composite systems addresses the need for composite interaction while in parallel enabling the non-destructive separation of the steel beam and the concrete floor as shown in Fig. 4.1. This separation can be achieved with the aid of demountable bolted connectors which have shown that they can provide the necessary shear connection that the traditional construction method with the welded shear studs can also achieve. Some types of demountable connectors are presented below.

Several connectors have been investigated (see Fig. 4.2) in the literature and include friction-based shear connectors [5, 6], bolted-headed studs with embedded nuts [7, 8], and embedded coupler shear connectors [9–11]. The former type of demountable shear connectors are bolts filled with a two-component epoxy resin from their head and has been researched in feasibility and large-scale application levels [9, 10]. These injection bolts are used in shear connections where slip should be avoided, as an alternative for fitted [12] or pretension bolts. In this paper, this technical practice will be discussed in Sect. 4.3, on component and structural system levels through research that has been performed at the Delft University of Technology and is part of the RFCS project "REDUCE" [22].

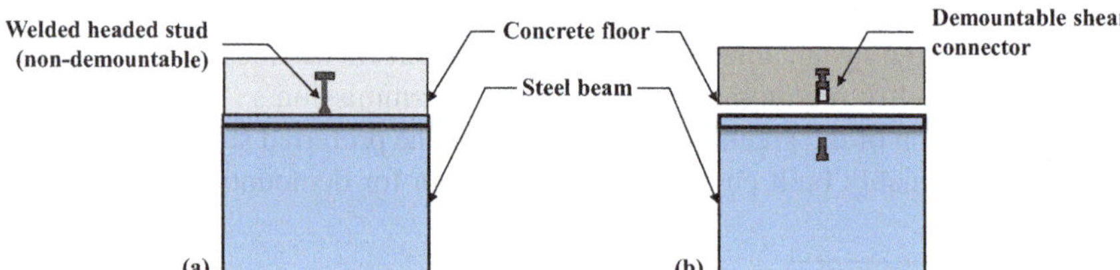

Fig. 4.1 The technical solution of the replacement of the **a** welded-headed studs with **b** demountable shear connectors in order to enable the reuse of the steel–concrete composite floors [13]

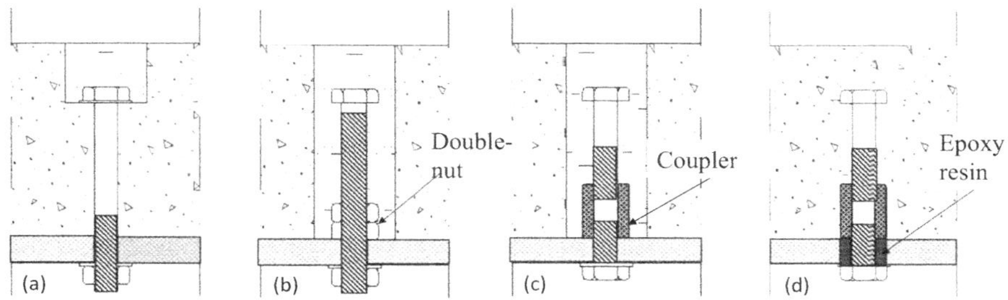

Fig. 4.2 Types of demountable bolted connectors **a** friction grip bolts, **b** double-nut, **c** coupler system connectors, **d** injected bolted shear connectors with a coupler system

4.2 Design Strategies for the Re-Use of Steel Structures

The approach for the reusability of reclaimed steel structural components is discussed in detail within the RFCS project Progress [14] and SCI [15] documents. They mainly focus on the design for deconstruction of one-story steel buildings where the in-situ and relocated reuse scenarios are examined thoroughly. According to the described processes, the procedure for the reuse of reclaimed steel structural elements is described in Fig. 4.3. The first phase includes the pre-deconstruction audit and assessment for reuse, which include a preliminary overall visual inspection (e.g. identifying problems such as excessive corrosion, excessive/plastic deformation of the structural elements, etc.) and quantitative/empirical evaluation and reporting (e.g. dimensions of joints and their connectors, etc.) of the existing steelwork. The second phase includes the sampling and testing techniques that quantify the material properties and verify the structure. The testing programs include a range of destructive (DT) (e.g. tensile testing, chemical composition analysis, charpy impact test) and non-destructive (NDT) tests (e.g. hardness tests, positive metal identification, and small punch testing). The third phase covers the design for reuse and the design considerations for achieving a reliable structure with reclaimed steel elements and structural analysis principles according to EN 1993 provisions.

4.2.1 Re-Used Steel Structural Elements in Practice

Real-life examples that address nowadays the reusability of buildings like "Donorskelet" [16], bridges like the "Nationale Bruggenbank" [17], or platforms that create digital databases on reclaimed steel structural elements like the "Circulair Bouwen in 2023" [18] are supported in the Netherlands. Also, several real-life case studies mainly in industrial one-story buildings are presented in detail in the RFCS project Progress [14] and illustrate the use of reclaimed steel structures in various EU countries. Through these case studies, the technical viability of the reuse of steel reclaimed elements is confirmed. The significant barriers across the supply chain have been identified as the additional time/program and cost of using reclaimed steel. However, several advantages in terms of environmental and economic benefits

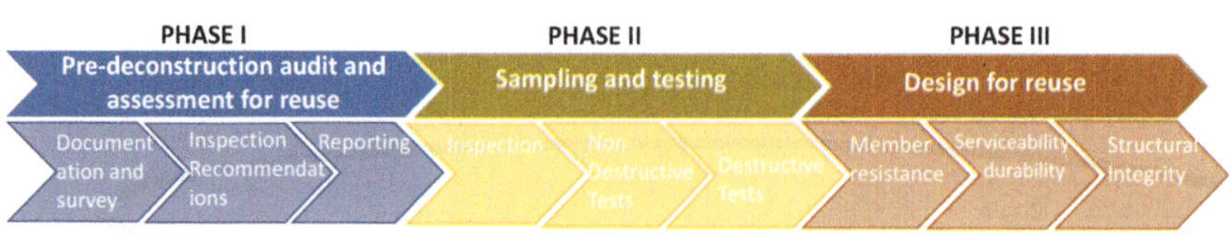

Fig. 4.3 Methodology of reusing a structural component into a new structure based on the circular building environment according to the RFCS project Progress [14]

have been underlined by the use of reclaimed steel structural elements [1, 15, 17]. Overall, as future goal is then that the required information about the reclaimed steel elements will be included in a material passport, which will be presented in digital databases for easy access [19].

4.3 Technical Solutions for Reusable Steel–Concrete Composite Structural Systems

The feasibility and the performance of demountable and reusable structural systems have been studied at Delft University of Technology through experiments on prefabricated composite floor systems which were designed to replicate a typical car park building [9] and experiments on cast-in-situ composite slabs [10]. The physical assemblies in the laboratory environment can be seen in Fig. 4.4. The demountable connectors used in both studies are the injected bolted shear connectors with a coupler system presented in Fig. 2d. The embedded and injection bolts were M20 grade 8.8 while the coupler was M20 and grade 10.9 and the injectant epoxy was the RenGel SW 404/2404. The load-bearing resistance of these injected bolts is substantially higher than its uniaxial compression strength due to the natural confinement provided by a bolt hole [20] caused by the injected epoxy resin acting as a load-bearing element. The benefit of the injected bolted shear connectors is that they allow for oversized holes which can account for the fabrication and execution tolerances and consequently improve the execution efficiency.

In the first phase of the experimental program (Fig. 4.4), a feasibility study was performed during the execution process and it was evaluated that Φ32 mm bolt holes would be appropriate to capture the deviations resulting from geometrical and dimensional imperfections. This led to a hole clearance of 12 mm which is much larger than the typical hole clearance of 1–3 mm for M20 bolts according to the

Fig. 4.4 Overview of experimental set-up [9, 13]

EN1993 [21] provisions. However, the injection of the epoxy resin in the bolt-to-hole clearance prevented the initial slip of the external bolt [20]. Six four-point bending experiments were performed with different shear connector arrangements. These different shear connector arrangements concluded the fact that concentrating the shear connectors near the supports reduced the deflection of the composite beam by 6% and thus led to increased composite interaction [20]. Regarding the inelastic tests, the experiment was terminated when the nominal resistance of the hydraulic actuators was reached (F = 550 kN) but up to that point the failure mechanism was yielding of the steel beam and plastic deformation of the shear connectors with no concrete damage.

The same injected shear connector was tested in a cast in-situ steel–concrete composite slab with profiled sheeting [10]. The composite slab in this case was tested in two life cycles under total working loads up to 200kN in a four-point bending set-up. During the first cycle, different arrangements of the shear connectors were tested within the serviceability limits. After the first cycle of experiments, the composite slab was cut through the timber joists which provided the cut edge of the slab. After the first cycle of testing the slab was demounted, reassembled, and tested again in a second life cycle. It was reported that the effective bending stiffness of the beam in the elastic area was decreased in the second cycle by an average of 10.5% [10]. This observation was attributed to the decrease in the bending stiffness of the steel section, of the concrete slab, and the initial stiffness of the shear connector. Regarding the bending stiffness of the slab, the possible decrease was inferred from the change in the longitudinal stress distribution across the slab width after cutting the composite slab. With respect to the decrease of the initial stiffness of the shear connectors, it was assumed due to possible damage which could have occurred during re-assembly.

4.4 Conclusions

This paper presents an overview of the most recent developments regarding the technical strategies and design recommendations for the reuse of reclaimed steel elements and steel–concrete composite structural components. It also discusses the types and properties of connections between these components which allow deconstruction and reuse. The main conclusions are summarized below:

- The idea of reusing composite structural elements is highly supported by guidelines and recommendations addressing the reusability of buildings, even if it is still at the initial stage of application in practice.
- The three main phases for the reuse of a structural component according to European guidelines [14] include (a) the pre-deconstruction audit and assessment for reuse, (b) sampling (in terms of grouping) and certification of structural components, and (c) design for reuse.

- Demountable connectors are the essential component that contributes to the demountability of a steel–concrete composite structural system. The injected bolted shear demountable connectors allow for oversized holes which can account for the fabrication and execution tolerances.
- The feasibility of assembly and disassembly of demountable prefabricated steel–concrete composite floor systems with the injected bolted shear connectors was confirmed through large-scale experiments.
- Beam tests with cast in-situ steel–concrete composite slab with profiled sheeting demonstrated a small decrease in the effective bending stiffness of the composite beam in the elastic area when tested in a second life cycle. It was also demonstrated that the system can be relatively easily demounted after the first life cycle and placed back to its original location for a second life cycle in laboratory conditions. However, more experiments will contribute to a better understanding of the second life cycle performance of the composite slabs.

References

1. Brambilla G, Lavagna M, Vasdravellis G, Castiglioni CA (2019) Environmental benefits arising from demountable steel-concrete composite floor systems in buildings. Resour Conserv Recycl 141(1):133–142. https://doi.org/10.1016/j.resconrec.2018.10.014
2. Organisation for Economic Co-operation and Development (2019) Global material resources outlook to 2060: economic drivers and environmental consequences. OECD Publishing
3. Cramer J (2015) Circulaire economie: van visie naar realisatie. Utrecht Sustainability Institute
4. Danatzko J, Sezen H (2011) Sustainable structural design methodologies. Pract Period Struct Design Constr 16(4):186
5. Liu X, Bradford MA, Lee MSS (2015) Behavior of high-strength friction-grip bolted shear connectors in sustainable composite beams. J Struct Eng 141(6):04014149
6. Ataei A, Bradford MA, Liu X (2016) Experimental study of composite beams having a precast geopolymer concrete slab and deconstructable bolted shear connectors. Eng Struct 114(1):1–13
7. Kwon G, Engelhardt MD, Klingner RE (2010) Behavior of postinstalled shear connectors under static and fatigue loading. J Constr Steel Res 66(4):532–541
8. Pavlovic M, Markovic Z, Veljkovic M, Budevac D (2013) Bolted shear connectors versus headed studs behaviour in push-out tests. J Constr Steel Res 88:134
9. Nijgh M.P, Gîrbacea I.A., and Veljkovic M.: Optimization of a composite (steel-concrete) floor system for fast execution and easy demolition, Proceedings of the Scientific Conference on Planning, Design, Construction and Building Renewal (iNDiS 2018), 45, (2018).
10. Gritsenko A (2018) Towards a demountable composite slab floor system. Master's thesis, Delft University of Technology
11. Odenbreit C, Kozma A (2019) Dismountable flooring systems for multiple use. In: IOP conference series: earth and environmental science, vol 225. IOP Publishing, p 012028
12. ECCS (1994) European recommendations for bolted connections with injection bolts, vol 79. ECCS
13. Kavoura F, Veljkovic M (2022) Technical practices of re-usable steel-concrete composite structural systems. In: 3rd coordinating engineering for sustainability and resilience CESARE 2022. Jordan University of Science and Technology
14. PROGRESS: Provisions for Greater Reuse of Steel Structures (2020) Grant agreement no: 747847, Research fund for coal and steel. European Commission

15. Brown D, Pimentel R, Sansom M (2019) Structural steel reuse. Steel Construction Institute SCI
16. Terwel MRC, Korthagen RC (2021) Voorbij de pioniersfase. Bouwen met Staal 54
17. Nationale Bruggenbank (2023) Nationale bruggenbank. https://www.nationalebruggenbank.nl/. Accessed 10 Feb 2023
18. Platform CB'23 (2018) Over platform cb'23. https://platformcb23.nl. Accessed 15 Feb 2023
19. van Lookeren Campagne F (2022) Efficiently including reclaimed steel elements in a truss bridge design: by performing a stock-constrained shape and topology optimization. MSc thesis, Delft University of Technology
20. Nijgh MP (2021) A multi-scale approach towards reusable steel-concrete composite floor systems. PhD thesis, Delft University of Technology
21. EN 1993-1-8 (2005) Eurocode 3: design of steel structures—part 1–8: design of joints
22. Research Fund for Coal and Steel (RFCS): Reuse and demountability using steel structures and the circular economy. REDUCE, project number RFCS-02-2015

Chapter 5
How Circular Economy Strategies Can Be Implemented in the Dwelling Renovation Design Phase

Ísis Figueirôa, Maria do Carmo Duarte Freitas, Sergio Fernando Tavares, and Luís Bragança

Abstract The world's largest consumer of raw materials is the construction sector, which mostly adopts the linear economy model. Several researchers make an effort to study how to realize a transition in the sector to a circular model of environmental development, applying strategies to preserve the raw resources, maintain materials in use as long as they can be, and reuse and recycle the building components. For increased circular efficiency of the construction, it is ideal to adopt strategies still in the design phase, however, most city buildings weren't built taking this into account and have a low possibility of adaptation of spaces and disassembly of their materials and components. Dwelling renovation has a crucial role in this scenario, and this research aims to figure out how these strategies can be applied to renovation projects since there is a lack of information on how to do this, as the studies are addressed more to new buildings. Through a systematic literature review, using relevant terms, this document identified the principal's circular economy strategies for the design phase, the dwelling spatial configuration models and the dissatisfaction of its inhabitants, and some design possibilities that can be applied to renovation projects. These findings contribute to the development of documents focused on architectural design practice.

Keywords Circular economy strategies · Dwelling renovation · Design

---

Í. Figueirôa (✉) · L. Bragança
University of Minho, 4800-058 Guimarães, Portugal

L. Bragança
e-mail: braganca@civil.uminho.pt

M. do Carmo Duarte Freitas · S. F. Tavares
Federal University of Parana, Curitiba 80230-901, Brazil

L. Bragança et al. (eds.), *Creating a Roadmap Towards Circularity in the Built Environment*, Springer Tracts in Civil Engineering,
https://doi.org/10.1007/978-3-031-45980-1_5

5.1 Introduction

The construction sector adopts practices and application of constructive techniques that make excessive and ineffective use of natural resources, typical of the linear economy model. The sector is the largest consumer of raw materials in the world [1] and one of the major emitters of greenhouse gases (GHG) [2].

There is an urgent need to revise the way the built environment is produced since the linear model has been shown to be inefficient. As an alternative to reduce the impacts, a model that preserves the raw resources, maintains the materials in use as long as they can be, and seeks to reuse and recycle the components of a building is more in line with sustainable development, and the circular economy (CE) emerges as a possibility for this pathway to the construction sector.

The concept of urban mining asserts that cities contain the essential materials for your own renovation, and some authors consider buildings to be material banks. In this way, adaptations of existing buildings can reduce waste, preserve natural resources, and extend the life cycle of buildings [3]. Cities have a massive stock of existing buildings that do not match increasing changes in market demand [4]. European Union data show that 85% of the existing building stock (more than 220 million) was constructed before 2001 and the majority (between 85 and 95%) of the buildings standing today will remain in 2050 [5].

Most cities are composed of residential buildings, and they store most of the construction materials [6]. A dwelling is estimated to have an average lifespan of around 100 years [7]. During the housing's life cycle, in the occupation and maintenance phase, modifications of various natures are requested. This is due to the dwelling meeting the basic needs of domestic activities according to the way of living of a person or household, which is dependent on the structure of values of a society. Therefore, as this structure changes, the needs and requirements for housing also change, generating the need for rehabilitation of its spaces [8].

A research verified, quantitatively, the construction and demolition waste (CDW) of three housing projects, considering the different stages of the building's life cycle, and concluded that the amount of waste at the end of life is 40% greater than the materials used in construction, due to the replacement of materials and components during the building's life [9]. These data show the importance of studying how the circular construction model can be applied to existing housing.

It is in the design phase that the main choices are made, and it includes circular decisions. This makes it possible to foresee some future scenarios for the buildings [10] and minimize or eliminate the need for using new materials [11]. Housing buildings are more susceptible than others to variables that require changes in spatial structure. Circular renovation can extend its useful lifespan, besides benefiting the surrounding area, such as economic and social development [7].

Some circular economy research focuses on design strategies, aiming to give concept professionals support in project decisions [12]. However, as the ideal is to apply the strategies in the early stages of designing a new building [12], it is not clear which parameters of the strategies apply when it comes to renovation projects.

This article aims to analyze how strategies can be applied in dwelling renovation projects. To achieve this purpose, the objective of this research is exploratory, for a better understanding of the subject, and carried out a bibliographic review, searching for publications about circular economy design strategies. In addition, the study uses a qualitative approach to provide context about the theme [13]. As a result, this study is expected to provide analysis that crosses data on the possibilities and characteristics of dwellings' renovation and what is suggested by design strategies of ecodesign methodologies. This can guide future research directions, as long as it will contribute to implementing CE for enhancing sustainability in refurbishing existing buildings.

The paper is organized into 3 parts. The first is the introduction, the second explains the method adopted for a systematic literature review, and the last one discusses and reports the research results gathering the circular design strategies found in the literature studied, the characteristics of the dwellings, and circular design possibilities that can be applied in renovation projects.

5.2 Literature Review Methodology

To a comprehensive and critical overview of academic studies on CE strategies applied to renovation design, this study developed a systematic literature review. It leads to a better understanding of the subject and identifies research gaps and potential directions.

Aiming to answer the article's question titled 'how circular economy strategies can be implemented in the dwelling renovation design phase', the literature review adopted a search (Fig. 5.1) with 6 steps: search terms, excluding duplicate articles, title analysis, abstract analysis, full-text analysis, and further research of relevant articles.

Using the Scopus database, a search was made for articles published in the last 5 years (2018–2022), using relevant theme terms in the title, abstract, and keywords. The terms 'circular economy', 'circular design', 'design for', and 'housing design' were searched as phrases.

The terms 'refurbishment' and 'renovation' were chosen based on Shahi et al. [3] research which concluded that 'renovation' is defined as 'the process of replacing or repairing outdated components or remodeling the interior spatial layout of existing buildings', and this term is a subcategory of 'refurbishment', a wider definition that is 'the process of improving the existing conditions of buildings and making improvements for the existing use'.

The 21 articles of the systematic literature review can be accessed through this link.

Search terms	AND building AND refurbishment	AND building AND renovation	AND building AND existing
"circular economy"	24	49	150
"circular design" AND strategies	1	2	2
"design for"	7	29	287
"ecodesign"	8	3	41

Search terms	AND flexib*	AND adaptab*	AND disassemb*	AND deconst*
"housing design"	19	15	0	1

Total	638 articles
Duplicates - 102 articles excluded	536 articles
Title analysis - 490 articles excluded	46 articles
Abstract analysis - 26 articles excluded	20 articles
Full-text analysis - 8 articles excluded	12 articles
Further research - 9 articles included	TOTAL 21 articles

Fig. 5.1 Literature review process (review date: February 2023)

5.3 Results and Discussion

5.3.1 Circular Economy Strategies

To achieve circularity in building projects, the literature already provided numerous circular economy strategies, also known as ecodesign methodologies, besides tools, and frameworks to apply the strategies in the design phase and support decision-making [14]. The strategies differ in circularity objective and can focus for example on the circularity of the materials, aiming at the reuse and recycling of materials, the adaptability of the buildings, analyzing their capacity for spatial modification, or the components disassembly, avoiding the common end-of-life demolition solution.

Some researchers studied different ecodesign terminologies often used, related to the design stage, and found many terms employed to address the changing needs of users and external factors all over the building life cycle. There are several similarities in terms' names and objectives, and this can be explained by the lack of standardization and interpretation of the terms, as reported by Munaro et al. [15], whose research presents the most distinct ecodesign terms. Table 5.1 shows the integration done in this research.

The terms can be summed up in two principal methods, Design for Adaptability and Design for Disassembly, to facilitate understanding and because they are considered in some studies as the main strategies to be adopted in design decisions [14, 16]. Both main methods encompass all the other strategies' objectives.

It is essential to understand how these strategies are presented in practice to identify which strategies can be applied during the renovation design phase. This

Table 5.1 Integration of CE strategies terms. Authors, adopted from [15]

Design for adaptability	Design for disassembly
Design for flexibility	Design for deconstruction
Design for durability	Design for dismantling
Design for change	Design for demountability
	Design out waste
	Design for recycling
	Design for reuse

can be seen in Figs. 5.3 and 5.4, on the next topic, where some design solutions were identified for adaptability or disassembly purposes.

Design for Adaptability (DfA) aims to apply in buildings, design characteristics of flexibility, reconfiguration, or change of function, structure, space, components, systems, services, and size as a response to accommodate change throughout time and minimize demolition risk [12, 15].

Design for Disassembly (DfD) seeks to make a reversible building, instead of demolishing it, by planning in design, the possibility of disassembly of the elements, and the reuse or recycling of its parts [12].

The next section reports on the problem of the rigid layout configuration of the dwellings, the desires for modifications on the part of the inhabitants, and the possible design options for a renovation project identified in the studies.

5.3.2 Dwellings and the Renovation Projects

Dwelling spatial configuration models and their consequences. In the literature review, some studies examined dwelling characteristics and the relation between them and the market and its inhabitants. The configuration of housing spaces has a logic depending on the period and local culture. This can be better understood using for example, the space syntax method, developed by Hillier and Hanson [17], which evaluates spatial configurations and, when applied to dwellings, assesses meaningful information about architectural spaces, their relationship, and the connectivity between them. In general, the construction market follows a spatial configuration model and builds dwellings without flexibility possibility. Therefore, the inhabitants show dissatisfaction with the model imposed on them [18]. Griz et al. [8, 19] report incompatibility between market offers and inhabitants' demands, causing modifications/customizations to be performed on the first resident as seen in Fig. 5.2. Gilani and Türker [20], in a survey with inhabitants, found that 64% of dwellers want to change space organizations, 62% the functions of interior spaces for functional flexibility and 54% want to change wall arrangements for privacy needs. Ollár et al. [18] report that the most frequent motivations for a renovation were dissatisfaction with the dwelling layout, a lack of workspace, insufficient floor area, obsolete furniture,

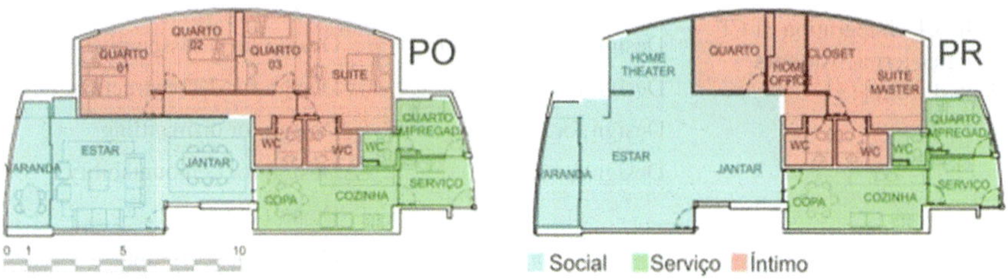

Fig. 5.2 Example of the original project and the customization [19]

Tarpio et al., 2022

- Fix wet spaces in a specific zone, others freely.

Rabeneck et al., 1974

- Use of moveable partitions.
- Moveable equipment and furniture.
- Propose lighting not in ceilings.
- Central heating/cooling.
- Fixed wet wall or stack to a wide choice of kitchen and bathroom layouts and relationship to adjacent spaces.
- Modular furniture.

Gilani & Türker, 2020

- Flexible arrangement of furniture.
- Movable division walls.

Fig. 5.3 Adaptability strategies possibilities for renovation projects

or appliances, improve the appearance of the kitchen, and an increase or decrease in household size.

Le Corbusier in 1914, with the Domino system, showed to architects the possibility to separate the interior from the structure, with the 'free floor plan' strategy to achieve space flexibility. The De Stijl movement's manifesto (1924) and Mies van der Rohe's ideas (1926) also talk about adaptability in spaces and defend movable and non-supporting walls [21]. But many housings built before and after these modern concepts were designed to have a specific function, interior dimensions, and space organization, with rigid structures and non-movable dividing walls, and considered low possibilities of adapting spaces to contemporary and future requirements of environments [16, 20].

This raises the question of the role of a renovation in minimizing the impact of future modifications by different residents who will still reside in the place.

Design possibilities in renovation projects. The literature review revealed some circular design strategies for new buildings that are also applicable to renovation design, to enhance circularity, more specifically, properties of adaptability, flexibility, disassembly, and reversibility in the modified housing area.

These possibilities were recognized based on modifications characteristic of a renovation in a dwelling, such as changes in layout, which lead to demolition and

Rabeneck et al., 1974

- Bathtub not fixed in walls.

Dams et al., 2021

- Use of steel frames gives the possibility to dismantle and reuse.
- Use of reusable material, previously used in the design life of a previous project.

Gilani & Türker, 2020

- Non-fixed finishing materials.

van Stijn et al., 2022

- Interior partitioning modular, demountable, and reusable wall.
- Kitchen with modular design to enable partial replacements, repair and prolonged use.

Wouterszoon Jansen et al., 2022

- Modular kitchen.

O'Grady et al., 2021

- Wall frames connected using screws enable disassembly.
- Internal cladding made of plywood sheets screwed to the steel frames.
- Use of connections such as screws, magnetic, and weld.
- Ceiling made of panels.

Fig. 5.4 Disassembly strategies possibilities for renovation projects [23–26]

construction of walls, kitchen and bathroom counter changes, furniture exchange, and a new heating/cooling system, among other services. Ollár et al. [18] report that in previous research from other authors, it is said that 'the kitchen is one of the functions of the home, which is most often subject to renovations and adaptations', and Stijn et al. [22] reports that domestic furniture and appliances account for 35% of building environmental impacts.

This section answers the article's question of 'how circular economy strategies can be implemented in the dwelling renovation design phase'. Figures 5.3 and 5.4 below show the circular design options for renovation projects, the reference where it was reported, and the relation to the specific principal strategy the design possibility belongs to.

As seen in the information presented above, there were just a few design options identified in studies that can also be applied to renovation projects, and many are common, such as modular kitchens and dry connections. It was noticed that minor-scale component modifications such as coatings, kitchen and bathroom pieces, lighting, and piping have minimal or no direction on how to be done, especially in renovation projects. Therefore, considering the literature reviewed, design decisions about other housing elements that are not addressed in the studies, are subject to the designer's interpretation through the broad concepts of strategies.

Also, for a more sustainable renovation proposal, it is imperative to analyze the quality and impact of the materials used and the economic viability of choices through a life-cycle assessment analysis [12].

Several researchers report the importance of material choice, focusing on reused materials, bio-based materials, and recyclability possibilities [12]. For 'Design for

Disassembly', it is also important that the material has a Material Passport to guide the path through reuse [11].

5.4 Conclusion

Through the literature review developed, and the definition of renovation given in Shahi et al. [3] research, this paper recognized design possibilities to be applied to renovation projects related to the principals' circular strategies studied. However, there were difficulties in gathering this information, as it was verified a lack of definitions regarding the practical application of circular strategies by design professionals, and, in general, those possibilities were described as an example in studies texts and tables.

This shows an opportunity for future research to increase the discussion that focuses on developing documents capable of guiding the application of circular economy strategies in projects, both for new buildings and for renovation. The development of research that addresses the investigation of executed architectural renovation projects that aim to apply circular characteristics, and the analysis of assessments and tools, such as Level(s) from the European Union, and certification guides, like the ones from the Green Building Council, are important in the definition of strategies for renovation projects.

It is essential to broaden this discussion and focus on its applicability, so that there is greater harmony between the market and inhabitants, offering greater possibilities for spatial, aesthetic, and functional modifications from day one of the housing, without major environmental damage.

References

1. Nairobi (2021) United Nations environment programme, global status report for buildings and construction: towards a zero-emissions, efficient and resilient buildings and construction sector. http://www.globalabc.org
2. Wang K et al (2022) Circular economy as a climate strategy: current knowledge and calls-to-action. Washington
3. Shahi S, Esnaashary Esfahani M, Bachmann C, Haas C (2020) A definition framework for building adaptation projects. Sustain Cities Soc 63. https://doi.org/10.1016/j.scs.2020.102345
4. Durmisevic E et al (2019) Circular economy in construction design strategies for reversible buildings
5. Bruxelas (2020) Comissão Europeia, Impulsionar uma Vaga de Renovação na Europa para tornar os edifícios mais ecológicos, criar emprego e melhorar as condições de vida
6. Soonsawad N, Martinez RM, Schandl H (2022) Material demand, and environmental and climate implications of Australia's building stock: current status and outlook to 2060. Resour Conserv Recycl 180:106143. https://doi.org/10.1016/J.RESCONREC.2021.106143
7. Cellucci C (2021) Circular economy strategies for adaptive reuse of residential building. Vitruvio 6(1):110–121. https://doi.org/10.4995/vitruvio-ijats.2021.15404

8. Griz C, Mendes L, Guedes V, Amorim L (2015) Reformar ou não reformar? Análise da influência da forma na customização de projetos de apartamentos, pp 677–685
9. Marrero M, Rivero-Camacho C, Alba-Rodríguez MD (2020) What are we discarding during the life cycle of a building? Case studies of social housing in Andalusia, Spain. Waste Manag 102:391–403. https://doi.org/10.1016/j.wasman.2019.11.002
10. Munaro MR (2022) The circular economy in the construction sector: existing trends, challenges, and tools towards buildings as material banks, Curitiba. https://www.prppg.ufpr.br/siga/visitante/autenticacaoassinaturas.jsp
11. Antwi-Afari P, Ng ST, Chen J (2022) Developing an integrative method and design guidelines for achieving systemic circularity in the construction industry. J Clean Prod 354. https://doi.org/10.1016/j.jclepro.2022.131752
12. Dams B et al (2021) A circular construction evaluation framework to promote designing for disassembly and adaptability. J Clean Prod 316. https://doi.org/10.1016/j.jclepro.2021.128122
13. Engel Gerhardt T, Tolfo Silveira D, Métodos de Pesquisa, Porto Alegre (2009)
14. Askar R, Bragança L, Gervásio H (2022) Design for Adaptability (DfA)—frameworks and assessment models for enhanced circularity in buildings. Appl Syst Innov 5(1). https://doi.org/10.3390/asi5010024
15. Munaro MR, Tavares SF, Bragança L (2022) The ecodesign methodologies to achieve buildings' deconstruction: a review and framework. Sustain Prod Consump Elsevier B.V. 30:566–583. https://doi.org/10.1016/j.spc.2021.12.032
16. Askar R, Bragança L, Gervásio H (2021) Adaptability of buildings: a critical review on the concept evolution. Appl Sci (Switz) 11(10). MDPI AG. https://doi.org/10.3390/app11104483
17. Hillier B, Hanson J (1984) The social logic of space. Cambridge University Press. https://doi.org/10.1017/CBO9780511597237
18. Ollár A, Granath K, Femenías P, Rahe U (2022) Is there a need for new kitchen design? Assessing the adaptative capacity of space to enable circularity in multiresidential buildings. Front Arch Res 11(5):891–916. https://doi.org/10.1016/j.foar.2022.03.009
19. Griz C, Amorim L, Loureiro C (2014) Entre a oferta e a demanda: a elite em busca do morar bem, in XV Encontro Nacional de Tecnologia do Ambiente Construído, Marketing Aumentado, pp 1863–1872. https://doi.org/10.17012/entac2014.107
20. Gilani G, Türker ÖO (2020) Assessing flexibility in real estate mass housing. Arquiteturarevista 16(1):154–175. https://doi.org/10.4013/arq.2020.161.09
21. Estaji H (2017) A review of flexibility and adaptability in housing design. Int J Contemp Arch 4(2). https://doi.org/10.14621/tna.20170204
22. van Stijn A, Eberhardt LCM, Wouterszoon Jansen B, Meijer A (2022) Environmental design guidelines for circular building components based on LCA and MFA: lessons from the circular kitchen and renovation façade. J Clean Prod 357. https://doi.org/10.1016/j.jclepro.2022.131375
23. Tarpio J, Huuhka S, Vestergaard I (2022) Barriers to implementing adaptable housing: architects' perceptions in Finland and Denmark. J Housing Built Environ 37(4):1859–1881. https://doi.org/10.1007/s10901-021-09913-1
24. Rabeneck A, Sheppard D, Town P (1974) Housing flexibility/adaptability? Arch Des 49:76–90
25. Wouterszoon Jansen B, van Stijn A, Gruis V, van Bortel G (2022) Cooking up a circular kitchen: a longitudinal study of stakeholder choices in the development of a circular building component, sustainability (Switzerland). 14(23). https://doi.org/10.3390/su142315761
26. O'Grady T, Minunno R, Chong HY, Morrison GM (2021) Design for disassembly, deconstruction and resilience: a circular economy index for the built environment. Resour Conserv Recycl 175. https://doi.org/10.1016/j.resconrec.2021.105847

Chapter 6
Design for Disassembly and Reuse of Timber in Construction: Identification of Trends and Knowledge Gaps

Rafael Novais Passarelli

Abstract The transition to a circular economy (CE) offers an alternative path to the current linear, high-polluting, and wasteful practices in construction. In this context, there is a growing interest in studying the potential environmental benefits of extending the lifespan of renewably-sourced wood-based building products through reuse. However, most publications still fail to present a conceptually integrated and comprehensive view of the topic that allows for a broader understanding of its possibilities and challenges. This paper assesses two decades of literature on DfD&R of timber in construction. It develops a comprehensive state-of-the-art framework about the topic, unveiling its most critical challenges, trends, and pressing knowledge gaps. The outcomes of this work contribute to determining more integrated strategies and decision-making tools that could point to further development in the field of timber construction from a DfD&R standpoint, thus facilitating the transition to a CE.

Keywords Circular economy · Wood construction · Timber · Design for circularity · Deconstruction · Disassembly reuse · End-of-life · Literature review

6.1 Introduction

The transition to a circular economy (CE) offers an alternative path to the current linear, high-polluting, and wasteful practices in construction [1]. According to the definition from the Ellen McArthur Foundation, a CE is "restorative and regenerative by design and aims to keep products, components, and materials at their highest utility and value at all times (...)" [2]. Through the implementation of a CE, the Circle Economy (2022) estimated that up to a 31% reduction in the emissions associated with the housing industry is possible if two conditions are met: (1) increased use of circular building materials and (2) more resource-efficient construction practices. In

R. N. Passarelli (✉)
Hasselt University, 3500 Hasselt, Belgium
e-mail: rafael.novaispassarelli@uhasselt.be

L. Bragança et al. (eds.), *Creating a Roadmap Towards Circularity in the Built Environment*, Springer Tracts in Civil Engineering,
https://doi.org/10.1007/978-3-031-45980-1_6

a nutshell, the CE message is that the inner circles of a product lifecycle demand fewer resources and energy than conventional recycling of materials for low-grade raw materials or their energy recovery [3].

Hence, the hierarchy for a CE indicates that recovering materials for reuse is a priority. Likewise, the European Directive on waste [4] urges the recovery of material waste streams through reuse and recycling as essential to limiting the extraction of virgin resources and, thus, reducing the environmental impact of construction. In this context, several publications studied the potential environmental benefits of extending the lifespan of renewably-sourced wood-based building products through reuse or recycling [5–7]. A number of studies also warn of the unforeseen challenges in pursuing an increased material cascading [8–12]. However, most publications still fail to present a conceptually integrated comprehensive view on the topic that allows for a broader understanding of its state of art, trends, challenges and opportunities, and pressing knowledge gaps that could support better decision-making and facilitate the transition towards a CE.

This paper aims to answer the following research question: what are the main trends, knowledge gaps, most critical challenges, and possibilities for DfD&R with wood in the current literature? To reach its goal, this paper assesses two decades of literature on the DfD&R of timber in construction, thus developing a comprehensive framework about the state-of-the-art on the topic. The results of this paper are relevant to both scholars and practitioners working in the field as they contribute to identifying significant gaps in scientific research and practical implementation of DfD&R with timber, which could contribute to developing more integrated strategies and decision-making tools, thus facilitating the transition to a CE.

6.2 Materials and Methods

This paper develops a comprehensive and systematic literature review method, followed by quantitative data analysis and qualitative interpretation. Firstly, the author retrieved all relevant scientific publications in English on the topic from 2002 to 2022. The publications were retrieved from the Scopus and Web of Science (WoS) databases, as the first contains the most relevant and high-impact publications in the field; the second comprises a larger database of indexed publications, including high-impact conference proceedings. Hence, the combination of the databases provides a comprehensive overview of the scientific developments in the field. Search string (1) was used for the papers sampling, excluding review papers and book chapters to focus on original scientific publications. Nonetheless, the search string was intentionally broad to yield the highest number of samples possible.

(1) *Design* AND *Deconstruction* OR *Disassembly* OR *Reuse* AND *Timber* OR *Wood*

In total, 843 publications were collected and checked for duplicates, resulting in an initial selection (IS) of 513 items. Then, the author refined the IS further through a manual screening of titles and abstracts to check for compliance with the main topic of DfD&R of timber in construction. The compliance check consisted of four sequential closed-ended questions (Yes/No). If the answer to any question was No, the publication was excluded from the final selection (FS). Conversely, all the publications included in the FS answered Yes to all four questions. After the content compliance check, the FS consisted of 83 publications. The four questions with their respective intents and number of publications filtered are displayed in Table 6.1.

Next, the author assessed the metadata of the publications in the FS quantitatively. The values analyzed were the number of publications, type (journal or conference paper), year published, the total number of citations, first-author country of affiliation, and keywords. The analyses aimed to identify general trends related to the publications on the topic of DfD&R of timber in construction in the last two decades. Then, the author developed a qualitative content analysis of the FS based on the abstracts and full-text reading when needed to categorize the publications into distinct primary and secondary content groups and approach types. The qualitative content assessment aimed to expose the main subjects already covered by the literature, the density of

Table 6.1 Compliance check questions, intents and number of filtered publications

Questions		Intent	513 (IS)
1	Is the publication about the DfD&R of timber in construction?	To filter publications that, despite containing all search terms, had no connection to the topic and were, for example, coincidental hits when one of the authors had the word wood as part of his/her surname	−274
2	Is the role of wood/timber construction significant in the publication?	To filter publications that, despite mentioning the words wood or timber, were focused on other materials. The role of wood/timber was considered significant when at least one fiber-based case study at the component or building level was present in the publication	−68
3	Is the publication about wood/timber as a building material?	To filter publications that investigated wood/timber at the fiber or molecule level or for uses such as formworks, scaffolding, interiors, and furniture	−51
4	Is the publication about the design, construction, or end-of-life phases?	To filter publications that did not deal with the DfD&R of timber in the phases mentioned above and focused on the laboratory testing or verification of assemblies and solutions pre-defined elsewhere	−37
			83 (FS)

knowledge in each category or subcategory, and its most relevant existing knowledge gaps.

The DfD&R approaches identified were downstream and upstream. According to Piccardo and Hughes [12], downstream approaches relate to activities occurring after the end of life concerning the salvage process of wood products from buildings being disassembled or demolished and their posterior reuse in new buildings, whereas upstream refers to strategies developed in the early stages of design to facilitate the future reuse of wood products at their end-of-life. The three primary content categories definitions were based on the working groups' division of the ongoing COST Action 21103 (Implementation of Circular Economy in the Built Environment). The 11 secondary content categories were defined iteratively during the in-depth reading of the abstracts of the publications in the FS. The content categories and their respective subcategories are displayed in Table 6.2.

In a subsequent study, all publications will be fully read. This last in-depth full-paper analysis aims to identify and theorize the main challenges, opportunities, insights, and pressing tasks yet to tackle related to the DfD&R of timber in construction. That will lead to a comprehensive framework for understanding the current state of the art in the topic, thus contributing to more integrated strategies and decision-making tools that could point to its further development.

6.3 Results and Discussion

6.3.1 Metadata Analysis

Figure 6.1 reveals that the first publication retrieved dates from 2006, despite searching for studies since 2002. Moreover, the number of publications remained low for ten years, averaging 1.8 per year. Next, the figure shows a short transition period from 2016 to 2018, with an increase in interest in the topic and an average of five publications per year. Then, we notice a sharp growth in the number of publications in the last four years, with an annual average of 12.5. That attests to the remarkably recent and yet booming interest in the subject. Also, the share of journal publications (J) outgrew the number of conference papers (C) in recent years. Assuming that journal publications frequently require more accurate data and in-depth studies, they indicate the ripening of the topic as a sound scientific field. The increasing number of citations in the same period also supports this argument.

Figure 6.2 shows that publications in the FS originated from 27 different countries, 19 in the European continent, 4 in Asia, 2 in North America, and 2 in Oceania. However, the share of publications per country is unequal, with more than half (44) of all publications coming from only one-fifth (6) of the countries. This unbalanced distribution indicates that the knowledge development on the topic is still highly concentrated. Specifically, Germany and the USA led the research on DfD&R of timber, accounting together for 18 publications (22%). It is perhaps no coincidence

Table 6.2 Definitions of content categories and subcategories

Categories		
1. Strategies and best practices: Includes publications that deal with innovative methods to apply DfD&R strategies in design and construction activities and identify best practices	**2. Stakeholder engagement**: Includes publications that analyze the value chain of DfD&R solutions from the standpoint of key stakeholders	**3. Performance indicators**: Includes publications that employ relevant, reliable, and replicable quantitative performance indicators to measure the benefits of timber DfD&R solutions in construction
Subcategories		
1a. Urban scale: Comprises publications dealing with strategies and best practices on the urban scale	**2a. Politic**: Comprises publications engaging political stakeholders, such as representatives from the public sector or legislators	**3a. Environmental assessment**: Comprises publications that developed a quantitative environmental impact assessment, such as LCA (lifecycle assessment)
1b. Building scale: Comprises publications dealing with strategies and best practices in the building scale	**2b. Economic**: Comprises publications engaging economic stakeholders, such as investors and customers	**3b. Cost assessment**: Comprises publications that developed a quantitative cost assessment, such as LCC (lifecycle costing)
1c. Component scale: Comprises publications dealing with strategies and best practices in the building component scale	**2c. Technical**: Comprises publications engaging technical stakeholders, such as designers or contractors	**3c. Circularity assessment**: Comprises publications that developed a quantitative circularity assessment, such as material input–output rate
1d. Other strategies: Comprises publications dealing with advanced strategies and best practices, such as BIM, computational design, automation processes, or robotics		**3d. Other assessments**: Comprises publications that developed other quantitative assessments, such as mechanical testing, thermal, or energy simulations

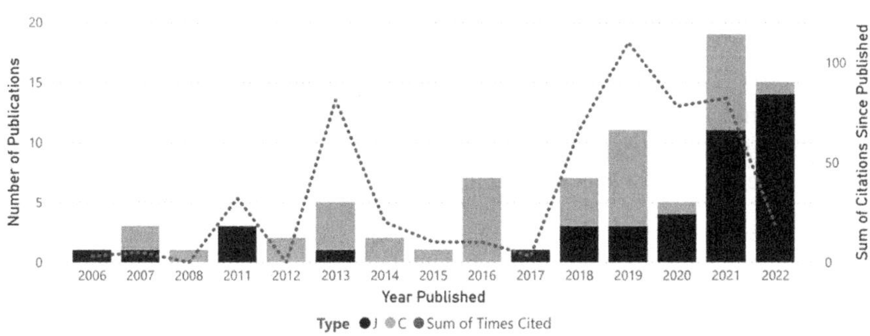

Fig. 6.1 Number of publications per year, per type and sum of times cited since published

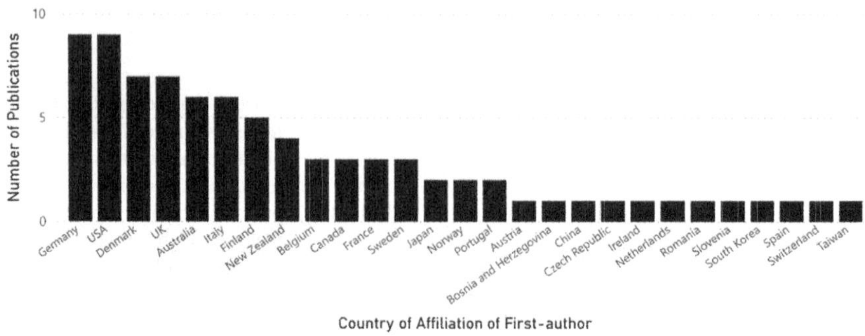

Fig. 6.2 Number of publications by country of affiliation of the first-author

that both countries hold the biggest GDP in their regions, with likely more access to funding, but also where the effects of a linear economy might be more visible. Next, we witness Denmark and UK tied in second place, with seven publications each. Australia and Italy follow closely with six publications each. It is also noteworthy that the list of origin countries in Fig. 6.2 only portrays so-called developed nations, except for one publication from China. Hence, developing countries in the global south are greatly underrepresented, despite sometimes hosting emergent timber construction industries such as Chile and Brazil. Although the scope of this study is insufficient to infer the reasons for this situation accurately, we can speculate they might include less availability of research funding, limited awareness or know-how on the topic, or even a language barrier for publishing in English.

Figure 6.3 shows the keywords assigned by the authors of the publications in the FS with an incidence higher than two. All keywords with identical meanings but different spellings were manually checked and adjusted to the same convention to ensure a fair assessment. For example, Building Information Modeling was rewritten as BIM. Not surprisingly, Circular Economy was the most popular keyword. The keyword Design for Disassembly (10) followed with a higher incidence than the keywords Reuse (6) and Material Reuse (3) altogether. This outcome suggests that researchers may be more prone to study the topic from an Upstream approach, as discussed in greater detail in item 3.2. The third most popular keyword was Life Cycle Assessment (9), which anticipates that this assessment method might be the prominent indicator to demonstrate the benefits of DfD&R quantitatively, as seen in item 3.2. It is also noteworthy that Cross Laminated Timber was the only product name to emerge on this list, hinting at the high relevance of mass timber construction in the topic.

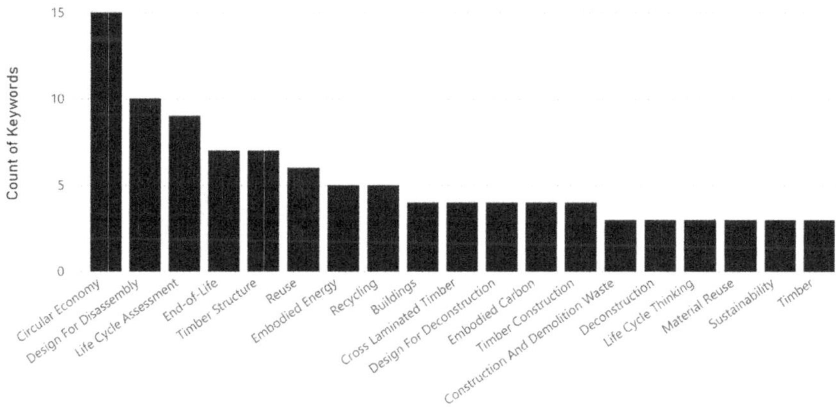

Fig. 6.3 Count of keywords assigned by the authors of the publications with incidence $n > 2$

6.3.2 Content Analysis

Overall, the content analysis revealed a balanced outcome in terms of approaches: 44 publications investigated DfD&R of timber from an upstream standpoint, 37 downstream, and three from both. It is important to note that publications in the last type analyzed both approaches as independent options. None of the publications in the FS investigated the link between upstream and downstream, thus confirming the knowledge gap previously uncovered by Piccardo and Hughes [12], mentioned in item 1.0. Nevertheless, we notice a new tendency when laying the approach distribution chronologically. Figure 6.4 shows a significant increase in the incidence of upstream-oriented publications in the latter half of the timeline. From 2006 to 2014, the share of upstream was 35%, whereas from 2015 to 2022, it almost doubled to 65%. That attests to a recent trend of considering DfD&R of wood/timber as a more integral part of the initial design task. On the other hand, that also indicates a diminishing interest in how to reuse the stock of materials already allotted to a building, thus neglecting to some extent the possibility of extending their service life span and trimming the need for extracting new resources, even if fiber-based renewed ones.

Figure 6.5 (left) shows that the primary content of publications in the FS deals predominantly with category 1 of Strategies and Best Practices (47), followed by studies on category 3 of Performance Indicators (29), whilst category 2 of Stakeholder Engagement was the last popular content (7). In category 1, publications tended to focus on subcategory C of the component scale (19), followed by the building scale subcategory B (15). Finally, the analysis found few publications that studied strategies and best practices related to the advanced topics in subcategory D, such as BIM and computational design, which might be a consequence of the relative novelty of such fields. Moreover, this study also revealed that a limited number of publications engaged with stakeholders as their main content. In category 2, subcategory C was predominant over the other subcategories, denoting a preference or convenience

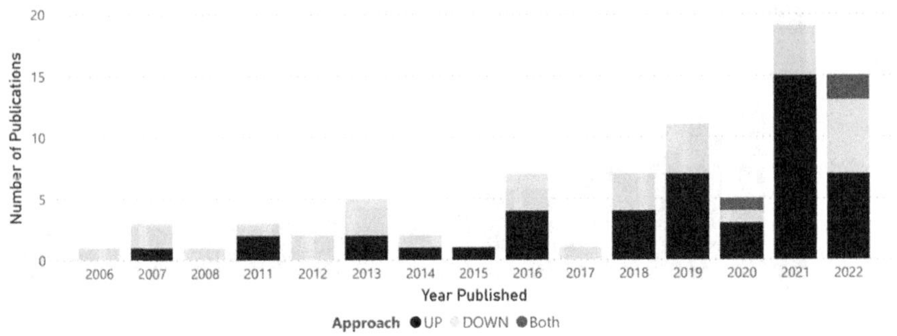

Fig. 6.4 Number of publications by year, by type of approach

for engagement with technical stakeholders. It is noteworthy that no publication engaged with economic stakeholders (subcategory B) and only one with a political stakeholder (subcategory C), which exposes a significant knowledge gap for the future implementation of these practices, for even when there are possible technical solutions, there still needs to be political will and economic feasibility. In category 3 of Performance Indicators, subcategory A of environmental assessment methods was dominant with 19 publications, often using the LCA method, characterizing it as the key indicator currently to measure the benefits of timber DfD&R. The predominant choice for LCA as an assessment method might be the consequence of a lack of alternative standard-defined methodologies to measure the circularity of buildings and components. Nonetheless, subcategory C of circularity assessment with six publications suggests that, although still limited, there is interest in experimenting with original quantitative methods to measure the circular benefits of timber DfD&R in construction.

Figure 6.5 (right) shows that the secondary content of publications in the FS is quantity-wise comparable between categories 1 (22) and 3 (18). Category 2 was once again the last popular content (3). Differently from the primary content outcome,

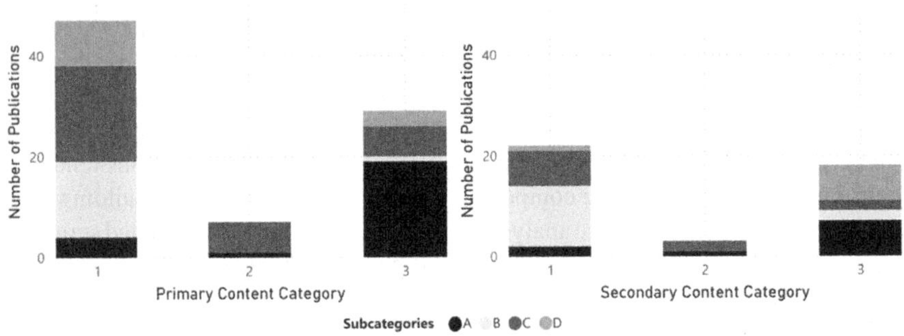

Fig. 6.5 Number of publications by content categories and subcategories. Primary publication content (left); Secondary publication content (right)

publications with secondary content in category 1 focused on subcategory B (12), followed by subcategory C (7). Furthermore, an even more limited number of publications engaged with stakeholders (category 2) as their secondary content, 2 in subcategory C and 1 in A. Once again, no publications engaged with economic stakeholders (subcategory B). Finally, the publications that introduced category 3 as secondary content were evenly distributed between subcategories A and C (7 each). Overall, the secondary content category of the publications in the FS maintained the same priorities as the primary content ones (Strategies and Best Practices and Performance Indicators) but presented a slightly more balanced concentration of the most prevalent subcategories (Building or Component Scale and Environmental or Circularity Assessment).

6.4 Conclusions

The metadata analysis showed growing interest and maturation of the topic, particularly in the last four years. Nevertheless, there is an accentuated tendency to focus on the upstream aspects of DfD&R of timber in construction. Although that is undoubtedly a critical task to produce more circular buildings in the future, there is a risk of neglecting enough attention to developing solutions to use the current stock of wood-based materials already trapped in our building stock. That would offer a counterpoint to the still linear mindset that we need to produce and build more, even if more sustainably, when, in fact, we need to do it less. Furthermore, this study confirmed a critical gap in publications linking upstream and downstream DfD&R approaches, for example, scrutinizing the true disassembly potential of allegedly demountable strategies. Finally, the metadata analysis also showed publications originated almost exclusively from so-called developed countries. Taking into account that developing countries are usually expected to experience faster growth, it is essential to ensure that it occurs with minimal impact. Furthermore, although the reasons for the lack of representation of developing countries are not evident in this study, the discussion on the decolonization of the narrative in the scientific field is imperative and long overdue. Even the simple inclusion of free-of-charge translation/revision services in big English language publishers could already contribute to giving voice to the countries absent in this narrative. Also, a more vigorous implementation of north–south co-developing research projects with specific grants for that purpose would be yet another way to empower underrepresented voices in the scientific field.

The content analysis uncovered that publications frequently seek to study innovative strategies and identify best practices for the DfD&R of timber in construction, particularly on the component scale. The choice for a smaller study scope may answer the academic need to minimize parameters to a managing point considering all the complexity in a building. Despite the fact this will surely yield in-depth results of critical individual aspects, that also indicates a significant knowledge gap related to the understanding of more holistic strategies on the building scale and an even more substantial gap on the urban scale. The second most covered content was on performance indicators to measure the benefits of DfD&R strategies. There was, however, a clear predominance of LCA as the preferred method to measure the outcomes of timber DfD&R strategies. It is highly plausible that this is the consequence of LCA being a well-known standardized method to assess the environmental impacts of products or buildings. At the time of writing, there was no established method to measure the circularity of constructions defined by a international association such as the ISO, creating an additional barrier to researchers trying to venture into systematically assessing the circularity index of given solutions. Yet, this study identified a growing number of publications implementing novel methods to quantify the benefits of DfD&R strategies as its primary or secondary content. Although still in their infancy and lacking in number, these isolated publications attest to the relevance of developing a scientific method that is fit to assess the circular benefits of timber DfD&R in construction. Hence, this gap will most likely be gradually filled, especially after the new ISO standards on circularity, currently under development, are published. Finally, this study unveiled a substantial knowledge gap related to stakeholder engagement as a publication content, with only a few studies reaching out to representatives in the technical field. It is worth highlighting that no publications in the FS investigated DfD&R of timber from the standpoint of an economic stakeholder. Therefore, new publications on the topic should actively pursue higher exchange with key decision-makers in the field, particularly the ones such as investors and customers (economic stakeholders) or legislators and government representatives (political stakeholders).

The results of this study also have some limitations that can be addressed in new studies on the topic. Firstly, as the systematic review only covered scientific publications, there may be additional know-how to investigate in technical reports and published material from companies and practitioners that were not within the scope of this paper. Secondly, although the sampling criteria were defined as objectively and clearly as possible, the final selection and classification process is, to some extent, subject to personal bias. Finally, the sampling in this study only included search terms and publications in English. Hence, further studies can employ a similar methodology to explore the state-of-the-art of DfD&R with timber developed and published in other languages.

References

1. Andrews D (2015) The circular economy, design thinking and education for sustainability. Local Econ 30(3):305–315
2. Ellen MacArthur Foundation (2015) Towards a circular economy: business rationale for an accelerated transition
3. Korhonen J, Honkasalo A, Seppälä J (2018) Circular economy: the concept and its limitations. Ecol Econ 143:37–46

4. European Parliament and Council (2018) Directive 2018/851 of the European Parliament and of the Council of 30 May 2018 amending Directive 2008/98/EC on waste (Text with EEA relevance). Luxemburg
5. Husgafvel R, Linkosalmi L, Hughes M, Kanerva J, Dahl O (2018) Forest sector circular economy development in Finland: a regional study on sustainability driven competitive advantage and an assessment of the potential for cascading recovered solid wood. J Clean Prod 181
6. Niu Y, Rasi K, Hughes M, Halme M, Fink G (2021) Prolonging life cycles of construction materials and combating climate change by cascading: the case of reusing timber in Finland. Resour Conserv Recycl 170
7. Risse M, Weber-Blaschke G, Richter K (2017) Resource efficiency of multifunctional wood cascade chains using LCA and exergy analysis, exemplified by a case study for Germany. Resour Conserv Recycl 126
8. Vis M, Mantau U, Allen B (2016) Study on the optimised cascading use of wood. No 394/PP/ENT/RCH/14/7689. Brussels
9. Höglmeier K, Weber-Blaschke G, Richter K (2017) Potentials for cascading of recovered wood from building deconstruction—a case study for south-east Germany. Resour Conserv Recycl 117:304–314
10. Ahn N, Dodoo A, Riggio M, Muszynski L, Schimleck L, Puettmann M (2022) Circular economy in mass timber construction: state-of-the-art, gaps and pressing research needs. J Build Eng 53
11. Sakaguchi D, Takano A, Hughes M (2017) The potential for cascading wood from demolished buildings: potential flows and possible applications through a case study in Finland. Int Wood Prod J 8(4)
12. Piccardo C, Hughes M (2022) Design strategies to increase the reuse of wood materials in buildings: lessons from architectural practice. J Clean Prod [Internet] 368. https://linkinghub.elsevier.com/retrieve/pii/S0959652622026737

Part III
Circular Materials and Building Products

Chapter 7
A Dynamic-Based Methodology for Optimising Insulation Retrofit to Reduce Total Carbon

D. M. Rakshit, Anthony Robinson, and Aimee Byrne

Abstract Optimisation is frequently mentioned in frameworks and assessments of design for a circular economy. Adopting circular economy principles in building retrofit can reduce the use of materials and minimise emissions embedded in building materials alongside reduced operational emissions. This paper presents the optimisation of retrofitted insulation thickness, using Ireland as a case study. Detailed and robust dynamic finite element models were developed based on in-situ boundary conditions and combined with economic and environmental considerations. It was determined that optimising insulation based on cost was considerably different to optimising based on carbon. Cost-optimised insulation can reduce overall cost which could expand the reach of retrofit, allowing for more existing homes to be used more efficiently. However, this approach can lead to significant increases in operational carbon and therefore a balanced decision must be made. The methodology presented can be adopted for different regions by inputting local data, which will facilitate the adoption of circular economy principals in European retrofit plans. The approach can benefit developing circular insulation materials of low embodied carbon by building a case for their use.

Keywords Optimisation · Retrofit · Material-efficiency · Circular insulation

D. M. Rakshit · A. Byrne (✉)
Civil and Structural Engineering, Technological University, Dublin, Ireland
e-mail: aimee.byrne@tudublin.ie

A. Robinson
Mechanical, Manufacturing & Biomedical Engineering, Trinity College Dublin, Dublin, Ireland

L. Bragança et al. (eds.), *Creating a Roadmap Towards Circularity in the Built Environment*, Springer Tracts in Civil Engineering,
https://doi.org/10.1007/978-3-031-45980-1_7

7.1 Introduction

7.1.1 Motivation

Energy used in buildings is primarily for space heating and cooling in order to provide and maintain thermal comfort [1–3]. Sustainability-focused energy-efficient design and operation of buildings has the potential for significant and positive environmental impact. In particular, energy-efficient retrofitting of existing buildings should be a primary focus since it is predicted that approximately 60% of the current building stock will still be in use in 2050 [4]. Optimising costs is one way of increasing the affordability of retrofit [5]. As detailed in the new EU Circular Economy Action Plan [6], the 'Renovation Wave' initiated by the European Green Deal [7] will be implemented in line with circular economy principles, notably optimised lifecycle performance, and longer life expectancy of build assets. In the Renovation Wave Strategy [8], the Commission aims to at least double renovation rates in the next 10 years and make sure renovations lead to higher energy and resource efficiency. Furthermore, the Green Deal identifies the twin challenge of addressing both energy efficiency and affordability [9].

The thermal performance of insulation material can be improved by simply increasing its thickness, since thermal resistance and thickness are proportional [10]. However, this results in the negative effect of increasing the material use and overall cost. In an attempt to resolve this push–pull tension between cost and thermal performance, researchers have developed optimisation methods that, for given thermal conductivity and material price by volume, requires minimum upfront cost and results in maximum long-term heat energy savings throughout its service life [11–13] referred to as *optimisation of insulation thickness* (OIT).

The standard approach to retrofitting insulation is based on U-Value, which is a steady-state thermal parameter that is the inverse of the total thermal resistance of the building element. This means that the lower the U-value, the better the insulator. It is estimated that around 67% of the buildings in Europe were built prior to 1980; prior to the introduction of building energy codes for housing sector in Europe [14]. In Ireland, 50% of buildings were built before the introduction of building wall thermal requirements into building regulations [15, 16]. The majority of walls built before 1980 are of a U-value above 1.3 W/m^{2}K [17]. In Ireland, government agencies offer grants for wall retrofit, aiming to achieve U-values of 0.55 W/m^{2}K for cavity walls and 0.35 W/m^{2}K for solid [18]. This, together with the assumption that typical heat loss through uninsulated walls is 30% of the total [3], forms the initial baseline in predicting current and potential heat transfer behaviour when optimising designs in the present work.

Despite this reliance on U-value, studies have shown that steady-state coefficients do not reflect the actual performance [19]. Thermal analyses under steady state tend to overestimate or underestimate performance when compared to more realistic, or in-situ, dynamic conditions [20, 21]. This can result in oversized building elements, increased upfront cost and associated carbon emissions. Therefore this study uses dynamic and in-situ environments for analysing the retrofit wall.

Ireland offers a unique testing ground for building energy analysis. Firstly, the Irish housing stock is recognised as amongst the least energy efficient in Northern Europe [22]. Second, Ireland is characterised as having a temperate oceanic climate, where the winter temperatures are mild and summer temperatures are moderate compared to the other European countries. This offers the opportunity to single-out home heating only, as air conditioning is rarely used or even installed in Irish homes.

7.2 Methodology

7.2.1 Finite Element Model (FEM) Development and Verification

Numerical Models were developed in Comsol® Multiphysics software of the three most common Irish wall types; Solid Wall (SW), Cavity Wall (CW) and Cavity Block Wall (CBW), the geometric domains are shown in Fig. 7.1 and Table 7.1 details thermal material properties. The inputs and assumptions were verified against experimental data under laboratory controlled conditions. The CW and SW type walls were verified against Hotbox data where one side of the wall air temperatures varied sinusoidally [23]. The CBW model was verified against published data [24]. The coefficient of variation (CV) and model efficiency factor (EF) were calculated to assess the capability of the models to reproduce real behavior. The coefficient of variation defines how well the model fits the experimental data by using offsetting errors between measured and simulated output. The EF compares the efficiency of the simulation model and the efficiency of describing the data as the mean of the experimental observations. The obtained simulation calibration index lies within the ASHRAE Guideline 14 [25] limits and therefore the models were considered calibrated. The following assumptions and simplifications were experimentally verified as acceptable for accurate models.

- Negligible thermal contact resistance between material layers.
- Isotropic thermal properties of the individual components, which are independent of temperature (Table 7.1).
- Mortar was not modelled since it has similar thermal properties to the solid component and represents a small percentage of the wall.
- One-dimensional conduction heat transfer model is used for a solid wall, a two-dimensional conjugate heat transfer model for cavity wall and a three-dimensional conjugate heat transfer model for the cavity block wall (Fig. 7.1).
- Workmanship, local variations, and thermal bridges were not considered.

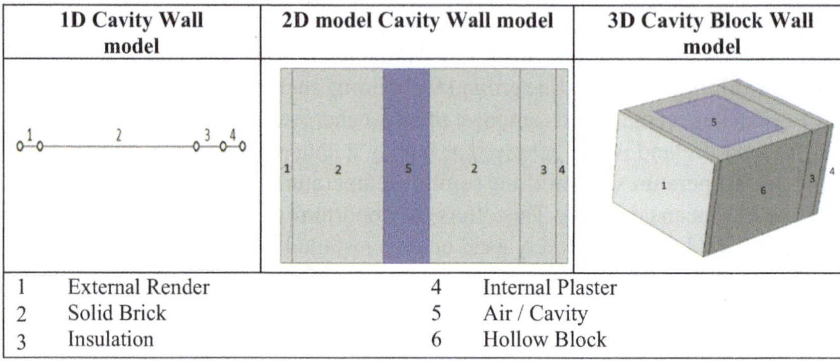

Fig. 7.1 Comsol multiphysics models for three wall types

Table 7.1 Thermal properties sourced from manufacturers and standards [3]

Material	Thermal conductivity (W/mK)	Density (kg/m^3)	Specific heat capacity (J/kg K)
EPS	0.031	30	Negligible
Brick	0.63	1970	950
Block	1.33	2100	1000
Internal plaster	0.2	850	850
External render	1.2	1200	850

7.2.2 Internal Boundary Condition Determination for FEM

An online self-completion survey was used to determine common home-heating setpoint temperature and heating patterns as the internal boundary condition for the Finite Element Models. A slider scale for setpoint temperature and dropdown menus for heating usage at different times of the day were used. A pilot study was completed initially with 25 participants followed by a survey with 205 responses, 71.1% of which were from Dublin. The interpretation and analysis of data was carried out only for Dublin.

Microsoft Excel™ and Minitab packages were used to analyse the data. The response data was directly exported to Excel before being imported into Minitab. Student's *t*-test and Descriptive statics were used for testing the means of two groups and measuring of central tendency such as mean and median.

7.2.3 External Boundary Condition Determination for FEM

Due to the limited weather parameters available from the Irish Meteorological Service, the data from Meteonorm was used to define external boundary conditions for the finite element models. Sophisticated interpolation models are used within Meteonorm, which allow for reliable calculation of solar radiation, temperature and additional parameters at any location in the world [26]. The data was further analysed in an ASHRAE sky model [25] to determine the solar air temperature for different orientations of the walls.

The wind data (speed and direction) obtained from Meteonorm for Dublin airport was analysed and wind direction and wind speed was determined using windrose diagrams and probability density distribution methods. The results were then used to calculate the external heat transfer coefficient with a developed formula of $h = 1.53\,V + 1.43$. To clearly understand the predominant wind speed in the west direction, a statistical analysis was carried out based on probability density function (DPD) of the yearly wind speed distribution, for which the Weibull model was used [27]. The probability density function represents the fraction of time the wind speed prevails at the location under investigation.

7.2.4 Numerical Models and OIT Determination

The most commonly used insulation material in Ireland, which is EPS, and most commonly used heating fuel, which is heating oil, were used as the example to explain the methodology and analysis approach. The three common walls types were evaluated for the influence of (a) heating pattern, (b) insulation location and (c) wall orientation. If determined to be significantly influential, the realistic heating pattern, optimal insulation position and the orientation which resulted in the worst-case heat loss were taken forward for OIT determination.

Economic and environmental models (EEM) were developed utilising the same equations as for the Heating Degree Day (HDD) approach detailed in this publication [28], but substituting the HDD determined heat loss value for the robust and verified FEM-determined heat loss. The total heat loss determined by FEM was directly linked to the heat loss (Q_{dhl}) in the EEM model through a Comsol® Multiphysics-Excel™ live link interface. Figure 7.2 illustrates the framework for FEM-EEM integrated model. This paper mainly focuses on the FEM-Environmental Model results.

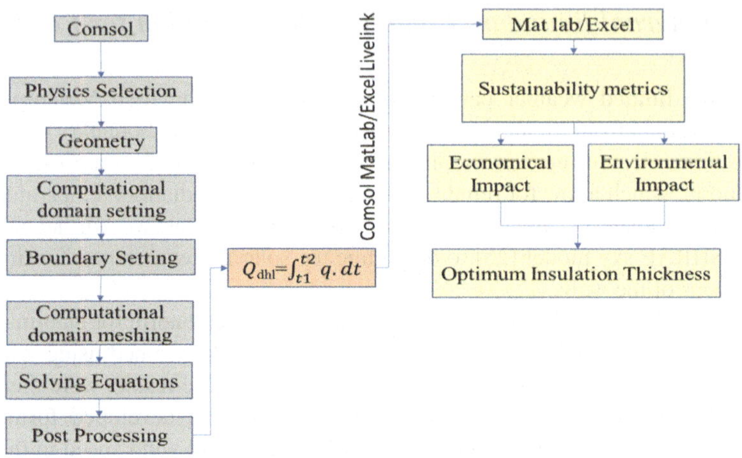

Fig. 7.2 Framework for integrated FEM-EEM model

7.3 Results

7.3.1 Internal Boundary Conditions

The mean set point temperature reported in the survey was 19.71 °C, which is 6.1% less than the comfortable indoor temperature recommended by World Health Organisation (WHO) [29]. The majority of the respondents came from working families and used the heaters only while they were at home resulting in a difference in heating pattern at weekends. The patterns were consistent with the literature [30–32]. Weekday and weekend heating patterns adopted for the numerical model input are shown in the below Fig. 7.3 for a single 24 h winter period.

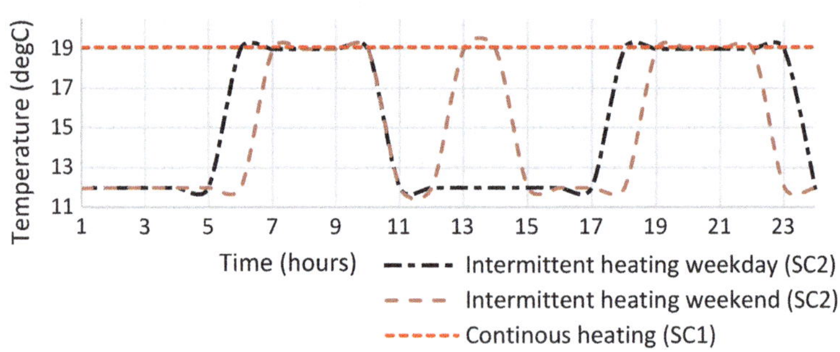

Fig. 7.3 Continuous and intermittent heating pattern for a 24 h period

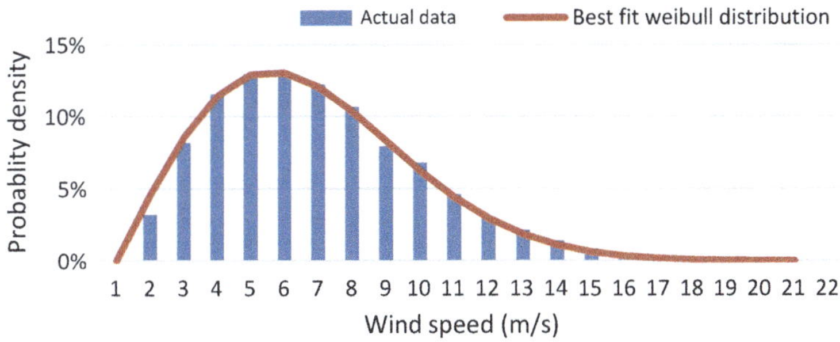

Fig. 7.4 Distribution of probability density (DPD) of annual wind speed for Dublin

7.3.2 External Boundary Conditions

For Dublin airport the wind blew from the west 14% of the time, from the north and south around 8% of the time and from the east 4–6% of the time. To clearly understand the predominant wind speed in the west direction, a statistical analysis was carried out based on DPD of the yearly wind speed distribution (Fig. 7.4), for which the Weibull model was used [27]. The probability density function represents the fraction of time the wind speed prevails at the location under investigation. The probability distribution curve for Dublin shows that the most frequently occurring wind speed is 6 m/s during winter, a value which corresponds to the peak of the probability density function.

7.3.3 Refined FEM Models

(a) Heating pattern: It was observed that, unlike continuously heated homes (SC1 in Fig. 7.5), under realistic heating patterns based on the survey responses (SC2) evidence of stored heat in thermal mass of the walls then returning to the internal space when the heating is turned off is identified by periods of positive internal heat flux (HFi) below (SC2). The overall impact on heat loss was very significant ranging from 20.9 to 27.5% higher (depending on wall type) in the continuous heating case compared to intermittent heating based on one week of weather data from January 2018.

(b) Insulation position: The above evidence of stored heat in thermal mass, re-released to the internal space is most evident in externally insulated walls. For externally insulated walls the living space takes longer to reach the target temperature, however, the heat stored in the wall during the heating period (brick/concrete layer) is prevented from fully dissipating into the outside environment. This stored heat is eventually released back into the indoor space during the non-heating period. Furthermore, in cavity walls and cavity block walls, the heat transfer coefficient

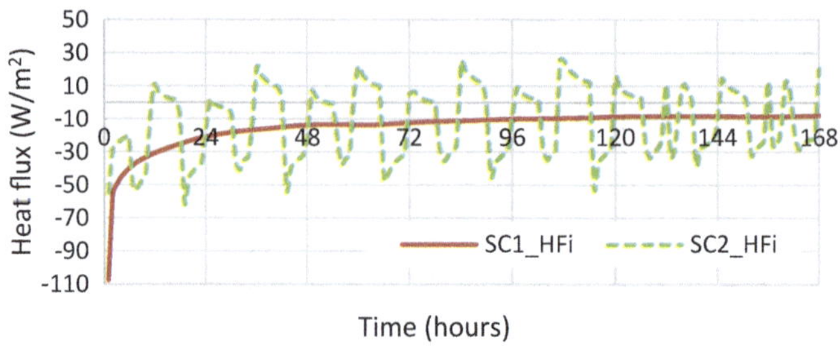

Fig. 7.5 Comparison of heatflux to solid walls for 1 week of constant internal heating (SC1), compared to common Irish heating pattern (SC2)

in the cavity is associated with the average temperature difference across it, which drives the buoyant flow of air. The higher the temperature difference, the greater the buoyancy force as will be the heat transfer coefficient. The heat transfer coefficient in the cavities of cavity block and cavity walls is lowest for internally insulated walls. These findings culminated in all wall typologies experiencing the lowest heat loss when internally insulated.

(c) Wall orientation: North facing walls in Ireland were found to experience significantly greater heat loss than other orientations and therefore should be used as the worst-case-scenario when designing for retrofit. For example north facing solid walls that are internally insulated experienced 29% greater heat loss than south facing. There is also greater conductivity and radiation across wall cavities due to the greater temperature difference across them during heating periods.

7.3.4 FEM-EM to Find OIT

Section 3.3 identified that the following should be included in the FEM portion of OIT determination; Intermittent heating pattern for internal boundary, defined solar air temperature and wind speed for six winter month heating period for external boundary, internally insulated walls as this results in lowest heat loss and north facing walls as they experience the worst-case heat loss in the building.

For determining economic and environmental impacts of changing insulation material thickness Table 7.2 values for calorific value, heating system efficiency and CO_2 emission factor and heating cost were used. For the economic analysis, costing for EPS insulation was taken as 78 €/m^3 based on the methodology published in [28]. Local recent and future projection data should be used where possible.

A large number of combinations were produced comparing the multiple variables which influence OIT values. This study preliminarily examined OIT based on carbon emissions related to cradle-to-gate embodied carbon estimates of 2.55 kg CO_2/kg for EPS [35] alongside the operational emissions for heating oil over 10 years based on

Table 7.2 Parameters for heating fuel type: heating oil [33, 34]

Hv[a] (kWh/unit)	Efficiency[b] (η)	CO_2 emission factor[c] (kg/kWh)	Cost (€/kWh)
10.55 (l)	0.9	0.2736	0.0788

[a] Heating fuel value, or calorific value, is the amount of heat released during the complete combustion process
[b] Efficiency is the amount of useful heat produced per unit of input energy (fuel)
[c] CO_2 emission factor: is an estimate of greenhouse gas emissions per unit of human activity (space heating)

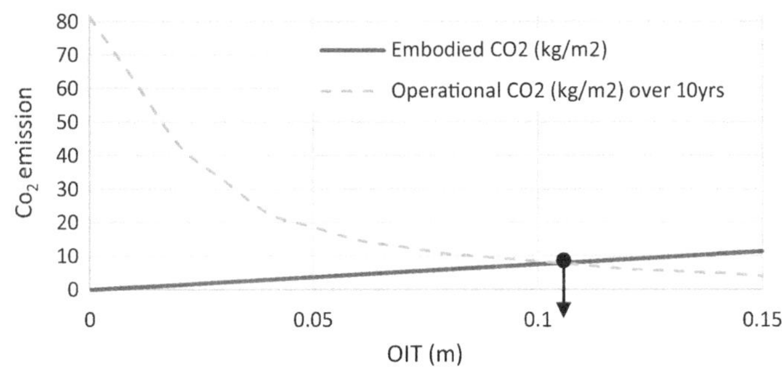

Fig. 7.6 OIT based on operational and embodied carbon for Internal EPS insulation on a Dublin Cavity Wall type using heating oil

the method outlined in [28]. To illustrate the findings, the example of EPS insulation, on cavity wall type, for Dublin, and heating oil as the heating fuel source is used in Fig. 7.6 below for environmental-OIT. OIT was determined to be 110 mm.

Conversely, when total cost is prioritised, 40 mm EPS was found to be optimal. The operational heating emissions at 40 mm EPS is considerably higher than for 110 mm. However, heating oil is associated with the greatest carbon emissions in Ireland when compared with the next most common alternatives of mains gas and electricity. As more efficient heating systems and renewable energy becomes more widespread resulting in lower operational-based CO_2, it is likely that reducing material production volumes will become more of a priority.

7.4 Conclusions and Recommendations

Using OIT insulation over a one-size-fits-all approach has the potential to save large quantities of carbon in a wholistic approach to circular design of buildings which includes both construction and operation. The method outlined here can be adopted for any region by substituting local data for: heating usage pattern, weather, wall typologies, insulation material and heating fuel. It is recommended that to ensure efficiency of design, dynamic thermal analyses are carried out and real-time boundary data (or approximations of such) are used where possible over simplified or steady-state approaches.

It is understood that limited computational power and time result in the need to simplify models to some extent, therefore future research can adopt the assumptions used (Sect. 2.1) in the calibrated models in this paper with confidence. Internal insulation was always identified as the optimal design for reducing heat loss in all wall types for intermittently heated spaces. The coldest or most shaded orientation of the building should be used as heat loss is significantly more due to the low solar gains.

The limitations of this work includes that cooling loads were not analysed, the embodied carbon value was based on a relatively old determination from the UK and that this only includes cradle-to-gate carbon which omits end of life reuse or disposal. These areas must be further examined, in particular when applying the methodology to other, potentially warmer, countries. Material volumes, costs and emissions were based on the insulating material alone, and did not include other costs such as labour, render, or fixings. This was so that the change in insulation material thickness could be isolated for comparison between options.

As bio-based and recycled insulation becomes more commonplace alongside the growing marked in renewable energy, the presented methodology can be used to build a case for adopting the most emission and cost efficient circular design combinations. Furthermore, this tool can be used as part of the toolkit in aligning the Renovation Wave with the Circular Economy Action Plan.

References

1. International Energy Agency (2013) Transition to sustainable buildings
2. Berardi U (2015) Building energy consumption in US, EU, and BRIC countries. Procedia Eng 118(Supplement C):128–136
3. NSAI (2014) Standard recommendation S.R 54
4. Gov. of Ireland (2022) Climate action plan 2023
5. SEAI (2020) Domestic technical standards and specification
6. European Commission Directorate-General for Environment (2020) A new circular economy action plan for a cleaner and more competitive Europe. 2020, communication from the Commission to the European Parliament, the Council, the European Economic and Social Committee and the Committee of the Regions
7. European Commission (2019) The European green deal, communication from the Commission to the European Parliament, the Council, the European Economic and Social Committee and the Committee of the Regions
8. European Commission (2020) A Renovation wave for Europe—greening our buildings, creating jobs, improving lives. Communication from the Commission to the European Parliament, the European Economic and Social Committee and the Committee of Regions, Brussels
9. EC (2019) The European green deal COM/2019/640. Brussels
10. Cengel YA (2002) Heat transfer, 2nd edn. McGraw-Hill, New York
11. Al-Homoud DMS (2005) Performance characteristics and practical applications of common building thermal insulation materials. Build Environ 40(3):353–366

12. Ozel M (2014) Effect of insulation location on dynamic heat-transfer characteristics of building external walls and optimization of insulation thickness. Energy Build 72:288–295
13. Mishra S, Usmani JA, Varshney S (2012) Energy saving analysis in building walls through thermal insulation system. Int J Eng Res Appl 2(5):128–135
14. Norris M, Shiels P (2004) Regular National Report on housing developments in European countries: synthesis report. Stationery Office
15. Ahern C, Norton B (2019) Thermal energy refurbishment status of the Irish housing stock. Energy Build 202:109348
16. Ahern C, Norton B (2020) A generalisable bottom-up methodology for deriving a residential stock model from large empirical databases. Energy Build 215:109886
17. SEAI (2019) National BER research tool. SEAI, Ireland
18. Department of Housing, Local Government and Heritge (2020) PART-L-Technical guidance document, Conversion of Fuel and Energy-Dwellings
19. Sassine E et al (2017) Experimental determination of thermal properties of brick wall for existing construction in the north of France. J Build Eng 14:15–23
20. Ramin H, Hanafizadeh P, Behabadi M (2015) Comparative study between dynamic transient and degree-hours methods to estimate heating and cooling loads of building's wall. Comput Appl Mech 46:153–165
21. Van der Veken J et al (2004) Comparison of steady-state and dynamic building energy simulation programs. Buildings IX
22. Ahern C, Griffiths P, O'Flaherty M (2013) State of the Irish housing stock—modelling the heat losses of Ireland's existing detached rural housing stock & estimating the benefit of thermal retrofit measures on this stock. 55:139–151
23. Muddu RD et al (2020) The use of three-dimensional conjugate CFD to enhance understanding of, and to verify, multi-modal heat transfer in dynamic laboratory test walls. In: Ruane K, Jaksic V (eds) Civil Engineering Research in Ireland (CERI), Civil Engineering Research Association of Ireland: Online, pp 360–365
24. Byrne A, Byrne G, Robinson A (2017) Compact facility for testing steady and transient thermal performance of building walls. Energy Build 152:602–614
25. ASHRAE Guideline 14-2014 for measurement of energy and demand savings, American Society of Heating. Refrigeration and Air Conditioning Engineers, Atlanta, GA
26. Meteonorm V8. https://meteonorm.com/
27. Axaopoulos I, Axaopoulos P, Gelegenis J (2014) Optimum insulation thickness for external walls on different orientations considering the speed and direction of the wind. Appl Energy 117:167–175
28. Muddu RD et al (2021) Optimisation of retrofit wall insulation: an Irish case study. Energy Build 235:110720
29. World Health Organisation (1987) Health impact of low indoor temperatures. In: Environmental health (WHO-EURO). World Health Organization. Regional Office for Europe.
30. Jones RV, Fuertes A, Lomas KJ (2015) The socio-economic, dwelling and appliance related factors affecting electricity consumption in domestic buildings. Renew Sustain Energy Rev 43:901–917
31. Jones RV et al (2016) Space heating preferences in UK social housing: a socio-technical household survey combined with building audits. Energy Build 127:382–398
32. Hanmer C et al (2019) How household thermal routines shape UK home heating demand patterns. Energ Effic 12(1):5–17
33. SEAI (2019) Domestic fuels comparison of energy costs. https://www.seai.ie/data-and-insights/seai-statistics/key-statistics/energy-data/. Accessed 20 Dec 2019
34. Badurek M, Hanratty M, Sheldrick W (2012) TABULA scientific report. IHER Energy Services, Ireland
35. Hammond G, Jones C (2011) Inventory of carbon & energy (ICE) V2.0. www.circularecology.com/ice-database.html

Chapter 8
Circular Strategies in Lightweight Steel-Framed (LSF) Buildings and the Research Project Tyre4BuildIns

Paulo Santos, Viorel Ungureanu, and Luisa Durães

Abstract According to the UN sustainable development goals, energy and waste management are two of the major concerns of humankind, these being related with the search for a more sustainable built environment. Innovative materials (e.g., recycled rubber-aerogel composite thermal insulation) and advanced construction systems, such as Lightweight Steel Framing (LSF), may contribute to a more sustainable built environment, due to their inherent advantages in comparison to traditional ones. Recently, the European Commission adopted the newly updated European Circular Economy Action Plan (ECEAP), which will be implemented not only to waste materials, such as End-of-Life Tyres (ELT), but also to the construction industry and the built environment. In the first part of the paper, the LSF construction system is described, including its suitability for circular strategies, such as modular construction. In the second part, the Tyre4BuildIns research project is described, which was conceived to develop a new eco-friendly and cost-effective thermal insulation material, made from a mixture of recycled tyre rubber (waste) and an advanced high-performance insulation material (aerogel). It was concluded that, compared to traditional construction, the LSF modular construction system can contribute significantly to a more sustainable built environment, being more suitable for the implementation of circular strategies. Furthermore, the Tyre4BuildIns project allowed the development of an eco-friendly high performance thermal insulation composite material based on recycled tyre rubber and aerogel, contributing to enhance the thermal performance of LSF building elements.

P. Santos (✉)
Department of Civil Engineering, University of Coimbra, ISISE, ARISE, Pólo II, Rua Luís Reis Santos, 3030-788 Coimbra, Portugal
e-mail: pfsantos@dec.uc.pt

V. Ungureanu
Politehnica University of Timisoara, Timisoara, Romania

Romanian Academy, Timisoara Branch, Timisoara, Romania

L. Durães
Department of Chemical Engineering, University of Coimbra, CIEPQPF, Pólo II, Rua Sílvio Lima, 3030-790 Coimbra, Portugal

L. Bragança et al. (eds.), *Creating a Roadmap Towards Circularity in the Built Environment*, Springer Tracts in Civil Engineering,
https://doi.org/10.1007/978-3-031-45980-1_8

Keywords Circular Strategies · Buildings · Lightweight Steel-Framed (LSF) system · Modular construction · Tyre4BuildIns research project · Recycled insulation materials · Recycled tyre rubber

8.1 Introduction

Today, energy and waste management are two of the main concerns of humanity [1]. In Europe, around 5 million tonnes of tyres are produced per year [2] and buildings account for 40% of the total energy consumption [3]. According to the same source, roughly 75% of the EU building stock is energy inefficient. Moreover, thermal bridges may be responsible for up to 30% of heat losses in buildings [4]. To achieve the ambitious goal of carbon-neutrality by 2050, improving energy efficiency in buildings is very relevant.

Over the last few years, alternatives to the traditional constructive method have emerged and proliferated worldwide, such as the lightweight steel-framed (LSF) system, which offers some advantages in fields such as production, transportation, durability, adaptability and construction economy [5]. However, if not correctly designed and built, the thermal bridges created by the high thermal conductivity of steel can penalize the thermal performance of the exterior envelope of buildings and, consequently, their overall energy efficiency [5].

Regarding end-of-life tyre (ELT) waste management, the Directive 1999/31/EC on waste landfill has prohibited the landfilling of used whole tyres since 2003 and shredded tyres since 2006. Therefore, with decreasing disposal options and increasing production, the volume of used tyres is becoming a major waste management issue.

In March 2020 the European Commission adopted the new updated European Circular Economy Action Plan (ECEAP), which will be implemented not only to waste materials (such as ELT), but also to the construction industry and built environment. Regarding this issue, a more recent and advanced construction system, such as the LSF system, may play a relevant role in this transition to a circular economy, due to some of their advantageous features, such as, high suitability for modular construction, easier reuse and adaptability [5].

The research project Tyre4BuildIns [6] was conceived to develop a new cost-effective eco-friendly thermal insulation material, made from a mixture of recycled tyre rubber (waste) and an advanced high-performance insulation material (aerogel) [7], which will be used mainly, but not exclusively, as a thermal break for LSF building structures [8].

In this study, modular construction and the LSF system are assessed to evaluate their suitability for implementing circular strategies in buildings. Furthermore, the Tyre4BuildIns research project is briefly described, with a focus on the main achievements in the management of waste from ELT (recycled tyre rubber and textile fibres), as well as in favour of improved thermal performance and improved energy efficiency of LSF building elements.

8.2 LSF Modular Construction

8.2.1 Modular Construction

Modular constructions are prefabricated buildings that consist of repeated components called modules [9]. One can identify two-dimensional (2D) panels, with or without openings, but also three-dimensional (3D) volumetric units, with or without full fixtures, much more complex than the two-dimensional ones. Timber, concrete, or steel can be used separately or in hybrid systems in different forms.

Modularity involves constructing units away from the building site and then delivering them to the site. The installation of the prefabricated unit is completed on site using cranes. The modules can be placed in various ways, i.e. side-by-side, end-to-end, or stacked, allowing for a variety of configurations and styles. Once the module is placed in position, they are joined using specific connections. These connections tie the individual modules together to form the overall structure of the building [10]. Modular buildings refer to the application of a variety of structural systems and building materials, rather than a single type of structure.

According to Rogan et al. [11], the attributes of modular construction are:

- reduced construction costs, especially when combined with economy-of-scale production (10%);
- shorter construction time on site (50–60%);
- increased profitability of the industry due to economy of manufacturing scale;
- increased site productivity (up to 50%) and reduced labour force on-site;
- greater certainty of completion on time and to budget;
- much reduced wastage in manufacture and on site—improved sustainability;
- greater reliability and quality.

Some other benefits can be identified, such as:

- lower weight, which provides lower foundation loads and expense, including transportation, and the ability for roof-top extensions of existing buildings;
- less disruption at construction sites due to multiple truck movements compared to conventional on-site construction;
- design for deconstruction, to encourage future reuse and recycling of products and materials;
- design for flexibility to extend building lifetimes and, where possible, further extend the life of buildings by renovation and refurbishment.

These benefits may be quantified in a holistic assessment of the costs and value of modular construction in relation to more traditional alternatives.

Lawson et al. [12] identified two different types of building modules according to their load transfer mechanisms. The first type is the corner supported module, where loads are transferred from edge beams to the supporting corner columns to the ground or a floor. The second type is the load bearing modules, where the loads are transferred from the side walls of the module to the ground.

Lusby-Taylor et al. [15] identified prefabricated construction in 5 categories according to the way it is manufactured and assembled on site. They can be (1) modular (volumetric) construction: production of three dimensional units in controlled factory conditions prior to transportation to site; (2) panelised construction, where flat panel units are produced in a factory and assembled on site to produce a three-dimensional structure; (3) hybrid (semi-volumetric) construction: combines both panelised and modular approaches; (4) off-site manufacturing (OSM) subassemblies and components: covers approaches that fall short of being classified as systemic OSM, but which utilise several factory fabricated innovative subassemblies or components in an otherwise traditionally built structural fabric, e.g. roof cassettes, pre-cast concrete foundation assemblies, but excluding window, door sets, roof trusses; and (5) non-OSM, where this category is intended to encompass schemes utilising innovative housing building techniques and structural systems that fall outside the OSM categories.

8.2.2 Lightweight Steel-Framed System

Light steel framing (LSF) is an integral part of modular construction as it is strong, lightweight, durable, accurate, free from long-term movement and has been well-tested in a wide range of applications.

LSF comprises galvanised cold-formed steel lipped channel sections of 70–150 mm depth in the wall panels, and 150–300 mm deep C-sections or lattice joists in the floors. The usual spans are of 3.6 m, but up to 6 m can be achieved, which can eliminate internal load-bearing walls and, therefore, leads to flexibility in internal space planning. The prefabricated wall panels are typically of the storey height and the length depending on transportation and lifting.

The following types of modules can be used Lawson et al. [12] in the design of buildings using either fully modular construction or mixed forms of steel construction: (1) 4-sided modules; (2) partially open-sided modules; (3) open-sided (corner-supported) modules; (4) modules supported by a primary structural frame; (5) non-load bearing modules; (6) mixed modules and planar floor cassettes, and; (7) special staircase or lift modules.

These systems may also have some drawbacks that could penalise their thermal behaviour and energy efficiency. Thermal bridges, originated from steel studs and reduced thermal inertia, are two major examples of these possible drawbacks [5]. Steel may be fire-resistant, but it also tends to attract extreme heat, due to its thermal conductivity, and home interiors can be affected during hot summers, given their usual reduced thermal inertia. Fortunately, this problem can be solved by adopting some strategies to increase their thermal mass.

Compared to timber, steel-framed homes need additional attention to insulate because steel framing could require a thermal break to be included as part of the insulation process. If a thermal break is not included, which can be an insulating strip along the steel flanges [13, 14] or an adequate External Thermal Insulation

Composite System (ETICS), then the cold steel can cause condensation within the walls of the home, and, over time, the moisture can compromise the integrity and durability of the wall.

The light steel frame is enhanced with the use of other materials such as cladding, insulation, and internal fittings, making it a truly sustainable option that is also energy efficient, when adequately designed. In addition, LSF supports creative and innovative design with a wide range of combinations with other materials.

Moreover, structural steel in LSF constructions is inherently recyclable but is also reusable. These and many other attributes, such as design for deconstruction and design for flexibility, are becoming increasingly significant in the context of the circular economy. The circular economy keeps resources in use for as long as possible, extracting the maximum value from them while in use, and then recovers and regenerates products and materials at the end of each service life through recycling and reuse.

8.3 The Tyre4BuildIns Research Project

8.3.1 Main Objectives

The Tyre4BuildIns research project [6] started on 26 July 2018 and had a total duration of four years. The main goals were as follows:

1. Develop a new eco-friendly and cost-effective insulation composite material based on recycled tyre rubber and other insulation materials;
2. Evaluate and optimize the performance of this new composite insulation material by characterizing its properties (hygrothermal, acoustic, fire reaction, mechanical resistance and durability);
3. Optimize the use of the new insulation material in building elements (e.g. walls) in order take maximum advantage of it regarding thermal and acoustic performance, and;
4. Assess the environmental impacts and cost of this new insulation material from a life-cycle perspective.

8.3.2 Main Tasks and Sub-Tasks

To achieve these objectives, the team make use of a research plan containing six main tasks and several sub-tasks, as illustrated in Fig. 8.1.

Task 1 (State-of-the-art) aimed to keep the team up to date on the research topics addressed in this project, given the frequent related new scientific advances. **Task 2** (Development of new thermal insulation composites) aimed to develop a new eco-friendly and cost-effective insulation composite material based on recycled tyre

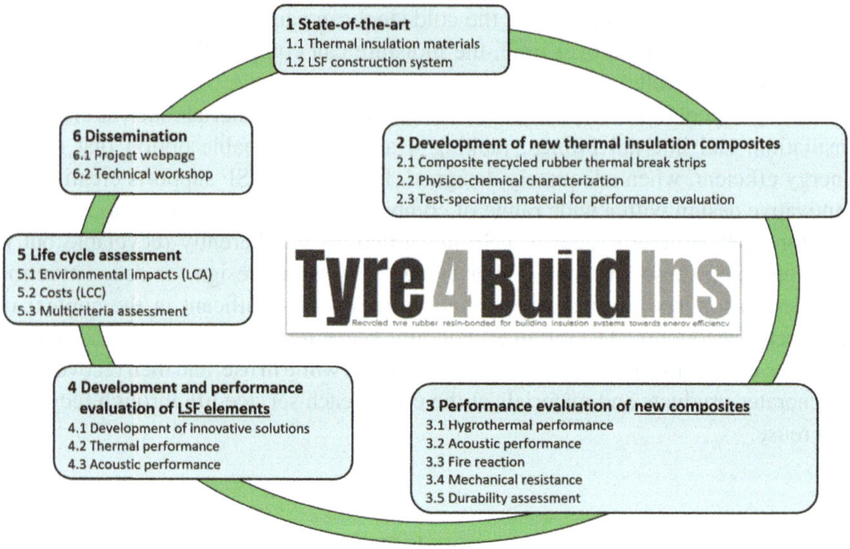

Fig. 8.1 Tasks and sub-tasks of the research project Tyre4BuildIns

rubber and aerogel materials, this being the most important task of this research project. **Task 3** (Performance evaluation of new composites) purposed to characterise the previously developed composite materials to evaluate their performance. **Task 4** (Development and performance evaluation of LSF elements) aimed to evaluate the thermal and acoustic performance and optimize the use of the material in LSF walls in order to take maximum advantage of it. **Task 5** (Life cycle assessment) focused on a life cycle assessment of the new insulation material in terms of environmental impacts and costs. **Task 6** (Dissemination) involves the dissemination of the results of the research project using a website [6] and the organisation of a technical workshop [16].

8.3.3 Research Team and Main Outputs

The research team included the participation of the Chemical and Civil Engineering Departments of the University of Coimbra. While the first department focused on the development of the innovative composite insulation material, the second aimed at its evaluation and optimization at the building element level (e.g., as thermal break strips (TBS) in LSF walls). The main outputs achieved until now [6], include: 21 scientific publications (e.g. books and book chapters, articles in international and national scientific journals); 14 communications in international and national scientific meetings; 9 PhD and Master thesis concluded; 1 computational application, and; 2 provisory patents applications (INPI and WIPO [17]).

8.3.4 Manufacturing Process of the New Rubber-Silica Aerogel Insulation

Among these outputs, it should be highlighted the new rubber-silica aerogel composite reinforced with fibres, where the rubber and the fibres can be obtained from recycled tyres [18], and for which the manufacturing process is now protected by an international patent application to WIPO—World Intellectual Property Organisation [17]. Figure 8.2 illustrates the main steps to reclaim the end-of-life tyres, from the process of recycling their components to the conversion into superinsulating thermal break strips (TBS) for LSF walls [19].

The manufacturing process to produce the superinsulation composite made of silica aerogel and recycled tyre rubber reinforced with fibres was described in Reference [18]. First, the recycled tyre rubber was dissolved using a Peracetic acid alcoholic solution, to obtain a sol of rubber which can be promptly mixed with the silica sol. The latter is obtained from hydrolysis and condensation of silica precursors (tetraethyl orthosilicate and organosilanes). Notice that a "sol" is a colloidal suspension made from tiny solid particles in a continuous liquid medium. Next, a gelification process occurs to obtain a rubber-silica gel. Just before gelification, this solution was poured over reinforcement fibres inside containers. Several reinforcement fibres were tested, including recycled tyre textiles, polyester fibres, silica fibres and glass wool.

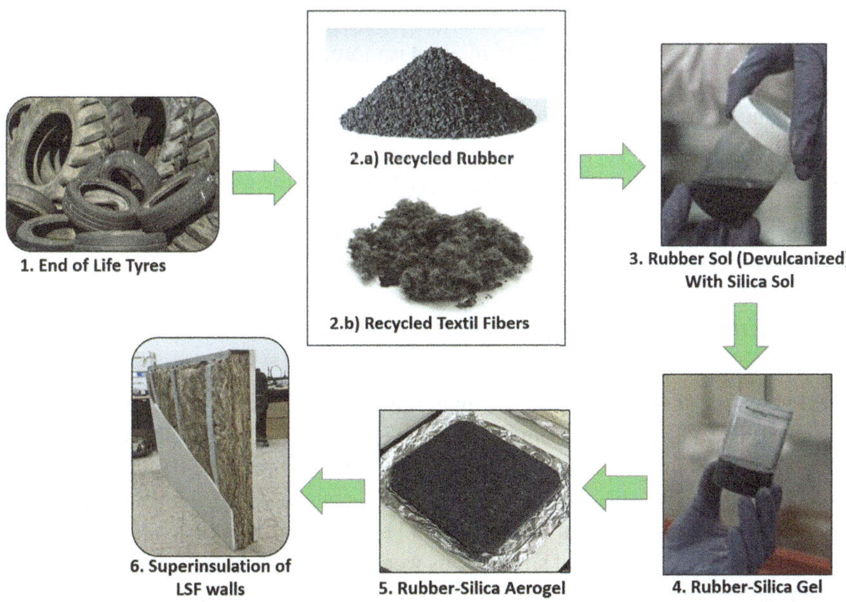

Fig. 8.2 From end-of-life tyres to rubber-silica aerogel and their application as super-insulating thermal break strips in LSF walls

Table 8.1 Measured thermal conductivities of the new fibre-reinforced rubber-silica aerogels (Adapted from Lamy-Mendes et al. [18])

Reinforcement fibres	Thermal conductivity [mW/m/K]	
	Hot Disk® TPS[a] method	HFM[b] method
Recycled tyre textiles	35.91 ± 0.43	28.1 ± 1.7
Polyester fibres	24.30 ± 0.25	16.4 ± 1.0
Silica fibres	25.37 ± 0.06	23.2 ± 1.4
Glass wool	58.05 ± 0.01	28.7 ± 1.7

[a]TPS—Transient Plane Source; [b]HFM—Heat Flow Meter

8.3.5 Thermal Performance of the New Insulation Composites

The thermal conductivity of these new composite insulation materials was measured using two different methods: (1) Transient Plane Source (TPS) method using a Thermal Constants Analyzer TPS 2500 S (Hot Disk, Göteborg, Sweden), with two similar samples maintained at 20 °C, and; (2) Heat Flow Meter (HFM) method using an HFM 436/3/1 Lambda (EN 1946-1:1999), from NETZSCH (Selb, Germany), at 23 °C.

The measured thermal conductivities for each measurement technique and for the different reinforcement fibres used, are listed in Table 8.1. The conductivities measured using the HFM method exhibit smaller values than those of the other method (TPS). Notice that the HFM method is more accurate for insulating materials and the test samples are larger: $21.5 \times 21.5 \times 1.6\ cm^3$ [18]. When the thermal performance of the several new composite insulation materials are compared, it can be concluded that the best performance was achieved when using the polyester fibres (16.4 ± 1.0 mW/m/K), while the worst was obtained by the glass wool reinforcement fibres (28.7 ± 1.7 mW/m/K).

8.3.6 Enhanced Thermal Performance of LSF Walls

These new superinsulation composite materials may have many applications in buildings, not only in the thermal domain, but also in acoustics given their good vibration attenuation, e.g., floating floors [20] or LSF walls [21]. In the Tyre4BuildIns research project [6] this new superinsulation composite material was mainly evaluated as thermal break strips (TBS) in LSF partition [14] and facade walls [13].

Moreover, some numerical simulations were performed, as well as measurements of thermal resistance under controlled lab conditions. Table 8.2 lists the measured conductive R-values for partition and facade load-bearing walls without TBS ("Reference"), as well as the thermal resistances when using a single aerogel TBS on the

Table 8.2 Measured thermal resistances of load-bearing LSF walls (Adapted from Santos et al. [13], Santos and Mateus [14])

	R-value [m^2 · K/W] (Δ R)		
	Reference[a]	1 TBS[b]	2 TBS[c]
Partition LSF wall	1.558	2.205 (+42%)	2.754 (+77%)
Facade LSF wall	3.200	3.695 (+15%)	4.093 (+28%)

[a]Without TBS; [b]Outer TBS; [c]Outer and inner TBS

outer steel stud flange ("1 TBS"), and two aerogel TBS (one in the outer flange and another on the inner flange). The thermal resistance improvement was for the LSF partition walls, ranging from +42% (1 TBS) up to +77% (2 TBS), relative to the R-value of the reference wall (1.558 m^2·K/W). The thermal performance improvement in the LSF facade walls was smaller, ranging from +15% up to +28%, relative to the corresponding reference R-value, i.e., 3.200 m^2·K/W. This reduced R-value increment was expected due to the existence of an external thermal insulation composite system (ETICS), which reduces the thermal bridge effect originated by the steel frame.

8.4 Final Remarks

In this paper, the lightweight steel-framed (LSF) system and the research project Tyre4BuildIns were presented within the scope of buildings, modular constructions, circular strategies, waste management and energy efficiency for a more sustainable built environment. First, a short review about modular construction and the main advantages of the LSF system was presented. After, the Tyre4BuildIns research project was described, including the main objectives, tasks and subtasks, the research team and the results, the manufacturing process of the new rubber-silica aerogel insulation material, thermal conductivities achieved and the enhanced thermal performance of LSF walls using thermal break strips (TBS).

It was concluded that, compared to traditional construction (e.g., reinforced concrete and masonry), LSF modular construction system can contribute significantly to a more sustainable built environment, being more suitable for the implementation of circular strategies (e.g., design for deconstruction and flexibility), due to its inherent advantages. Moreover, the Tyre4BuildIns research project allowed to develop a high-performance thermal insulation composite material based on recycled tyre rubber and aerogel. This new eco-friendly thermal insulation has a thermal conductivity similar to the state-of-the-art commercial aerogels available in the market. When used as TBS along the steel stud flanges in LSF building elements, allowed to significantly enhance their thermal performance. This increase in thermal resistance reached +42% and +77% compared to a reference partition LSF wall, when using one and two TBSs, respectively. For LSF facade walls, the increase in R-value was more modest (+15% and +28%, respectively), due to the influence of ETICS, which mitigated the importance of the steel frame thermal bridge effect.

Regarding the eventual reuse of the new TBS, it is facilitated by the way these TBS are fixed to the steel stud flanges, using screwed mechanical connections. Thus, the TBS could be easily removed after unscrewing the bolts and removing the adjacent sheathing panels (e.g., gypsum plasterboard or OSB).

8.5 Funding and Acknowledgements

This research was funded by FEDER funds through the Competitivity Factors Operational Programme–COMPETE, and by national funds through FCT, Foundation for Science and Technology, within the scope of the project POCI-01-0145-FEDER-032061. The authors also want to thank the following companies, partners of the Tyre4BuildIns project: Pertecno, Gyptec Ibéria, Volcalis, Sotinco, Kronospan, Hulkseflux, Hilti, and Metabo.

References

1. United Nations (2015) Sustainable development goals (SDGs). The 2030 agenda for sustainable development, https://sdgs.un.org/goals. Last Accessed 15 Jan 2021
2. ETRMA (2021) European tyre & rubber industry, statistics, edition 2021, N. 11. European Tyre and Rubber Manufactures Association, p 48
3. European Commission (2020) In focus: energy efficiency in buildings. https://commission.europa.eu/news/focus-energy-efficiency-buildings-2020-02-17_en. Last Accessed 20 Jan 2023
4. Erhorn-Klutting H, Erhorn H (2009) ASIEPI-impact of thermal bridges on the energy performance of buildings. Fraunhofer Institute for Building Physics, Germany, pp 1–8
5. Santos P, Simões da Silva L, Ungureanu V (2012) Energy efficiency of light-weight steel-framed buildings, 1st ed. In: Technical committee 14-sustainability & eco-efficiency of steel construction. European Convention for Constructional Steelwork (ECCS), N. 129. ISBN 978-92-9147-105-8
6. Tyre4BuildIns (2018) Research project Tyre4BuildIns-recycled tyre rubber resin-bonded for building insulation systems towards energy efficiency, funded by FEDER European and FCT National funds. Reference: POCI-01-0145-FEDER-032061. University of Coimbra, PT, www.tyre4buildins.dec.uc.pt. Last Accessed 19 Jan 2023
7. Lamy-Mendes A, Rodrigues Pontinha AD, Alves P, Santos P, Durães L (2021) Progress in silica aerogel-containing materials for buildings' thermal insulation. Constr Build Mater 286:122815
8. Santos P (2019) Kick-off presentation of the project 'Tyre4BuildIns'–recycled tyre rubber resin-bonded for building insulation systems towards energy efficiency. In: 4th Energy for sustainability international conference-designing a sustainable future, 24–26 July. Turin (IT), pp 1–5. ISBN: 978-989-54499-0-3
9. Lacey AW, Chen W, Hao H, Bi K (2018) Structural response of modular buildings–an overview. J Build Eng 16:45–56
10. Lacey AW, Chen W, Hao H, Bi K (2019) Review of bolted inter-module connections in modular steel buildings. J Build Eng 23:207–219
11. Rogan AL, Lawson RM, Bates-Brkljac N (2020) Better value in steel. Value and benefits assessment of modular construction. The Steel Construction Institute
12. Lawson RM, Ogden RG, Bergin R (2012) Application of modular construction in high-rise buildings. J Arch Eng, Am Soc Civ Eng 148–154

13. Santos P, Mateus D, Ferrandez D, Verdu A (2022) Numerical simulation and experimental validation of thermal break strips' improvement in facade LSF walls. Energies 15(21):8169
14. Santos P, Mateus D (2020) Experimental assessment of thermal break strips performance in load-bearing and non-load-bearing LSF walls. J Build Eng 32:101693
15. Lusby-Taylor P, Morrison S, Ainger C, Ogden R (2012) Design and modern methods of construction. The Commission for Architecture and the Built Environment (CABE), London, UK
16. Webinar Tyre4BuildIns: sustainable silica based nanomaterials for thermal protection systems online, 12 January 2022, http://www.dec.uc.pt/~tyre4buildins/Webinar.pdf. Last Accessed 19 Jan 2023
17. Durães L, Mendes AL, Pontinha AD, Santos P (2022) Fibre-reinforced aerogel composites from mixed silica and rubber sols and a method to produce the rubber-silica aerogel composites. Patent cooperation treaty international application, PCT/IB2022/050094. WIPO-World Intellectual Property Organization
18. Lamy-Mendes A, Dora A, Pontinha R, Santos P, Durães L (2022) Aerogel composites produced from silica and recycled rubber sols for thermal insulation. Materials 15(22):7897
19. University of Coimbra (2022) Tyre4BuildIns, NextLap accelerator 2022 winner, Notícias UC. https://noticias.uc.pt/artigos/equipa-de-investigadores-da-universidade-de-coimbra-vence-desafio-nextlap-accelerator-2022/. Last Accessed 19 Jan 2023
20. Pereira A, Mateus D, Lopes P (2022) Desempenho acústico de um material desenvolvido a partir do reciclado de borracha e aerogel para utilização em pisos flutuantes. In: TecniAcustica 2022-XII Congresso Ibérico de Acústica, 2–4 November. Elche (SP), pp 1–10
21. Pereira A, Mateus D (2022) Avaliação do isolamento acústico de divisórias e paredes de fachada com estrutura em LSF. In: TecniAcustica 2022-XII Congresso Ibérico de Acústica, 2–4 November. Elche (SP), pp 1–9

Chapter 9
The Recycling of Construction Foams: An Overview

Nuno Gama, Ana Barros-Timmons, and Artur Ferreira

Abstract In 1987, the United Nations Brundtland Commission defined sustainability as "meeting the needs of the present without compromising the ability of future generations to meet their own needs." Yet, after all these years, the humankind is dealing with catastrophic environmental problems which may jeopardize the future generations wellbeing. One cause of such issue is pollution associated to polymer's disposal. Polymers are mainly produced using petroleum derivatives and/or non-degradable. In addition, after their use, they are normally disposed in land fields or burned for energy. Yet, due to environmental problems, these solutions are not valid options, so plastic wastes must be recycled and used to produce new materials. This circular economy concept is not only a requirement for preventing pollution but is also a need for the reduction of the costs associated with their production and for the enhancement of the eco-efficiency of materials. Furthermore, this approach also addresses the risk of shortage of raw materials in the medium future. With this in mind, this document intends to give an overview of the recycling of construction foams with special focus on polyurethane (PU) and polystyrene (PS) foams. It aims to highlight the possible routes to recycle construction foams, presenting the differences and challenges of recycling different types of polymers. In that perspective, chemical and mechanical recycling routes are discussed, as well as energy recover alternatives. Finally, life cycle analysis (LCA) reports of these products are presented.

Keywords Circular economy · Building's circularity · Sustainability · Polyurethane foam · Expanded polystyrene

N. Gama (✉) · A. Barros-Timmons
Chemistry Department of University of Aveiro (UA), Aveiro, Portugal
e-mail: nuno.gama@ua.pt

N. Gama · A. Barros-Timmons · A. Ferreira
Aveiro Institute of Materials (CICECO), Aveiro, Portugal

A. Ferreira
Águeda School of Technology and Management (ESTGA-UA), Águeda, Portugal

L. Bragança et al. (eds.), *Creating a Roadmap Towards Circularity in the Built Environment*, Springer Tracts in Civil Engineering,
https://doi.org/10.1007/978-3-031-45980-1_9

9.1 Polymer Foams

Polymer foams are porous structures, therefore extremely lightweight. Furthermore, they are versatile, since their mechanical properties can be adjusted, are highly durable, and have excellent sound absorption and thermal insulation properties, among others [1]. All these characteristics, together with their low cost, make them the primary choice to be used in protective packaging, thermal and sound insulation, or seat cushioning applications. Hence, they are indispensable i.e. for the construction industry, automotive or packaging sectors [1].

Theoretically, all polymers can be foamed, in addition, different process of foaming can be used. Extruded foams based on polystyrene (PS), poly(vinyl chloride) (PVC), polyethylene (PE), polypropylene (PP) or poly(ethylene terephthalate) (PET) are used in food, construction, decoration, packaging, medical application; injected foams-based PS, PP or PE are used in automotive, mounded beads based on PS or PP are used in food and packaging sectors, among many others. This results in materials with different properties and different applications, as already mentioned [2].

Due to the wide variety of applications, the global market of polymer foams was valued at USD 123.1 billion in 2021 and is expected to expand at a compound annual growth rate (CAGR) of 3.6% from 2022 to 2030 [3]. In what concerns the foams used in construction, they are mainly produced from PU or PS (representing almost 50% of the market [3]), but phenolic resins and PVC foams can be used as well. Independent of the type of polymer, they can be classified as flexible which are mainly used for dumping and sound absorption, or classified as rigid which are mainly used as structural or thermal insulation materials, as presented in Fig. 9.1.

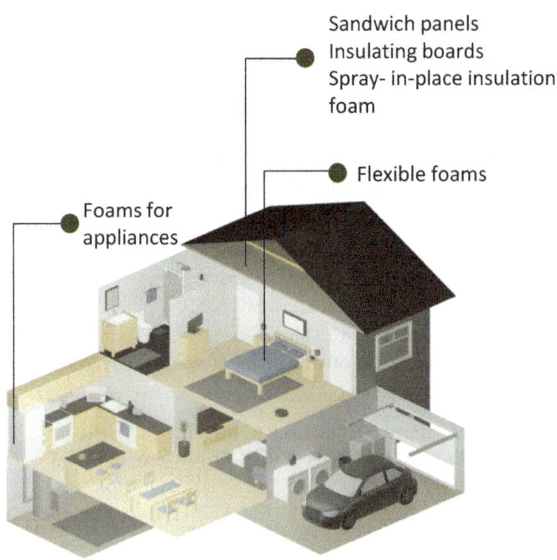

Fig. 9.1 Types of foams used in construction (Adapted from Metgen [4])

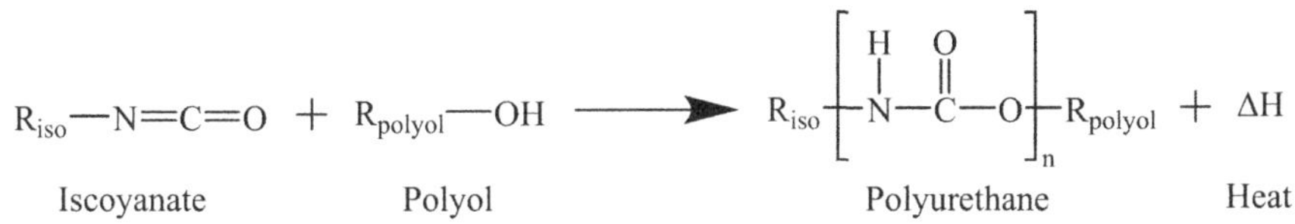

Scheme 9.1 Reaction scheme of polyurethane production [7]

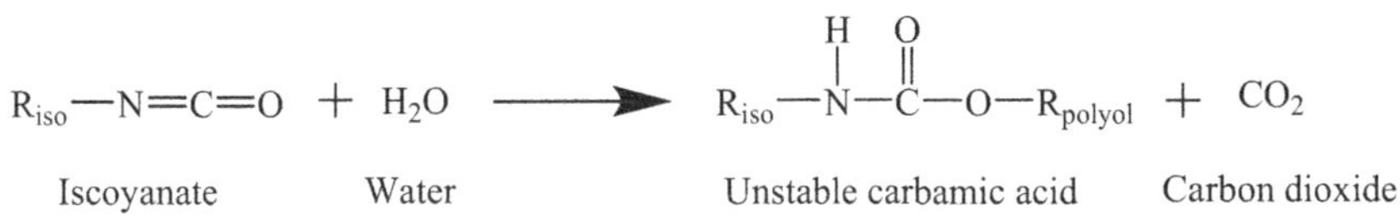

Scheme 9.2 Reaction scheme of the isocyanate with water [7]

9.1.1 Polyurethane Foams

PU foams correspond to 50% of global PU consumption, being mainly classified as flexible foams or rigid foams. They are the main type of foam used in construction, since they can be used to produce thermal insulation boards, sandwich panels, spray in-place foam, structure panels, pillows, mattress, etc. as illustrated in Fig. 9.1 [5].

All of these types of foams follow similar chemistry, being synthetized by the reaction between the OH (hydroxyl) groups of a polyol with the NCO (isocyanate functional groups) of an isocyanate, as illustrated in Scheme 9.1 [6].

Where R_{iso} and R_{polyol} are isocyanate and polyol moieties, respectively. Normally, the polyol has an average functionally ~3 and the isocyanate have an average ~2, therefore, the ensuing polymer is highly crosslinked. This means that PU foams are thermoset polymers.

Besides the polymer reaction, the cellular structure of PU foams results from a parallel reaction between the isocyanate groups and water which releases CO_2 as presented in Scheme 9.2, which is trapped within the cells.

The PU foams expanded using water as blowing agent do not present thermal conductivity suitable for thermal insulation applications, hence, these type of materials requires the use of physical blowing agents (such as solvents with low boiling point such as *n*-pentane, acetone, or hexane) which expand the polymer by vaporization, are frequently used [8].

9.1.2 Expanded Polystyrene

Expanded PS (EPS) is ideal to be used as lightweight filler, damping and insulation materials, hence, it is mainly used in building applications to insulate wall structures (in the cavity, internally or externally), roofs and floors [9].

Scheme 9.3 Polymerization reaction of styrene

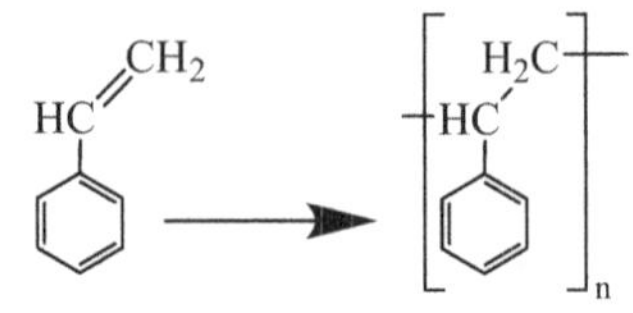

It is a thermoplastic polymer based on the styrene monomer (as presented in Scheme 9.3) and to produce EPS products, the PS resin is blended with 4–7% (by weight) of a hydrocarbon blowing agent (usually *n*-pentane) to form an expandable PS commonly referred in the industry as 'bead'. Next, the beads are expanded to about 40 times their original size using a flow of steam. This causes the blowing agent to boil and a honeycomb of closed cells is formed. In this form EPS, consists up to 98% of gas [9].

Similarly, boards of extruded PS (XPS) are commonly used in construction, which are manufactured using extrusion. These materials presents a closed-cell structure which provides low thermal conductivity, prevents the penetration of water and ensure strength and durability [10].

9.2 Recycling of Polymer Foams

As mentioned, polymer foams are mainly produced using petroleum derivatives. In addition, not all polymers can be easily reprocessed therefore, after their use they are normally disposed in land fields or burned for energy [11, 12]. As a consequence, the pollution associated with polymers has become one of the most important environmental issues, since the disposal of these products overwhelms the world's ability to deal with them.

Following this issue, the European Community is implementing measures for the prevention of the disposal of plastic urban residues, following the circular economy principles [13].

Hence, the European Union has become a pioneer against plastic pollution. In fact, the Directive 2019/904 of the European Parliament which aims to reduce the volume and impact of single-use plastics products on the environment is a major example of the measures being taken [14].

There are two major methods for recycling polymers: mechanical recycling and chemical recycling. The type of the recycling process to adopt is dependent on the type of polymer since thermoset and thermoplastics behave differently when exposed to heat. While thermoplastics melt under heat, thermoset can degrade at elevated temperatures [9]. In Fig. 9.2, the schematic representation of mechanical and chemical recycling of foam wastes, to produce recycled foams is represented.

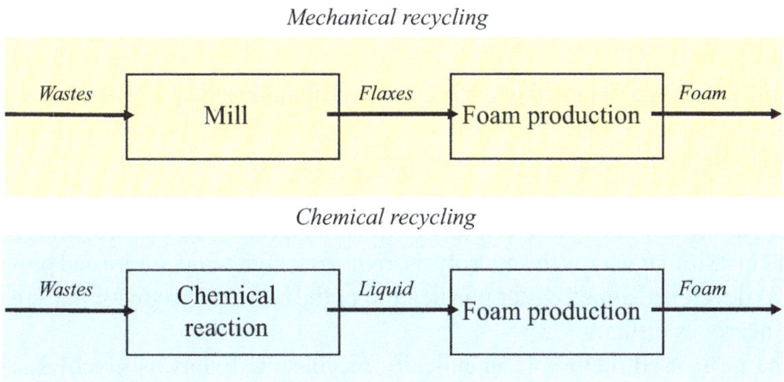

Fig. 9.2 Schematic of mechanical and chemical recycling

In mechanical recycling, polymer scraps are milled and reintroduced in the production process, while the chemical recycling follows the degradation principle. In this case, the polymer wastes are depolymerized via a chemical reaction into oligomers, which can be used as raw-materials to produce new polymers [15]. Nonetheless, it must be highlighted that the life cycle assessment (LCA) is critical to evaluate whether these strategies are in fact suitable to ensure eco-friendly products, since this methodology evaluates all the environmental impacts associated to the life cycle of the products [16–18].

9.2.1 PU Foams

The majority of PU foams are thermoset polymers and are not easily biodegradable [19]. Consequently, only 29.7% is recycled, while 39.5% is recovered through energy recovery and 30.8% is disposal of in landfills [20].

The use of mechanical recycling is a simple and cheap process which can be an opportunity to increase the eco-efficiency of the polymers. Using this methodology, PU foam wastes can be used to produce agglomerates, however the market is very limited and so is the added value [21–24]. In turn, in chemical recycling, polymer wastes are depolymerized, being the recycled products used to produce new materials [7]. Despite of the necessity of equipment's, energy and reactants, this option affords chemicals that can be used in the production of several/different materials. Hence, this route can be economically advantageous. In fact, there are several thermo-chemical methods that can be used, such as hydrolysis, aminolysis, alcoholysis or glycolysis [25]:

i. Hydrolysis of PU uses overheated steam which hydrolyzes urethane bonds leading to the formation of polyols and amines [26]. Afterwards, purification is required, being the recycled products used as raw materials for PU production.

However, the purification step is costly, making the process not economically attractive;

ii. Aminolysis is associated to the breaking urethane bonds using amines (e.g. with dibutylamine or ethanolamine). The final products are disubstituted ureas and polyols which can be used to produce new PU [27];
iii. The alcoholysis used alcohols as decomposition reactants and is based on the substitution of one hydrogen atom in water by an alcohol. The reaction is carried out in a similar way as the hydrolysis, requiring high temperature and pressure. It was developed aimed easier purification of the recycled products, but separation is nearly as difficult [28];
iv. The most used method to chemically recycle PU foams, is glycolysis, which uses high boiling glycols as cleavage agent [29]. Yet, similar to the methods described previously, it requires purification, high energy demands and long reaction times.
v. Nowadays, a new route for recycling PU scraps is emerging—the acidolysis - which converts PU wastes into a recovered polyol (RP), using dicarboxylic acid(s) (DA) [30–33]. As opposed to glycolysis, in acidolysis only one phase and no residues are obtained.

Despite of the method used, after the chemical reaction, the RP can be used as partial substitute of the petroleum-based counterparts to produce both flexible [31] and rigid [30] PU foams. Yet, the LCA results regarding chemical recycling of PUF indicate that optimization of the recycling conditions are still necessary [16–18].

9.2.2 EPS

Thermoplastics can melt under heat, hence the waste generated in the process or that resulted from the end-of-life of products can be easily recycled. In simple terms, thermoplastic wastes such as those from EPS products can be milled in to a powder and blended with cement, sand and wire mesh to produce sandwich blocks [34]. Yet, this causes a large drop in the polymer molecular weight due to thermal degradation. In addition, the ensuing material has very different characteristics when compared to the original one. Nonetheless, similar to the thermoset polymers, thermoplastic materials such as EPS can be recycled using chemical approaches.

One easy way of recovering polymers is by using appropriate solvents. These solvents must have the ability to dissolve the polymer with negligible degradation of their properties. As example, toluene was used as solvent to recycle EPS and it was proved to be more effective than benzene, ethylbenzene or *p*-xylene, at 360 °C for 20 min [34]. Another option is to use metal-based catalysts at 300–450 °C. Using this process styrene monomer is obtained with a high selectivity using relative low temperature. Furthermore, oxides such as MgO, CaO or BaO can be used as catalysts. In fact, using BaO it was observed a 90 wt. % conversion of PS into styrene [34].

Similarly to the PU recycling reactions, in all these chemical approaches, purification may be required.

Afterwards, the recycled products can be used as partial substitute of the petroleum-based styrene monomer to produce PS products. However, during these approaches, the material loses its cellular structure. Thus, to produce recycled EPS, it is necessary to blend the recycled PS resin with the expanding agent. This can be achieved using i.e. co-extrusion. The ensuing recycled beads can be molded into new and recycled EPS products [34].

Similar to the approach used in PUF, LCA is essential to evaluate the benefits of the recycling EPS. From the results available in literature, it can be observed that the manufacturing process of EPS is the major contributor of the overall environmental impacts while recycling has the lower impact. From these observations, it is concluded that recycling is in fact a suitable alternative to the disposal of EPS wastes [35, 36].

9.3 Other Alternatives

Whilst recycling tends to be environmentally preferable, energy recovery via incineration can provide valuable energy while hazardous additives are safely destroyed. Incineration for energy recovery can be in fact a viable alternative to recycling because it enables to generate substantial amounts of energy, since polymers have higher calorific value than coal and approximately the same value as fuel oil [9]. In addition, energy recovery can be the only suitable disposal method for wastes which do not have market. In fact, waste-to-energy and other thermal processing routes, such as gasification, pyrolysis and combustion has contributed for the disposal of significant amounts of PU and PS foam scraps [25, 37]. There are several contributions on literature regarding the energy recover from construction foams, however it is not used to produce new materials. For that reason, since it falls out the scope of this review, this alternative will not be further discussed.

9.4 Life Cycle Analysis

Environmental assessments are critical to evaluate the viability of the circular economy strategies. This tool is crucial to evaluate if the circular economy strategies contribute to the production of more eco-friendly products or not. This can be achieved by using the life cycle assessment (LCA) methodology, which complies and evaluates the inputs, outputs and environmental impacts of a product throughout its life cycle.

Lindstrom et al. [38] evaluated the EPS handling system and several disposal alternatives using LCA. While a closed-loop reuse system for EPS was most environmentally desirable, the study results indicated that conventional recycling of EPS is the only disposal scenario that generates net environmental benefits while also being logistically feasible. In a similar manner, Quinteiro et al. [16] conducted the LCA of different strategies to produce rigid PUF, including the use of recovered polyol as partial substitute of the conventional petroleum-based counterparts. It was claimed that the environmental impacts from the polyol recovery via acidolysis exceeded the environmental benefits of PUF produced using partial replacement of virgin polyol by the recovered polyol. In turn, Marson et al. [39] conducted the LCA of PUF from polyols obtained through chemical recycling, reporting that the use of recycled polyol obtained via glycolysis can contribute to the reduction of the potential environmental impacts of PUF when compared to the use of virgin polyol, provided that physical and thermal characteristics are guaranteed.

The discrepancy of results highlights the need to carry out further studies to improve the environmental performance related to recycling processes.

9.5 Challenges and Future Perspectives

Despite recycling is not the top option in the waste hierarchy, efficient recycling is widely acknowledged to mitigate the negative effects of plastic wastes. Yet, the recycling faces different challenges, and it must be continuously improved.

As previously shown, the recycling methods are very sensitive to the type of polymer, hence, separation of the different type of plastics make its recycling easier. In addition, the presence of contaminants in the waste streams, such paper, metals, glasses, or organic matter difficult the plastic recycling. For these reasons, sorting is a current challenge that foam recycling is facing. To resolve this problem, recycling facilities are being implemented at production sites to manufactures recycle their production wastes. Yet, this is not a solution to the end of life of products, which can be minimized by legislation on the waste management. Another issue associated to the recycling of foams, specially via chemical methods is the fact that these processes require high energy demand and chemical reactants. Hence, in the future the use of renewable energy and bio-based chemicals will improve the sustainability of the recycling of plastics.

9.6 Conclusions

Polymeric foams used in construction on them are mainly produced using PU or PS resins and different recycling methodologies can be used. The simplest way to recycle these materials is using mechanical methods i.e. mill them and reintroduce the ensuing powder in the production process. However, using this method, the ensuing materials has very different properties when compared with the original foam. In turn, both PU and PS foams can be recycled using chemical methods. Different routes can be found in literature, being the reaction products used to synthetize recycled PU and PS. Nonetheless, further developments are still required to ensure the sustainability of these materials. As example, mechanical recycling is not suitable to obtain value-added products and the environmental impacts associated with chemical recycling can exceeded the environmental benefits of producing materials from recovered raw materials. Following the ambition for a sustainable future, this overview aimed to highlight the benefits and challenges of the recycling of construction foams, contributing that way to the knowledge available about the recycling construction foams.

References

1. Lee ST, Ramesh NS (2004) Polymeric foams: mechanisms and materials. CRC Press, London and New York
2. David E (2001) Polymer foams: trends in use and technology. Rapra Technology Ltd., Shawbury
3. Bulk Chemicals, https://www.grandviewresearch.com/industry-analysis/polymer-foam-market. Last Accessed 01 Apr 2023
4. Metgen, https://www.metgen.com/wp-content/uploads/2020/10/MetGen-Company-Presentation.pdf. Last Accessed 01 Apr 2023
5. Ashida K (2007) Polyurethane and related foams chemistry and technology. Taylor & Francis Group, Florida
6. Szycher M (2006) Szycher's handbook of polyurethanes, 2nd edn. CRC Press, London and New York
7. Gama N, Ferreira A, Barros-Timmons A (2018) Polyurethane foams: past, present, and future. Mater (Basel) 11:–35
8. Singh SN (2002) Blowing agents for polyurethane foams. Rapra Technology Ltd., Shawbury
9. Eaves D (2004) Handbook of polymer foams. Rapra Technology Ltd., Shawbury
10. Aksit M, Zhao C, Klose B, Kreger K, Schmidt HW, Altstädt V (2019) Extruded polystyrene foams with enhanced insulation and mechanical properties by a benzene-trisamide-based additive. Polym (Basel) 11:268
11. Gardiner F, Garmson E (2010) Plastics and the environment. iSmithers Rapra Publishing, Shawbury
12. Harrison RM, Hester RE (2018) Plastics and the environment. Royal Society of Chemistry, Cambridge
13. Cregut M, Bedas M, Durand M, Thouand G (2013) New insights into polyurethane biodegradation and realistic prospects for the development of a sustainable waste recycling process. Biotechnol Adv 31:1634–1647
14. Directive (EU) 2019/904 of the European Parliament and of the Council, 2019
15. Yang W, Dong Q, Liu S, Xie H, Liu L, Li J (2012) Recycling and disposal methods for polyurethane foam wastes. Procedia Environ Sci 16:167–175
16. Quinteiro P, Gama NV, Ferreira A, Dias AC, Barros-Timmons A (2022) Environmental assessment of different strategies to produce rigid polyurethane foams using unrefined crude glycerol. J Clean Prod 371:133554
17. Manzardo AL, Marson A, Roso M, Boaretti C, Modesti M, Scipioni A (2019) Life cycle assessment framework to support the design of biobased rigid polyurethane foams. ACS Omega 4:14114–14123
18. Marson A, Masiero M, Modesti M, Scipioni A (2021) Life cycle assessment of polyurethane foams from polyols obtained through chemical recycling. ACS Omega 6:1718–1724
19. Yadav D (2019) Sustainable waste management of polyurethane polymers. Int J Sci Res Chem 4:1–5

20. Gadhave RV, Srivastava S, Mahanwar PA, Gadekar PT, Ravindra SS, Gadhave V, Mahanwar PA, Gadekar PT (2019) Recycling and disposal methods for polyurethane wastes: a review. Open J Polym Chem 9:39–51
21. Gama N, Godinho B, Barros-Timmons A, Ferreira A (2021) Insights into PU/EVA blends produced using industrial residues towards eco-efficient materials. J Polym Environ 30:1–11
22. Gama N, Godinho B, Barros-Timmons A, Ferreira A (2021) PU/lignocellulosic composites produced from recycled raw materials. J Polym Environ 30:1451–1461
23. Gama N, Godinho B, Barros-Timmons A, Ferreira A (2021) PU composites based on different types of textile fibers. J Compos Mater 30:194–205
24. Gama N, Ferreira A, Barros-Timmons A (2020) 3D Printed thermoplastic polyurethane filled with polyurethane foams residues. J Polym Environ 28:1560–1570
25. Zia KM, Bhatti HN, Bhatti I (2007) Methods for polyurethane and polyurethane composites, recycling and recovery: a review. React Funct Polym 67:675–692
26. Johnson OB (1977) US 4025559 A method for continuous hydrolysis of polyurethane foam in restricted tubular reaction zone and recovery
27. Kanaya K, Takahashi S (1994) Decomposition of polyurethane foams by alkanolamines. J Appl Polym Sci 51:675–682
28. Behrendt G, Naber BW (2009) The chemical recycling of polyurethanes (Review). J Chem Technol Metall 44:3–23
29. Machado RM, Farrell BE (1994) US 5300530 A process for modifying the glycolysis reaction product of polyurethane scrap
30. Gama N, Godinho B, Marques G, Silva R, Barros-Timmons A, Ferreira A (2021) Recycling of polyurethane by acidolysis: the effect of reaction conditions on the properties of the recovered polyol. Polym (Guildf) 219:123561
31. Gama N, Godinho B, Marques G, Silva R, Barros-Timmons A, Ferreira A (2020) Recycling of polyurethane scraps via acidolysis. Chem Eng J 395:125102
32. Godinho B, Gama N, Barros-Timmons A, Ferreira A (2021) Recycling of different types of polyurethane foam wastes via acidolysis to produce polyurethane coatings. Sustain Mater Technol 29:e00330
33. Ferreira A, Barros-Timmons A, Godinho B, Gama N, Silva R, Marques G, Teixeira S (2018) P449.8 WO Methods for recycling polyurethane using dicarboxylic acids
34. Uttaravalli AN, Dinda S, Gidla BR (2020) Scientific and engineering aspects of potential applications of post-consumer (waste) expanded polystyrene: a review. Process Saf Environ Prot 137:140–148
35. Lim YS, Izhar TNT, Zakarya IA, Yusuf SY, Zaaba SK, Mohamad MA (2021) Life cycle assessment of expanded polystyrene. IOP Conf Ser Earth Environ Sci 920:012030
36. Junior HRA, Dantas TET, Zanghelini GM, Cherubini E, Soares SR (2020) Measuring the environmental performance of a circular system: emergy and LCA approach on a recycle polystyrene system. Sci Total Environ 726:13811
37. Maafa IM (2021) Pyrolysis of polystyrene waste: a review. Polym (Basel) 13:1–30
38. Lindstrom T, Hickers AL (2022) Life cycle assessment of expanded polystyrene shipping boxes at a public research institution: insights for infrastructure at the end of life. Environ Res Infrastruct Sustain 2:031001
39. Marson A, Masiero M, Modesti M, Scipioni A, Manzardo A (2021) Life cycle assessment of polyurethane foams from polyols obtained through chemical recycling. ACS Omega 6(2):1718–1724

Chapter 10
By-product Phosphogypsum Valorisation Possibilities in the Context of Circular Economy of Building Materials

Ignacio Villalón Fornés, Danute Vaiciukyniene, Dalia Nizeviciene, Diana Bajare, Ruben Paul Borg, and Reda Bistrickaite

Abstract Phosphogypsum (PG) is the most significant solid waste in the world. However, only 15% of it is recycled, and the rest is stored in useless and environmentally-damaging stockpiles, which are continually growing. Therefore circular economy of PG through its utilisation in useful applications, is relevant and critical from an ecological perspective. In this context, its utilisation in building materials is among the most promising and attractive recycling possibilities, since PG typically exhibits excellent binding properties. However the applicability of PG is often restrained by its radiological risks and economic non-competitivity in comparison to traditional building materials, such as natural gypsum. In some cases, these obstacles refer to cultural prejudices and misinformation rather than real issues. Therefore, through a comprehensive literature review, this article considers the real potential of PG in building materials, providing useful information to interested stakeholders.

Keywords Phosphogypsum · Waste valorisation · Building materials · Circular economy

---

I. V. Fornés · D. Vaiciukyniene (✉) · R. Bistrickaite
Faculty of Civil Engineering and Architecture, Kaunas University of Technology, Studentu 48, 50270 Kaunas, Lithuania
e-mail: danute.vaiciukyniene@ktu.lt

D. Nizeviciene
Faculty of Electrical and Electronics Engineering, Kaunas University of Technology, Studentu St. 48, 51367 Kaunas, Lithuania

D. Bajare
Institute of Materials and Structures, Riga Technical University, Kalku 1, Riga 1658, Latvia

R. P. Borg
Faculty for the Built Environment, University of Malta, Msida 2080, MSD, Malta

L. Bragança et al. (eds.), *Creating a Roadmap Towards Circularity in the Built Environment*, Springer Tracts in Civil Engineering,
https://doi.org/10.1007/978-3-031-45980-1_10

10.1 Introduction

In the context of global warming and sustainable use of energy and natural resources, the need to develop real circular economies (CE) is becoming of critical importance. The main goals of CE are: '(1) eliminate waste and pollution; (2) circulate products and materials; (3) and regenerate nature' [1]. In this sense, the recycling of phosphogypsum (PG), the "world's most significant solid waste [2], is crucial for CE on a global scale since it directly relates to the first and second CE goals and, indirectly, to the third one".

PG is a by-product of the industry of mineral fertilisers, which, nowadays, are irreplaceable in agricultural activities. The main nutrients needed for plant growth are nitrogen, potassium, and phosphorus [3]. However, the production of these fertilisers carries notorious secondary effects, some of which are hazardous to the environment. The case of phosphorus-based fertilisers deserves special attention, as they are responsible for the production of the leading solid residue in the world. In fact, every year the PG global production increases and currently it is about $1.8–2.8 \times 10^8$ t [2]. However, until now, most of the produced by-product, approximately 85%, has been stored in open dumps or discharged in water bodies, and only with 15% of it being recycled.

The most problematic issue related to long-term storage is not only the fact that humungous stockpiles occupy extensive area but also cause other kinds of ecological damage: radionuclides, waste-soluble phosphates, fluorides, sulphates, and many heavy metals leach into underground waters and end up being absorbed by plants, thus seriously harming the entire ecological environment [4, 5]. Heavy metals migrate along the food chain and enter the human body [6].

Considering the global relevance of this problem, it is critical for CE to find effective ways to utilise PG. Abundant research is carried out in this area, but the obtained results do not necessarily reach lead to implementation due to economic, regulatory, or even cultural reasons [7]. Thus, the current publication aims to review the PG utilisation possibilities in the field of building materials, thus providing PG use cases which might be of interest to stakeholders of both the public and private sector stakeholders.

To address this objective, a comprehensive literature review has been carried out, within the *Google Scholar* database, using keywords "phosphogypsum utilisation", "building materials", "binding materials", "press-forming", "agriculture", "radioactivity", and pertinent synonyms.

10.2 Characterisation of Industrial By-product Phosphogypsum

PG is an undesirable by-product resulting from the production process of phosphate fertilisers. The raw material used during the process consists mainly of phosphate ores of igneous or sedimentary nature [8]. According to the mineral composition, impurities and microstructure, the main minerals of these rocks belong to the apatite or phosphorite group, for which the general chemical formula is $Ca_{10}(PO_4,\pm CO_3,\pm OH)_6(OH,F,Cl)_2$ [7]. To produce fertilisers, phosphate ores are processed into orthophosphoric acid (H_3PO_4), through the so-called "wet process", during which the phosphate rock is digested by concentrated sulfuric acid [7]. Furthermore, it can be clarified that H_3PO_4 is not the final product but an intermediate feedstock to manufacture the final phosphate fertilisers [8]. The general reaction is provided by the International Atomic Energy Agency (IAEA) [7] in Eq. (10.1). This reaction occurs in two stages: first, the phosphate raw material reacts with a mixture of sulfuric and phosphoric acids (the latter formed during the process); then, subsequently, calcium sulphate compounds crystallise.

$$Ca_5(PO_4)_3F + 5H_2SO_4 + 5nH_2O \rightarrow 5CaSO_4 \times nH_2O + HF + 3H_3PO_4 \quad (10.1)$$

Equation (10.1) reveals that in addition to H_3PO_4, a by-product based on calcium sulphate crystals ($CaSO_4$) and known as PG is produced. For each tonne of orthophosphoric acid, up to 4.5–5.5 t of PG by-product are generated [9, 10]. Depending on the process conditions (sulfuric acid concentration, temperature, phosphate raw material composition, and impurities), the solid phase of calcium sulphate can present three alternative forms: dihydrate (DH): $CaSO_4{\cdot}2H_2O$; hemihydrate (HH): $CaSO_4{\cdot}0.5H_2O$; anhydrous: $CaSO_4$.

According to the mineral composition, PG belongs to the group of gypsum raw materials, as it contains 80% to 98% calcium sulphate [4]. Furthermore, PG is typically characterised by a certain content of undesirable impurities, such as unreacted apatite or phosphorite and H_3PO_4, soluble and insoluble phosphate and fluoride compounds, salts of iron, aluminium and other heavy metals, REE and quartz [6, 11, 12]. PG usually contains an elevated radionuclide content [7, 13, 14], which significantly hinders its applicability.

10.3 A Sustainable Approach to the Utilisation of PG

Currently, the main applications of PG, according to IAEA [7], are soil stabilisation and road base construction [15–18], agriculture and building materials. There are other promising PG valorisation opportunities and fields of activity, such as is the biochemical approach presented by Chernysh et al. [5], consisting in the exploitation of various bioprocesses, with PG valorisation from existing stockpiles

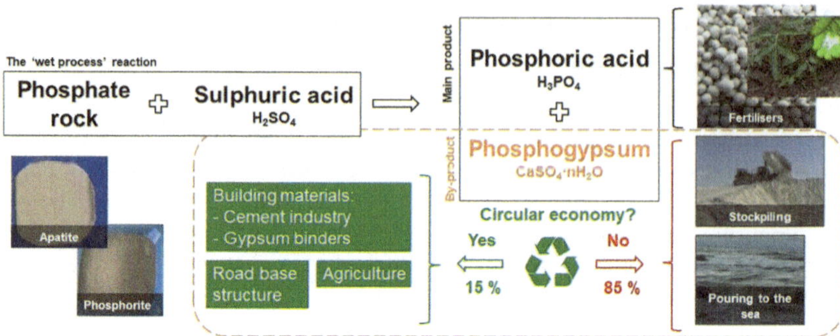

Fig. 10.2 Scheme of PG production, disposal, and main ways of utilisation

and implementing innovative biotechnological solutions for phosphorus raw materials processing. A schematic view of the PG production and main options for its utilisation is presented in Fig. 10.2. Although the current article focuses on the field of building materials, the utilisation of PG in agriculture is also commented on due to its relevance.

10.3.1 PG in Agriculture: Acidic and Saline Soils Improvement and Other Uses

Agricultural activity is one of the main ways to use PG on large scales, as it has shown good performance in soil, improving it in various ways.

First, PG can be used to improve acidic soils. As explained by Qi et al. [2], PG is an excellent alternative to more expensive inorganic improvements for acidic soils, such as natural gypsum (which is scarcely available) or lime, which exhibits poor solubility, so its nutrients within the soil profile are limited to a thin layer [19]. Acidic soils appear due to long-term fertilisation. The problems of acidic soils are related to toxic metals, especially exchangeable aluminium ions (Al^{3+}), and the lack of calcium [20], which is necessary for plants to grow. The addition of PG to the soil determines the reduction of the exchangeable aluminium ions of $Al(OH)_3$, binding them to $Al(OH)SO_4$ compounds. This process was named the "self-limiting effect" by Reeve and Sumner [21]. Moreover, PG is a suitable amendment for saline soils, effectively neutralising Na^+ ions typical of their composition [2].

PG treatment can also improve agricultural soils by enhancing their resistance to crusting [22] and rainwater runoff-induced erosion [23] or immobilising toxic heavy metals in the soil by binding them in complex and non-soluble metal-fluorine compounds [2]. In addition, PG can be used as a source material for calcium and sulphur, which are necessary elements for the growth of crops [24, 25]. In summary, PG can be used effectively to improve agricultural activities.

10.3.2 PG in Building Materials

According to the content of the primary substance, calcium sulphate, PG could be exploited as a raw material for cement and gypsum binders. However, compared to natural gypsum, the quality of PG is usually lower because of the acid-soluble impurities (phosphate and fluoride), which limit its use. Soluble phosphates prolong gypsum hydration and reduce gypsum specimens' compressive strength (*CS*) [26, 27]. It is also the main cause of a long setting time and a low early cement strength [26]. Therefore, extensive research has been conducted on neutralising or removing soluble phosphate and fluoride compounds from the PG. The principal methods are the following: washing of PG with water; neutralisation of impurities in citric, nitric, or sulphuric acid solutions with subsequent washing with water [26–28]; application of various neutralising, mineralising and crystallisation-regulating additives before dehydration of PG [29, 30]; thermal treatment and neutralisation in lime suspension [31, 32].

10.3.2.1 The Problem of Radioactivity

Radioactivity is the key factor that hinders the application of PG in building products. PG generally contains a relatively high amount of ^{226}Ra radionuclide, which decays to ^{222}Rn [33, 34], a toxic gas to which exposure in indoor environments causes between 3 and 14% of global lung cancer cases, according to the World Health Organisation [35]. This issue restricts the use of PG in building materials, as the competent authorities limit the radioactivity levels of building products. For example, the European Commission established a limit gamma emission rate of 1 mSv/year [36]. Usually, this limit is not depicted directly but is evaluated through the radioactivity concentration index I, calculated from the activity concentration of radionuclides ^{226}Ra, ^{232}Th, and ^{40}K. A material is considered safe from a radiological perspective when $I \leq 1.0$, if employed in bulk amounts (slabs, wall blocks, etc.) or $I \leq 6.0$, if used in superficial applications (wall coatings, ceilings, etc.) [37]. If this condition is not met, the material could be hazardous to human health, so additional investigations must be performed.

According to Kovacs et al. [38], the PG produced in most countries presented I value between 0.7 and 2.0, although in some cases (UK and Morocco), the values were much higher. This variety depends on the nature of the phosphate rock from which PG is processed [39–41]. For instance, the PG produced from igneous phosphate ores tends to exhibit a lower radioactivity level than those manufactured from sedimentary rocks. It must be considered that the imposed limits apply to the final building material, not to each ingredient. This is the main reason why PG is mainly utilised in small amounts as a supplementary material of OPC or asphalt (the final I index will not exceed the limit) and not so frequently as the main binding material since, in this case, the I limit could easily be surpassed. In the latter case, employing a PG with $\mathrm{I} \leq 1.0$ index is crucial.

10.3.2.2 PG as a Cement Retarder

One of the most important uses of natural gypsum is the manufacture of ordinary Portland cement (OPC), which is included in amounts of 2.5–5%, acting as an OPC setting time retarder [42, 43]. If gypsum was not present during mixing OPC with water, tricalcium aluminate compounds (C3A) would immediately hydrate into calcium hydro aluminate compounds (CAH), which would not allow further mixing. When there is gypsum, a layer of ettringite and calcium hydro silicates (CSH) precipitate on the surfaces of the cement particles, and so the start of hydration is postponed (by about 2 h), allowing workability for a longer time [42]. Therefore, a retarder such as gypsum is crucial for suitable OPC hydration. In addition, gypsum is a retarder of the setting time and acts as an accelerator of the early strength of OPC mixtures.

In the same way, PG could be utilised as an OPC retarder. The retarding effect could be attributed not only to the ettringite coating but also to the soluble content of phosphorus and fluorine in PG, which, during OPC mixing with water, precipitates in the form of phosphate and fluoride compounds ($Ca_3(PO_4)_2$ and CaF_2), forming a layer that hinders further cement hydration and determines a later setting time than OPC with gypsum [26]. However, this additional retarding effect caused by the fluoride and phosphate impurities is usually too long. Therefore, the purification of the PG from the soluble phosphate and fluoride compounds is necessary to obtain a good performance as an OPC retarder.

10.3.2.3 PG as a Binding Material

Most of the research on using PG in building materials attempts to explore the potential of PG as a secondary material that acts as an additive or modifier of the main binding material, such as OPC or asphalt, in building products or roads. The problem with this approach is self-evident: PG is used in rather small amounts. In contrast, it must be considered that PG, in essence, is the same material ($CaSO_4$) as natural gypsum, which is one of the most widely used binding materials in the world. Therefore, in the same way, PG can be used as the primary binding material for building products. Its utilisation rate would be much higher, allowing it to solve the environmental problems derived from PG stockpiling more efficiently. It must be observed that PG (and natural gypsum) exhibit binding properties only in its HH form. When mixed with water, it hydrates and hardens, achieving its DH phase [44].

Researchers are exploring ways to increase the applicability of PG as a binding material by improving its main properties: mechanical strength, degree of hydration, water resistance, thermal and acoustic insulation performance, etc. The main focus is on improving mechanical properties. The mechanical performance of PG mostly depends on its microstructure, chemical composition, and processing method through which the final products are obtained.

First, abundant research has been conducted on the influence of microstructure and chemical composition. Valančius et al. [45] created a laboratory imitation of the "wet process" to determine the influence of reaction conditions on the microstructure and properties of the produced PG. Applying wet process conditions of 8 h duration, 97 °C temperature, and 1.03 apatite/sulfuric acid ratio, the α-type HH PG crystals were regularly orientated and presented needle shapes. Later, the HH PG powder was ground and mixed with water. The suitable microstructure of the HH crystals determined the fast hydration of the pastes and the excellent mechanical properties of the hardened specimens, even reaching a *CS* of 27 MPa. In a similar way, Leškevičienė et al. [46] studied the influence of the mineralogical composition of HH-PG on the physical and mechanical properties of hardened specimens.

Moreover, the mechanical properties of DH PG can also be improved by processing the specimens in certain ways, such as adding fibre, press-forming, autoclaving, or treating with a constant magnetic field. Press-forming has proven to be a very effective method for improving the mechanical properties of specimens. Zhou et al. [47] suggested a processing method called 'the two-step hydration process', including press-forming and subsequent immersion in water, with which high-strength non-fired PG mortar bricks were manufactured. During the investigation, it was found that the higher the press-forming pressure determined higher *CS* values, even reaching a 7-day *CS* value of 32 MPa when the pressure was 20 MPa. However, for economic reasons, the author recommends using a lower press-forming pressure of 10 MPa, with which the specimens still present a satisfactory *CS* value of 29 MPa. Similarly, Villalon Fornes et al. [48] explored this method using plain PG (instead of mortar) and obtained high *CS* values. In addition, Zhou et al. employed intermittent press-forming hydration to produce load-bearing tiles [49] and plasterboards [50]. Therefore, strong PG products, such as tiles, plasterboards, bricks or blocks, could be created by applying a suitable processing method and a pertinent type of PG.

10.4 Discussion

10.4.1 Insights and Risks of PG Utilisation in Agriculture

As described in Sect. 10.3.1, PG exhibits suitable properties to be utilised to amend and improve certain soil, render the types to be suitable for agricultural purposes. However, certain aspects must be considered. PG contains noticeable amounts of P, F, and heavy metals, so it can contaminate the soil or leach water, creating even more serious problems than those it is intended to solve. Phosphorus dissolved in leaching water can lead to the eutrophication of rivers and lakes [51], and plants with large amounts of accumulated fluorine can led to dental or skeletal fluorosis if plants containing high amounts of it are digested [52]. Similarly, heavy metals (especially Cd) and radionuclides in the PG composition can also cause health problems when ingested. Furthermore, an excessive amount of PG in the soil can reduce microbial activity in the soil, reducing agricultural yield [2]. Hence, the chemical composition of PG and the extent of its utilisation in the soil must be evaluated and monitored.

10.4.2 Possibilities in the Field of Building Materials

The potential of PG to be used in building materials as a substitute for natural gypsum is great. However, there are some objective constraints for PG usage in building materials: aside from the mentioned radioactivity, the economic factor is also relevant, since the additional costs derived from PG processing (purification, drying, and slightly different processing methods) may make PG non-competitive when compared to natural gypsum. However, in places where natural gypsum is scarcely available (for example, Japan, the Netherlands, and France), PG could be a competitive solution [40].

Some representative examples of PG applications in building materials are given in Table 10.1.

10.5 Conclusions

Valorisation of by-product PG is a non-trivial subject for circular economy. The real constraints to effective PG application in building materials are high radioactivity and economic non competitively (regarding natural gypsum). However, in some cases these risks do not apply. If that is the case, stakeholders must be open to the incorporation of PG to their products (OPC retarder, tiles, blocks, bricks, ceiling, partition boards, etc.), not behaving according to prejudices, cultural obstacles or simply, inertia, but making a step forward towards a more real concrete circular economy.

Table 10.1 PG processing methods to make it suitable for building materials

Processing method	Final product	References
Cement retarder was prepared from PG. It was dried and ground	Belite-ferroaluminate cement	Yang et al. [53]
Calcined at 500 °C and neutralisation with 4.00% of lime	Super sulphated cement	Liu et al. [12]
Addition of aqueous citric acid solution (1–5%)	Gypsum plaster, Portland and Portland/slag cements	Singh et al. [26]
Bitumen modified with 10% phosphogypsum	Asphaltic bitumen	Cuadri et al. [10]
Use of lime modified PG as a partial filler of white paints	White textured paints	Valančius et al. [54]
Neutralisation in 1% lime suspension at 60 and 80 °C temperature	PG binding material	Nizevičienė [4]
Addition of waste fluid catalytic cracking (0.5–10%), with and without sonication	PG binding material	Vaičiukynienė et al. [55]
Addition of carbide slag (1–5%)	PG binding material	Wu et al. [6]
Addition of $Ca(OH)_2$ (0.5–1.5%)	PG binding material	Villalón Fornés et al. [41]
Calcination and further addition of fly ash and lime	PG plasterboards	Kumar [56]
Granulation with water, press-forming and hydration with intermittent pressing	Paper-free and fiber-free plasterboards	Zhou et al. [50]
Intermittent press-forming hydration	PG load-bearing tiles	Zhou et al. [49]
Press-forming with two-step hydration process	PG load bearing non-fired bricks and blocks	Zhou et al. [47] Villalon Forncs ct al. [48]
Partial substitution 10–30% of natural gypsum in the gypsum binder	Gypsum-based biocomposite	Bumanis et al. [57]

References

1. Ellen MacArthur Foundation. How to build a circular economy. https://ellenmacarthurfoundation.org/. Accessed 16 March 2023
2. Qi J et al (2023) Application of phosphogypsum in soilization: a review. Int J Environ Sci Technol 1–16
3. Kirkby EA (2023) Chapter 1—Introduction, definition, and classification of nutrients. In: Rengel Z, Cakmak I, White PJ (eds) Marschner's mineral nutrition of plants, 4th edn. Academic Press, San Diego, pp 3–9
4. Nizevičienė D (2001) Pushidratinio fosfogipso susidarymas, savybės ir panaudojimas: daktaro disertacija (EN—Formation of semihydrate phosphogypsum, properties and use. Doctoral thesis). Kaunas University of Technology, Kaunas

5. Chernysh Y, Yakhnenko O, Chubur V, Roubík H (2021) Phosphogypsum recycling: a review of environmental issues, current trends, and prospects. Appl Sci 11(4):1575
6. Wu F et al (2022) Highly targeted solidification behavior of hazardous components in phosphogypsum. Chem Eng J Adv 9:100227
7. IAEA (2013) Radiation protection and management of NORM residues in the phosphate industry, safety reports series no. 78. International Atomic Energy Agency (IAEA), Vienna
8. U.S. Department of the Interior & U.S. Geological Survey (2023) Mineral commodity summaries 2023. Reston, VA
9. Nedelciu CE, Ragnarsdottir KV, Schlyter P, Stjernquist I (2020) Global phosphorus supply chain dynamics: assessing regional impact to 2050. Glob Food Sec 26:100426
10. Cuadri AA, Navarro FJ, García-Morales M, Bolívar JP (2014) Valorization of phosphogypsum waste as asphaltic bitumen modifier. J Hazard Mater 279:11–16
11. Sun ZL (2016) Outlook and thinking of phosphogypsum comprehensive utilization in China. Sulphuric Acid Ind 1:55–58
12. Liu S, Wang L, Yu B (2019) Effect of modified phosphogypsum on the hydration properties of the phosphogypsum-based supersulfated cement. Constr Build Mater 214:9–16
13. Sahu SK, Ajmal PY, Bhangare RC, Tiwari M, Pandit GG (2014) Natural radioactivity assessment of a phosphate fertilizer plant area. J Radiat Res Appl Sci 7(1):123–128
14. Campos MP, Costa LJP, Nisti MB, Mazzilli BP (2017) Phosphogypsum recycling in the building materials industry: assessment of the radon exhalation rate. J Environ Radioact 172:232–236
15. Paige-Green P, Gerber S (2000) An evaluation of the use of by-product phosphogypsum as a pavement material for roads. In: South Africa transport conference "action in transport for the new millennium"
16. Degirmenci N, Okucu A, Turabi A (2007) Application of phosphogypsum in soil stabilization. Build Environ 42(9):3393–3398
17. Folek S, Walawska B, Wilczek B, Miśkiewicz J (2011) Use of phosphogypsum in road construction. Pol J Chem Technol 13(2):18–22
18. Sarkka A (2001) Disposal management system for utilisation of industrial phosphogypsum and fly ash. In: Life environment in action: 56 new success stories for Europe's environment. Office for Official Publications of the European Communities, Luxembourg, pp 98–99
19. Crusciol CAC et al (2016) Soil fertility, plant nutrition, and grain yield of upland rice affected by surface application of lime, silicate, and phosphogypsum in a tropical no-till system. CATENA 137:87–99
20. Alva AK, Sumner ME (1990) Amelioration of acid soil infertility by phosphogypsum. Plant Soil 128(2):127–134
21. Reeve NG, Sumner MF (1972) Amelioration of subsoil acidity in Natal Oxisols by leaching of surface-applied amendments. Agrochemophysica 4(1):1–5
22. Al-Naser Y (2019) Effect of phosphogypsum on formation and development of soil surface crust and wheat crop growth. Tikrit J Agric Sci (English Part) 18(3):90–95
23. Cochrane BHW, Reichert JM, Eltz FLF, Norton LD (2005) Controlling soil erosion and runoff with polyacrylamide and phosphogypsum on subtropical soil. Trans ASAE 48(1):149–154
24. Mi N et al (2022) Synergistic effects of microorganisms and passivation materials on the growth and Cd uptake of coriander (Coriandrum sativum L.) in Cd-contaminated soils. Rhizosphere 24:100604
25. Wang Y, Ying Y, Lu S (2020) Si-Ca-K-Mg amendment reduces the phytoavailability and transfer of Cd from acidic soil to rice grain. Environ Sci Pollut Res 27(26):33248–33258
26. Singh M (2002) Treating waste phosphogypsum for cement and plaster manufacture. Cem Concr Res 32(7):1033–1038
27. Aliedeh MA, Jarrah NA (2012) Application of full factorial design to optimize phosphogypsum beneficiation process (P2O5 Reduction) by using sulfuric and nitric acid solutions. In: Sixth Jordanian international chemical engineering conference, Amman, Jordan, pp 1–10
28. Ennaciri Y, Zdah I, El Alaoui-Belghiti H, Bettach M (2020) Characterization and purification of waste phosphogypsum to make it suitable for use in the plaster and the cement industry. Chem Eng Commun 207(3):382–392

29. Nizevičienė D, Vaičiukynienė D, Vaitkevičius V, Rudžionis Ž (2016) Effects of waste fluid catalytic cracking on the properties of semi-hydrate phosphogypsum. J Clean Prod 137(Supplement C):150–156
30. Li Z et al (2018) Investigation of phosphate adsorption from an aqueous solution using spent fluid catalytic cracking catalyst containing lanthanum. Front Environ Sci Eng 12(6):15
31. Kaziliunas A, Leskeviciene V, Vektaris B, Valancius Z (2006) The study of neutralization of the dihydrate phosphogypsum impurities. Ceram Silik 50(3):178–184
32. Leškevičienė V, Nizevičienė D, Valančius Z (2003) The effect of the processing way of technogenic raw materials on the properties of gypsum binding materials. J Civ Eng Manag 9(1):76–81
33. Kovler K (2009) Radiological constraints of using building materials and industrial by-products in construction. Constr Build Mater 23(1):246–253
34. Cánovas CR, Macías F, Pérez-López R, Basallote MD, Millán-Becerro R (2018) Valorization of wastes from the fertilizer industry: current status and future trends. J Clean Prod 174:678–690
35. World Health Organization (2009) WHO handbook on indoor radon: a public health perspective. World Health Organization. In: Zeeb H, Shannoun F (eds). Geneva, Switzerland
36. The Council of the European Union (2014) Council Directive 2013/59/Euratom of 5 December 2013 laying down basic safety standards for protection against the dangers arising from exposure to ionising radiation, and repealing Directives 89/618/Euratom, 90/641/Euratom, 96/29/Euratom, 97/43/Euratom and 2003/122/Euratom. Off J Eur Union L13:1–73
37. European Commission (1999) Radiological protection principles concerning the natural radioactivity of building materials. OPOCE, Luxembourg
38. Kovacs T et al (2017) From raw materials to NORM by-products. In: Schroeyers W (ed) Naturally occurring radioactive materials in construction. Woodhead Publishing, pp 135–182
39. Nizevičienė D et al (2018) The treatment of phosphogypsum with zeolite to use it in binding material. Constr Build Mater 180:134–142
40. Kovler K, Somin M (2004) Producing environment-conscious building materials from contaminated phosphogypsum. In: Proceedings of the RILEM international symposium on environment-conscious materials and systems for sustainable development (ECM 2004), Koriyama, Japan, pp 241–250
41. Villalón Fornés I, Vaičiukynienė D, Nizevičienė D, Doroševas V, Michalik B (2021) A comparative assessment of the suitability of phosphogypsum from different origins to be utilised as the binding material of construction products. J Build Eng 44:102995
42. Bhanumathidas N, Kalidas N (2004) Dual role of gypsum: set retarder and strength accelerator. Indian Concrete J 78:1–4
43. Neville AM, Brooks JJ (1987) Concrete technology, 2nd edn. Longman Scientific & Technical England, Harlow
44. Singh NB, Middendorf B (2007) Calcium sulphate hemihydrate hydration leading to gypsum crystallization. Prog Cryst Growth Charact Mater 53(1):57–77
45. Valančius Z, Nizevičienė D, Leškevičienė V, Kybartienė N (2005) Influence of the technological parameters on the structure and properties of hemihydrate phosphogypsum. Ceram Silik 49(2):120–125
46. Leškevičienė V, Kybartienė N, Nizevičienė D, Valančius Z (2006) The influence of mineralogical composition and crystal structure of phosphogypsum on its physical-mechanical properties. IBAUSIL: 16. Internationale Baustofftagung, September 20–23, Weimar, Germany, pp 883–889
47. Zhou J et al (2014) A novel two-step hydration process of preparing cement-free non-fired bricks from waste phosphogypsum. Constr Build Mater 73(Supplement C):222–228
48. Villalón Fornés I, Doroševas V, Vaičiukynienė D, Nizevičienė D (2021) The investigation of phosphogypsum specimens processed by press-forming method. Waste Biomass Valoriz 12(3):1539–1551
49. Zhou J et al (2016) Preparation of hardened tiles from waste phosphogypsum by a new intermittent pressing hydration. Ceram Int 42(6):7237–7245

50. Zhou J et al (2020) Preparation of paper-free and fiber-free plasterboard with high strength using phosphogypsum. Constr Build Mater 243:118091
51. Peng X et al (2020) The addition of biochar as a fertilizer supplement for the attenuation of potentially toxic elements in phosphogypsum-amended soil. J Clean Prod 277:124052
52. Ropelewska E, Dziejowski J, Zapotoczny P (2016) Changes in the microbial activity and thermal properties of soil treated with sodium fluoride. Appl Soil Ecol 98:159–165
53. Yang L, Yan Y, Hu Z, Xie X (2013) Utilization of phosphate fertilizer industry waste for belite–ferroaluminate cement production. Constr Build Mater 38:8–13
54. Valančius Z, Vaickelionienė R, Vaickelionis G, Makčinskas P (2022) Use of an industrial by-product phosphogypsum in the production of white textured paints. J Clean Prod 380:134888
55. Vaičiukynienė D, Nizevičienė D, Šeduikytė L (2017) Sustainable approach of the utilization of production waste: the use of phosphogypsum and AlF3 production waste in building materials. In: Izabela H (ed) Energy efficient, sustainable building materials and products. Politechnika Krakowska, Krakow, pp 183–197
56. Kumar S (2003) Fly ash–lime–phosphogypsum hollow blocks for walls and partitions. Build Environ 38(2):291–295
57. Bumanis G, Irbe I, Sinka M, Bajare D (2022) Biodeterioration of sustainable hemp shive biocomposite based on gypsum and phosphogypsum. J Nat Fibers 19(15):10550–10563

Part IV
Adaptive Reuse of Existing Buildings

Chapter 11
Identifying Abandoned Industrial Zones in Skopje and Evaluating the Potential of Extending Their Useful Lifespan by Adaptive Reuse

Teodora Mihajlovska, Ana Trombeva - Gavriloska, Meri Cvetkovska, and Liljana Dimevska Sofronievska

Abstract The built environment of cities can play a key role in the transition to a circular economy, especially considering the large resource consumption and waste generation of building construction. This paper explores the potential for adaptive reuse of abandoned industrial buildings through the lens of the sustainability within the context of circular economy in Skopje, North Macedonia. The adaptive reuse of vacant industrial buildings can bring environmental, social, and economic benefits by employing an urban strategy based on circular economy principles and innovative approaches. Considering the transformation of Skopje's industrial zones through the city's development, this paper identifies potential abandoned industrial sites in the city where the adaptive reuse of buildings, could improve environmental, social, and economic quality of the built environment. The identified abandoned industrial sites were analyzed in terms of their functional transformation and their vulnerability to current demolishing practices.

Keywords Circular economy · Adaptive reuse · Industrial buildings

11.1 Introduction

The built environment of cities can play a key role in the transition to a circular economy (CE), especially considering the large resource consumption and waste generation of building construction. CE is a model of production and consumption, which involves sharing, leasing, reusing, repairing, refurbishing and recycling

T. Mihajlovska (✉) · A. Trombeva - Gavriloska · L. Dimevska Sofronievska
Faculty of Architecture, Ss. Cyril and Methodium University, Skopje, North Macedonia
e-mail: mihajlovska.teodora@arh.ukim.edu.mk

M. Cvetkovska
Faculty of Civil Engineering, Ss. Cyril and Methodium University, Skopje, North Macedonia

L. Bragança et al. (eds.), *Creating a Roadmap Towards Circularity in the Built Environment*, Springer Tracts in Civil Engineering,
https://doi.org/10.1007/978-3-031-45980-1_11

existing materials and products as long as possible [1]. This is in contrast to the approach of linear economy where natural resources are turned into products that are ultimately intended to become waste because of the way they have been designed and manufactured; a process that is often summarized by "take, make, waste" [2]. This process has been the dominant approach in the build environment where the tendency is to regard buildings as products with a limited useful life that have to be eventually discarded and demolished. In this context, the goal of the implementation of CE is to achieve the principles of circularity relating to the entire life cycle of buildings. Since much of the existing building stock will still be in use for another 100 years [3], there is a need to develop not only strategies to design durable and adaptable new construction projects, but also develop sustainable CE approaches for existing buildings.

Adaptive reuse (AR) is a strategy to improve the environmental, social, and financial performance of a building, site, or area by transforming them from unused buildings to ones with a new function [3–5]. AR thus plays a significant role in the transition towards CE as it transforms buildings into a regenerative and reusable resource. Adaptability is among the general principles for the design of circular buildings in the perspective of extending their life, preventing their premature demolition, through the provision of transformations and adaptations for new uses; however, AR approaches can also be developed for existing buildings that were not designed with CE principles in mind.

In support of the development of CE initiatives, this paper focuses on exploring the potential of applying an AR approach on underused or abandoned industrial sites in Skopje, North Macedonia as a useful practice to generate new values by supporting innovative approaches. Through a historical analysis of past and current urban plans, the development of industrial sites in Skopje is overviewed. The identified abandoned industrial sites were analyzed in terms of their functional transformation and their vulnerability to current demolishing practices.

11.2 Industrial Buildings in Skopje

11.2.1 Historical Development of Industrial Sites in Skopje

The first enterprises in Skopje where machine production was established appeared at the end of the nineteenth century. After the Second World War, North Macedonia, as a part of SFR Yugoslavia, began intensive economic and industrial development. The main focus for growth and development in Skopje was heavy industrialization, through which the economic and social modernization of the city and the country was meant to be achieved.

This process was disrupted in 1963 when an earthquake destroyed 80% of the city's building stock, prompting the creation of a new General Urban Plan for reconstruction and development of the city. With this Plan, adopted in 1965, the intent for industrialization was continued; the borders of the city expanded and the area of the city territory increased to 11,156 ha, with about 1,215 ha or 10.9% of the city area intended for industrial purposes. The industry was rounded in planned industrial zones: (1) industrial zone Zhelezara formed by the construction of the industrial complex "Mines and Ironworks Skopje", (2) Eastern industrial in which the largest number of enterprises were located, (3) Southern industrial zone, which was made up of the existing capacities of the industry, and (4) a new Western industrial zone. Due to major changes in the basic planning parameters (disproportionate population increase and unfulfilled planning goals) in 1985, a new General Urban Plan was adopted. However, this Plan was a continuation of the previous and it only confirmed the existing industrial zones in the city determined by the 1965 plan: (1) Southeast zone (by the 1965 Plan named as the Southern zone), (2) Northeast zone (by the 1965 Plan named as the Eastern zone and Ironworks) and (3) Western Zone.

An important shift was made in 2002, with the new General Urban Plan, which emphasized the need for rational and efficient use of space and land. Starting from the postulate of a 'sustainable city', a city that is integrated into the natural environment, takes from nature as much as it gives back, a move away from industrialization was made. This was illustrated by the planned reduction of the industrial area in the city from 1,269.6 ha foreseen by the 1985 Plan to 1,050.29 ha planned by the 2002 Plan, a reduction of 219.31 ha. The tendency of planned reduction of industrial areas in the urban area of the city was confirmed with the General Urban Plan of 2012; that reduced the area to 782.19 ha in relation to the area determined by the 2002 Plan of 1050.29 ha, a decrease of 268.1 ha.

11.2.2 Industrial Sites in Skopje

As shown by the historical development of the urban plans for Skopje, during the socialist period, the plans determined oversized territories of city land for industrial purposes that surpassed the real needs and possibilities of the city. The change of the industrial enterprises in the post-socialist period, conditioned by the transition to the new socio-economic and political system, caused the abandonment or transformation of a large number of these sites. For the purposes of this research, 163 individual industrial sites in Skopje with a total area of 849.79 ha were taken into account.

The data for these sites was obtained by analyzing the studies made for the new General Urban Plan for Skopje (2022–2032) which is currently being made, as well as any previous relevant urban plans. The research carried out on the former industrial sites shows that the industrial zones in the city are fragmenting and adapting to new functions. The former exclusively industrial areas are transforming under the influence of the processes of deindustrialization, functional transformation and reconstruction of the industry. The sites were classified in three groups: (1) Abandoned—sites that are completely inactive or severely underused; (2) Restored—sites that are actively used for industrial purposes; and (3) Transformed—sites that were completely changed and are used for non-industrial purposes [6].

From the data in the Table 11.1 it can be concluded that the total area of abandoned industry (382.25 ha) is 44.98% of the total analyzed industrial area in the city (849.79 ha); while a similar area or 45.07% has been restored and is still functional as an industrial complex. Only 9.94% have been successfully transformed and repurposed. This paper will look into the abandoned sites, because they show the greatest potential and need for the application of AR, and the transformed sites in order to determine what kind of functional transformations have been made so far (Fig. 11.1).

By analyzing the spatial distribution of the sites of abandoned industry, we can conclude that they are spread over the entire territory of the city, both within the three industrial zones and in the isolated locations outside the industrial zones. The status of the abandoned sites was determined according to the degree of use: (1) inactive—completely deserted sites, (2) underused—sites where the degree of use does not correspond to the size and potential of the site and (3) occasionally used sites (Table 11.2).

A small percentage of the abandoned sites has already undergone a functional transformation. From the data presented in Table 11.3, we can conclude that dominant function to which industrial sites are transformed is commercial, 46.38%, closely followed by housing, 40.25%. Smaller percentage of sites, 10.42%, have been cleared and transformed into green surfaces and parks.

The change of function is a result of the changes in the structure of the economy and the tendency is to perform the most profitable transformation of sites within the city territory, which is the construction of residential buildings, as well as buildings for commercial functions. These types of changes in function are also supported by the location of the sites within the urban area of the city. As the city grew and the industrial zones fragmented, certain sites became surrounded by incompatible residential and commercial functions. As a result, sites near the city center tend to be

Table 11.1 Overview of transformation of researched industrial sites in Skopje, 2022 [7]

Zone	Abandoned		Restored		Transformed		Total	
	Area (ha)	No. of sites	Area (ha)	No. of sites	Area (ha)	No. of sites	Area (ha)	No. of sites
Not defined	19.17	6	20.04	15	46.62	20	85.83	41
Western zone	9.36	4	44.87	9	13.13	1	67.36	14
Southeastern zone	107.57	6	109.63	19	7.22	5	274.42	30
Mines and iron works	166.11	8	108.38	6	/	/	274.49	14
Northeastern zone	80.05	18	100.11	35	17.54	11	197.70	64
Total	382.25	42	383.04	84	84.50	37	849.79	163
Percentage (%)	44.98		45.07		9.94		100	

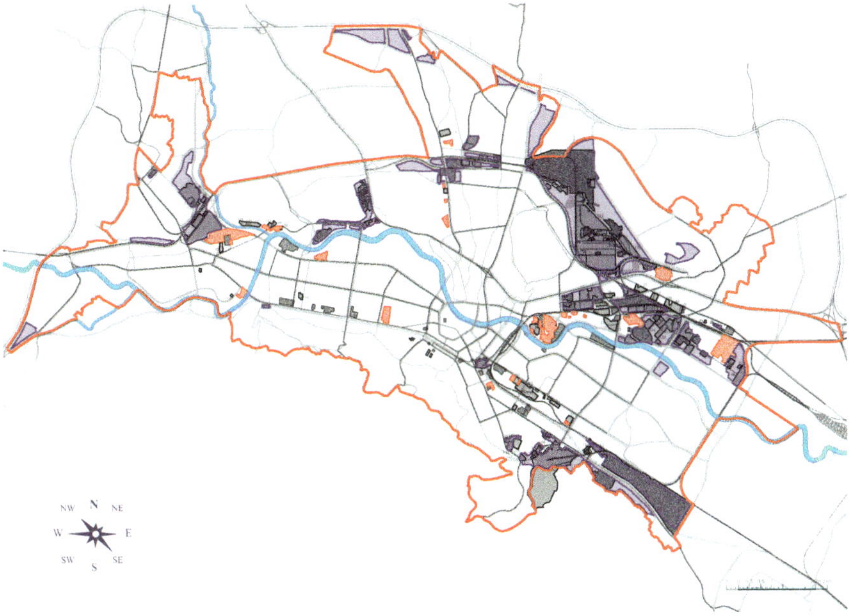

Fig. 11.1 Map of industry sites in Skopje in 2022; current industrial zones (purple), abandoned industrial sites out of the current industrial zones (red)

Table 11.2 Overview of abandoned industrial sites in Skopje, 2022 [7]

Zone	Inactive		Underused		Occasionally used		Total	
	Area (ha)	No. of sites	Area (ha)	No. of sites	Area (ha)	No. of sites	Area (ha)	No. of sites
Not defined	18.11	5	1.06	1	/	/	19.17	6
Western zone	9.36	4	/	/	/	/	9.36	4
Southeastern zone	104.42	3	3.15	3	/	/	107.57	6
Mines and iron works	96.77	3	61.25	4	8.08	1	166.11	8
Northeastern zone	27.78	11	51.51	6	0.76	1	80.05	18
Total	256.44	26	116.97	14	8.84	2	382.25	42
Percentage (%)	67.09		30.60		2.31		100	

Table 11.3 Overview of transformed sites in Skopje according to new purposes in 2022 [7]

Zone	Housing		Greenery		Commercial		Education		Total	
	Area (ha)	No. of sites	Area (ha)	No. of sites	Area (ha)	No. of sites	Area (ha)	No. of sites	Area (ha)	No. of sites
Not defined	11.99	6	8.81	3	24.48	9	1.34	2	46.62	20
Western zone	13.13	1	/	/	/	/	/	/	13.13	1
Southeastern zone	5.58	3	/	/	0.48	1	1.16	1	7.22	5
Mines and iron works	/	/	/	/	/	/	/	/	/	/
Northeastern zone	3.31	2	/	/	14.22	9	/	/	17.54	11
Total	34.01	12	8.81	3	39.19	19	2.50	3	84.50	37
Percentage (%)	40.25		10.42		46.38		2.96		100	

developed for commercial purposes, and sites located in residential areas are mostly used for construction of new residential buildings.

Although the transformation in terms of function is necessary, the current approach for transforming the sites from their industrial functions has been by demolishing the industrial buildings and construction of new ones, resulting in both waste generation and further resource consumption (Fig. 11.2). Developing an AR approach for the remaining locations would bring environmental, social, and economic benefits.

Fig. 11.2 Transformed site—form an industrial soap factory (demolished) to residential building

11.3 Potential for AR

However not all abandoned sites necessarily need to undergo a functional transformation by means of AR. Certain abandoned sites are situated with in current industrial zones and as such can still be used for their original purpose. The most vulnerable sites are the ones that according to the General Urban Plan of the city are no longer located in industrial zones. The buildings on these sites cannot be used for their original function, and as observed by the analysis of the already transformed sites are at risk of demolition. In order to identify these sites, the map of abandoned industry sites is overlapped with the industrial zones planned with the most recent and current General urban plan for Skopje. A number of abandoned industrial sites, which are located near the central city area and in parts of the city where the dominant function (mostly housing) is in collision with industry, are marked for a change in function. In order to explore the potential of the mapped out sites, analysis in context of other urban studies made for the master plan was carried out. The potential of AR was analyzed by taking into consideration the study for the economic perspectives of the space of the city and the study for environmental protection.

The study for environmental protection identified two important aspects; the presence of brownfield sites and pollution [8]. During the measurements and monitoring of the soils, a high degree of contamination of the soil with heavy metals around the organic-chemical industry "Ohis"; the industry for production and processing of iron and steel "Makstil" and tannery "Godel" (Fig. 11.3). These localities are marked as industrially contaminated localities—"hotspots" and were counted among the 14 largest hotspots in the Republic, given the results of the recent soil quality monitoring in Skopje, which confirmed the risk that these localities have on the environment due to the high contamination of the soil with heavy metals. The presence of hazardous substances and contaminants at all three sites is linked to the previous industrial activities carried out at the locations and is especially concerning near the Ohis site where hazardous chlorinated organic substances (a technical mixture of HCH-hexachlorocyclohexane isomers, known as lindane) have been stored in certain concrete storage areas for more than 20 years. Any plans for AR of these locations would have to take into consideration the need to remedy the soil pollution before any other activities can take place.

The second important issue that could affect the application of AR is the air pollution. The air quality is measured on five locations throughout the city, the most concerning is the suspended particulate matter pollution, especially PM_{10} particles, a city wide problem. However, the highest concentrations have been measured at the Lisiche station as well as in the city center. This data must be taken into account when determining the future functions that could be housed at the sites.

In the study for the economic perspectives of the city, the analysis of the distribution of facilities showed that there is no adequate policy for even urbanization of the entire area of the City [9]. A lack of polycentric dispersion of business buildings and commercial facilities in the City, as well as dispersion of economic and social contents in the wider area were also identified. From a spatial point of view, the

Fig. 11.3 Map of brownfield sites (purple) and air pollution (red)

largest number of buildings classified as commercial, are located in the city center or in the neighborhoods that gravitate and are part of the wider city core. The study also pointed out that the industrial zones can be developed by establishing business incubators and technology parks that will represent support for entrepreneurship. The expansion and enrichment of the contents in the peripheral zones, especially with the establishment of commercial centers can also have a double effect of business development and of relieving the burden on the city center, allowing better accessibility and mobility, less pollution in the city center and a more even distribution of commercial and business facilities.

However, in order to ensure the polycentric development of the City, the study proposes that the new urban plan should foresee the construction of commercial centers in the peripheral city settlements, well connected with appropriate approaches and suitable infrastructure. The study does not take into consideration the opportunity for AR of certain buildings and misses out on the opportunity for a sustainable CE approach in favor of the more conventional linear approach where introducing new functions in any area of the city is achieved by the construction of new buildings. The analyzed sites have great potential to offer an alternative solution to the problem, because most of the industrial sites were historically located on the outskirts of the city. With the expansion of the urban area these parts are now integrated in the city but still represent its most peripheral areas (Fig. 11.4).

Fig. 11.4 Potential for polycentric dispersion of business buildings and commercial facilities

11.4 Further Evaluation the Potential for AR

The evaluation of the potential of adaptive reuse is extremely complex and dynamic because the process involves different stakeholders [10] and a number of aspects need to be taken into consideration for each individual building. So far, this research has been focused on the city level, and exploring the potential benefits for an AR approach at this scale. However, in order to evaluate whether or not the specific buildings have the physical capacity to adapt to new functional needs, a number of aspects at the building scale need to be examined. Their compliance with health and safety requirements needs to be checked; their energy performance needs to be assessed in comparison to current standards; their infrastructures such as electricity, drainage, mechanical systems evaluated; the presence of hazardous materials assessed etc. [11].

Ball [12, 13] investigated the industrial property stock in Stoke-on-Trent in the UK and identified the characteristics of buildings that were reused or reoccupied in comparison to the vacant ones. He argued that the characteristics suggested the potential of a building's adaptive reuse. This approach could be implemented in Skopje, since a number of the mentioned sites have been transformed, an analysis of the type of function and location in the city of these sites could help identify which ones could be next. The redevelopment of brownfield and reuse of industrial buildings has attracted considerable research interest but only a handful of researchers attempted to find solutions for the evaluation of adaptive reuse potential [14]. Two groups of Australian researchers have proposed evaluation models for buildings. The first group, Langston and Shen [15, 16] developed an Adaptive Reuse Potential (ARP) model which calculates the adaptive reuse potential at any point in a building's life cycle and helps not only determine the potential in physical sense but also determine the right timing for intervention. It takes into account the buildings' expected physical life, the current age and the effect of physical, economic, functional, technological, social and legal obsolescence, factors that the researchers believe have reduced the building's useful life [17]. The other group, Bullen and Love [11, 18] identified variables which affected adaptive reuse decision-making and categorised them into three groups, namely capital investment, asset condition and regulation. The applicability of these models in other contexts needs further research, however the development of evaluation models for evaluation of the potential of AR in the context of these research is needed.

11.5 Conclusion

This paper analyzed the abandoned industrial sites in Skopje, North Macedonia in order to explore the potential for AR of underused or abandoned buildings in order to help improve current practices that are not aligned with CE. From the analyzed data it can be concluded that the city of Skopje is undergoing a transformation of its industrial zones. Currently a large number of former industrial sites are underused or completely abandoned, with a number of sites located in areas where according to the current urban plans no industrial functions are planned. These sites will inevitably undergo a functional transformation in the future. The small percentage of sites that adapted to a new function, have been transformed by means of demolition and new construction, which results in waste creation and depletion of resources. This research identifies the sites where by developing a proper AR strategy the buildings can be repurposed, rather than demolished. The AR of vacant industrial buildings can bring environmental, social, and economic and help change the perspective of buildings as products with a limited useful life. The development and implementation of such strategies is crucial in the transition towards CE in the built environment. Further research should look into the specific conditions of the sites identified by this paper and focus on developing a clear plan for AR of the buildings. This type of applicative use of adaptability on existing industrial buildings can be helpful in developing a more general strategy for AR on buildings with different functions. Finally, an in-depth analysis of the application of more sustainable approach from the perspective of CE and AR would provide a set of useful guidelines for practitioners.

References

1. European Commission. Directorate General for Communication (2020) Circular economy action plan: for a cleaner and more competitive Europe. Publications Office, LU. https://data.europa.eu/doi/10.2779/05068. Accessed 13 Feb 2023 [Online]
2. Ellen MacArthur Foundation (2017) A new textiles economy: redesigning fashion's future [Online]. http://www.ellenmacarthurfoundation.org/publications
3. Bullen P, Love P (2011) Factors influencing the adaptive re-use of buildings. J Eng Des Technol 9(1):32–46. https://doi.org/10.1108/17260531111121459
4. Abastante F, Corrente S, Greco S, Lami IM, Mecca B (2022) The introduction of the SRF-II method to compare hypothesis of adaptive reuse for an iconic historical building. Oper Res Int J 22(3):2397–2436. https://doi.org/10.1007/s12351-020-00611-4
5. Lami IM (2020) Shapes, rules and value. In: Lami IM (ed) Abandoned buildings in contemporary cities: smart conditions for actions, vol 168. Springer International Publishing, Cham, pp 149–162. https://doi.org/10.1007/978-3-030-35550-0_9
6. Mickovski G (2015) Transformation of industrial space in the context of post-socialist transition of the city of Skopje. University of Belgrade, Belgrade
7. Faculty of Architecture—UKIM (2022) Study of the development of the industry of the city of Skopje for general urban plan 2022–2032
8. Faculty of Architecture—UKIM (2022) Study for the environmental protection of the city of Skopje for general urban plan 2022–2032
9. Faculty of Architecture—UKIM (2022) Study for the economic development of the city of Skopje for general urban plan 2022–2032
10. Kurul E (2007) A qualitative approach to exploring adaptive re-use processes. Facilities 25(13):554–570
11. Bullen P, Love P (2011) A new future for the past: a model for adaptive reuse decision-making. Built Environ Project Asset Manag 1(1):32–44
12. Ball RM (1989) Vacant industrial premises and local development: a survey, analysis and policy assessment of the problem in Stoke-on-Trent. Land Dev Stud 6:105–128
13. Ball RM (2002) Re-use potential and vacant industrial premises: revisiting the regeneration issue in Stoke-on-Trent. J Prop Res 19(2):93–110
14. Urban Task Force (1999) Towards a strong urban renaissance. Urban Task Force, London
15. Langston C (2012) Validation of the adaptive reuse potential (ARP) model using iconCUR. Facilities 30(3/4):105–123
16. Langston C, Shen LY (2007) Application of the adaptive reuse potential model in Hong Kong: a case study of Lui Seng Chun. Int J Strateg Prop Manag 11(4):193–207
17. Shen LY, Langston C (2010) Adaptive reuse potential: an examination of differences between urban and non-urban projects. Facilities 28(1/2):6–16
18. Bullen P, Love P (2010) The rhetoric of adaptive reuse or reality of demolition: views from the field. Cities 27(4):215–224

Part V
Recovery and Reuse of Salvaged Building Materials and Products from Existing Structures

Chapter 12
The Materials Bank of the City of Porto: Flow of Processes to Recover Tiles in Urban Operations of Historic Buildings

Mayara Regina Munaro, Maria do Carmo Duarte Freitas, Sergio Fernando Tavares, and Luís Bragança

Abstract The recovery and reuse of materials are one of the main objectives of the circular economy (CE) to minimize the environmental impacts of the construction industry. Public policies, citizen cultural awareness, urban mining practices, and analysis of material flows are being increasingly investigated seeking the reinsertion of materials into the economy. These practices are particularly important when they aim to protect the history and culture of a country, as in the case of Portuguese tiles, which constitute a significant part of Portuguese cultural heritage and must be safeguarded and protected. This article presents a case study of the Materials Bank (MB) of the City of Porto, Portugal, which aims to recover and protect the tile heritage of historic buildings in the municipality. The study sought to map the flow of processes that enable the formation of the MB, due to the joint efforts of local governance and public policies. The flowchart obtained guides managers to create public policies associated with actions for the maintenance and reuse of building materials in buildings related to historical and cultural heritage. It is observed that from legislation and engaged governance, it is possible to accelerate the operation of circular actions in the built environment and promote greater awareness about the reuse of secondary materials and the protection of historical heritage.

Keywords Portuguese tiles · Circular economy · Materials bank · Public policies · Cultural heritage

---

M. R. Munaro (✉)
São Paulo University, São Paulo, Brazil
e-mail: mayara.munaro@lme.pcc.usp.br

M. do Carmo Duarte Freitas · S. F. Tavares
Federal University of Paraná, Curitiba, Brazil

L. Bragança
Minho University, Braga, Portugal

L. Bragança et al. (eds.), *Creating a Roadmap Towards Circularity in the Built Environment*, Springer Tracts in Civil Engineering,
https://doi.org/10.1007/978-3-031-45980-1_12

12.1 Introduction

The global economy is only 7.2% circular due to rising material extraction and use, leaving fewer materials to cycle back into the economy [1]. In the last fifteen years, the annual global extraction of materials grew from 27 to 92 billion tons [2]. 70% of global greenhouse gas (GHG) emissions, over 90% of global biodiversity loss, and water stress are tied to material handling and use [1]. The highlight is in the category of non-metallic minerals, which includes most building materials, such as sand, gravel, and clay. Globally, the construction sector accounts for almost 50% of resource extraction [3], and 40% of global waste generation [4].

The continuous accumulation of materials in cities in the form of built environment stocks has led to the prospects of urban mining for secondary materials that can be looped back into the economy [5]. The concept of circular economy (CE) has shown promise toward resource sustainability while material flow assessment support in estimating the scale at which a circularity potential exists [6, 7]. Using a lifecycle approach, it has been estimated that a 10–20% reduction in GHG emissions could be achieved through waste prevention, landfill mitigation, and other types of solid waste management [4].

Sustainability in the construction industry requires moving beyond waste management and focusing on building components management [8]. Yet, urban mining of the built environment remains less explored, driven by long lifetimes, and bulky and heavy materials of buildings and infrastructures [5]. However, there has been an increase in the studies analyzing material stock and flows and producing insights on the scale of in-use materials and waste outflows [6, 7]. In addition, public authorities' involvement is seen as an effective mechanism to promote a circular built environment [9]. If materials embodied in demolition waste flows are reused, it can help reduce the environmental pressure related to the extraction of virgin resources, which relies on the proper management of waste flows and supply chains [10].

The European Commission and several national governments have formulated strategies exalting circularity. In Portugal, the regard and protection of historical and artistic glazed tiles (called "azulejo") became mandatory in Portuguese cities with the Portuguese Law No. 79/2017, representing an advance in terms of the protection of the Portuguese glazed tiles' heritage [11]. In the city of Porto, for example, the Materials Bank of the Porto City Council (MB) proposes mechanisms to protect the tile heritage in urban operations. The MB is intended for the storage, preservation, restoration, and return of tile items to the city in the rehabilitation of façades of historic buildings [12].

However, the inventory of traditional Portuguese glazed tiles is scarce and not integrated with other heritage inventories [11]. In addition, the resources for carrying out the Bank's activities are restricted, and common citizens and institutions do generally not value artistic glazed tiles [13]. The result is needless glazed tile removal, demolitions of azulejo-covered buildings, vandalism, and glazed tiles conservation needs [11].

This study seeks to analyze how the flow of processes occurs in the MB, with guidance on procedures for the recovery and reuse of historic Porto tiles and other identity elements. The study seeks to clarify how the MB organizes and manages the realization of its value proposition recovering and reusing non-structural demolition waste. For the authors, it is the first time that this type of flow analysis has been carried out, and the research is the first step in mapping what already exists. It also seeks to understand the opportunity to expand the collaboration and scope of the Materials Bank, to optimize a methodology to coordinate and accelerate recovering of traditional glazed tiles, which will enable stakeholders to constructively engage and play their part in enabling the reused of secondary materials and in the preservation of Porto cultural heritage.

12.2 Background

12.2.1 Reuse of Buildings' Materials Stock

Component reuse in buildings is the process where the building parts are dismantled during deconstruction and are reused in new lifecycles and contexts, without any type of requalification or reprocessing operation [14]. Despite environmental superiority, the systemic reuse of building components is still far from reality [7]. Most construction and demolition waste is either recycled off-site, landfilled, or used for road construction, which is more useful than landfilling but has a higher risk of environmental leakage [10]. In this way, urban mining is a strategic component of circularity or sustainability aiming the resource efficiency use [15].

Different approaches and deconstruction processes to recover buildings' materials and components can be followed in urban mining [16]. Technical, regulatory, and policy considerations, including labor requirements, as well as the market demand needed to be considered in this practice [17]. However, information on deconstruction projects and the own deconstruction process are limited, mainly because is not attractive in terms of cost and time and by the uncertainties surrounding the markets for secondary materials [16]. With the absence of building components reuse and the lack of a building components reuse market, contractors prefer traditional demolition to the deconstruction process [17].

Experimentally, deconstruction costs could be 17–25% higher than demolition costs, considering the prospects of selling salvaged building components, avoided disposal costs, and additional labor costs [18]. Tatiya et al. [19] proposed that lower net deconstruction costs than the costs of demolition could be achieved primarily based on the sale of recovered materials. Arora et al. [7] developed a methodology to estimate the potential of urban mining of public housing developments, evidencing regulatory requirements for deconstruction and the need to accommodate the uprising of circular business models and define old, yet reusable products/materials or building components out of the concept of waste. Yet, the ease of recovering the same building component varies from site to site within a building itself and hence the combined recovery and reuse potential is based on many onsite experimental exercises of urban mining [7].

To minimize costs and time effects on the deconstruction process, it is important to know the number and features of materials that are stocked in buildings. The Urban Mining and Recycling unit project is a temporary storage of materials and a laboratory that monitors and evaluates the circular potential of materials through an online platform [20]. Cai and Waldmann [21] proposed a database/bank of materials and components based on Building Information Modelling (BIM) to promote the reusing and recycling of materials. Although these initiatives, information on the material composition of older construction projects often lacking or incomplete, and those projects were not designed for disassembly [5]. A detailed understanding of the composition, quantity, and spatial distribution of materials stored in buildings in the built environment is required [5].

The development of material passports for information on the type and quantity of materials used in a building aims at increasing the circularity potential of buildings [22]. For historical buildings, it is essential to develop digital inventories to allow the heritage management process, including identification; research and analysis; heritage impact assessment; planning for conservation, and management activities; and providing information to the public, authorities, and decision-makers [11]. However, the existing Portuguese digital heritage inventory has an incomplete database or lacks the georeferenced location of the building or structure where the glazed tiles are applied [11].

12.2.2 Protection and Safeguarding of Porto Glazed Tile Cultural Heritage

The traditional glazed tile is a secular element of material culture and a heritage rightly considered one of the country's most distinctive art forms [11, 13]. Introduced by the Arabs to the Iberian Peninsula and in Portugal in the sixteenth century, azulejos can be found on both the interior and exterior of churches, palaces, houses, and bars around Portugal, playing a decorative role, serving to control the temperature, or depicting stories about Portuguese history, religion, and culture [13]. In the nineteenth century, ceramic factories opened in Porto and Lisbon to mass-produce several patterns of glazed tiles [11]

In Portugal started in 2007 the "SOS Azulejo" project mentored by the Portuguese Judiciary Police Museum. The project was created with a global approach and a strategic line for the effective protection of Portuguese historic and artistic glazed tiles, preventing art and antique burglary and trafficking [13]. In 2017, the initiative managed to create the "National Day of Azulejo" to raise awareness of the importance of the Portuguese historical and artistic tile heritage [13].

(a) (b)

Fig. 12.1 **a** Façade with stonework, iron, wood, and toponymic plaques on a public road in the city of Porto; **b** building artifacts shelves—tiles and roof tiles. *Source* Authors, 2023

In 2017, after requests from the "SOS Azulejo" project, Law No. 79/2017 was instituted, which prohibits the demolition of buildings covered with tiles and/or the removal of tiles from the façades. In 2009, based on these movements, the Porto Materials Bank (MB), implemented nearly two decades ago, was restructured, and underwent numerous institutional changes as a public policy to protect the tile heritage of the city of Porto. The MB's principle is to preserve Porto's buildings with their historic tiles—Hispano-Arabic—from patterns from the 17th to the twentieth centuries (Fig. 12.1). Open to the public since 2010, the MB is a pioneer at the national level and has received awards for promoting the safeguarding of materials that characterize the city's image. In addition to tiles, other decorative and constructive elements from buildings in Porto are recovered, including wooden artifacts, iron, and stonework [12].

The importance of protecting and restoring Porto's tile heritage was emphasized by changes to the Municipality's Urbanization and Building Legal Regime, with emphasis on Service Order No. I/194083/19/CMP, which determines the obligation to include in the licensing of urban operations with demolition work to descriptive memory, photographic record, architectural survey, or any other method that allows identifying the framework of the demolition proposal [23]. Projects that provide for the demolition of buildings with façades covered in tiles require a technical assessment to grant or reject the demolition and assess the pertinence of removing the tiles for the MB of the demolished building elements [23]. Law No. 79/2017 states that urban operations where the removal of tiles is carried out only for restoration and/or cleaning, and subsequent replacement, will not require a license [24]. Compliance with this law implies, however, the existence of technical teams, protocols, and an integrated national heritage inventory system, which does not exist yet [11].

12.3 Research Strategy

This study has a qualitative-descriptive approach and uses the case study as a strategy, with bibliographic and documentary research techniques, interviews, and on-site visits. Case studies allow the researcher to develop an in-depth analysis of a case, with a variety of data collection procedures over an extended period [25]. On-site data collection started in August 2022. The case study was carried out at the Materials Bank of the city of Porto, located in the palace of the Viscondes de Balsemão, De Carlos Alberto square, Porto, Portugal. The interviewees were carried out in a non-structured way in the search to know the processes and activities carried out by the unit. Qualitative interviews are intended to elicit views and opinions from the participants [25].

The study aims to analyze how the flow of processes and policy considerations can leverage urban mining in historical buildings/cities. Mapping a process flow can identify activities that need to be improved, deleted, or added, in addition to serving as a parameter for the creation of other banks of recovered materials. The MB is, hierarchically, subordinate to the Municipal Department of Cultural Heritage Management of the city of Porto. The unit has a physical space like an open museum and shares with other chamber units spaces for storing its collection of recovered building materials.

12.4 Process Flow in the Operation of the Porto Materials Bank

The flow of information for approval of a building rehabilitation project considers Decree-Law No. 555/1999, which provides for the Legal Regime for Urbanization and Building, with amendments to Laws No. 79/2017 and No. 118/2019 that act on "legal regimes for the municipal licensing of urban subdivisions and urbanization works and private works [...]" in the search to meet the requirements of safeguarding the public interest with the administrative efficiency to which citizens legitimately aspire [26].

Figure 12.2 presents the process flowchart for the recovery of tiles from urban demolition works used by the MB in Porto. The flowchart was developed according to the information obtained from the MB, following current legislation. The recovery processes of historic artifacts are concentrated in three main agents: (1) the client, a representative of the historic building; (2) the Municipal Directorate of Urban Development (MDUD), which receives the request for urban intervention; and (3) the MB that is an advisory unit in case there are tiles or other historical artifacts for recovery.

Normally, the recovery of construction elements occurs when the client requests a license for an urban operation involving demolition services. The license is requested from the Legal Regime for Urbanization and Building, following Service Order No.

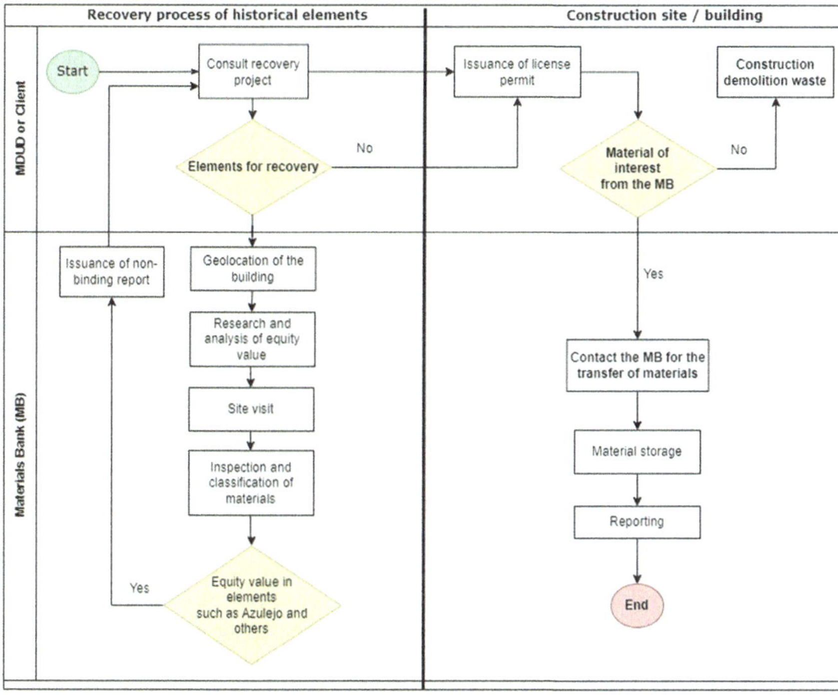

Fig. 12.2 Simplified process flowchart of the Porto Materials Bank for the recovery of tiles from historic buildings

I/194083/19/CMP. With due, attention to tile façades, it is important to highlight the decision to hire a service to survey the tile values identified in Law No. 79/2017.

The technical opinion issued by the MB will guide decision-makers. In the analysis, it is identified whether the dwelling is among the 2860 buildings mapped since 2007, 623 of which are in the urban perimeter. Next, an analysis of the documentation and attached photographs are carried out to identify the patrimonial value of the constructive elements (catalog of types of ceramics, plaster, wood, paper, metal, and other decorative elements—Table 12.1). In the identification of tiles or other elements related to Porto's historical heritage, a georeferenced mapping of the site is carried out, with a subsequent visit to the site. The visit aims to inspect and characterize the possible recovery and restoration elements. The MB team prepares a technical report that is forwarded to the requesting unit. After analysis of the MDUD, technical guidelines are sent to the customer to proceed with the demolition or recovery of the elements.

The recovery of historical artifacts can also occur when a degraded building or one in the process of demolition is reported, an activity carried out in partnership with the civil protection of the municipality. It will be up to the MB to guard, preserve, restore, and return the items to the city in the rehabilitation of façades with the same topology, according to the Architectural Heritage Catalog (Table 12.1).

Table 12.1 Cataloging of items by material category and registered buildings

Element catalog type	Ceramics	Plaster	Wood	Paper	Metal	Stone
LOCVS	4672	982	89	4	79	2
RM	1038	981	90	4	79	2

LOCVS: Management of Architectural and Archaeological Heritage of the Municipality of Porto
RM: Museum of Materials Reserve [12]

When donating building materials of historic, patrimonial, and cultural value, the MB team evaluates the item for incorporation into the collection, emphasizing historical aspects and productive elements. Next, the artifact is registered with its characteristics and cataloging, design, photographic record, and a label that provides information recorded in the database. Physically, each tile is stored with its proper identification label in custom-made wooden boxes and organized in chronological order on shelves.

It is a critical part of the process of the amount of building elements to be organized and structured for physical and virtual storage in the database. Database planning allows content management with photographic files of the location of the property and the material with its drawings, among other details for retrieving future information. The historical value and the analysis of the data flow behavior in the rescued pieces indicate the structure to organize the information for the team's decision-making.

In addition, MB operates in other activities such as workshops and support for historic building restoration projects; guidance to architecture and construction engineering professionals; mapping of buildings and works that are part of the city's memory and identity; registration, organization, and archive of the constructive elements that tell the history of the city; between others. It should be added that the MB offers educational workshops for young people and children who visit the site.

12.5 Discussion

The historical and architectural heritage valued in this study is a public policy associated with the CE. The recovery of materials encourages the dismantling of buildings and structures to store or guide reuse, requalification, remanufacturing, or recycling. The tiled façade with its different details represents a tourist asset that adds economic and environmental value to the buildings. The preservation of the historical heritage of cities is a practice of urban sustainability, reducing the extraction of natural resources, the production of GEE, and construction waste; by improving economic models of recovery and reusing materials, and allowing the salvage of building social–historical values.

The presented flowchart allows the MB process to be constantly revisited, enabling critical analysis, identification of failures, and opportunities for improvement [27].

It is possible to keep it updated and optimized for the reality of the organization. In addition, whenever there is a doubt, it is possible to consult the flowchart, ensuring that this information is not lost, facilitating the understanding and standardization of procedures among those involved in the operations of recovery and storage of historical elements.

In the last 10 years, the requalification of historic buildings has changed the economy and revitalized the city, contributing to the culture of preserving existing buildings, instead of demolition. However, the study found the need to enhance proposed projects with solutions to problems related to the abandonment of buildings with historic potential for the city. Partnerships with public and private institutions that support the generation of circular business and favor the MB's priority activity would be necessary. To this end, the knowledge of Law No. 79/2017 on the corresponding value of the tiled façade elements needs to be communicated to the citizens.

The informational flow of the MB's projects, processes, and interests is inventoried as a set of data, files, documents, photographic records, drawings, and notes that allow for guidance in decision-making in the process of managing the city's assets. The technical and legal information delivered by the MB is a consultative action, but not mandatory in the internal processes of the municipality. As for the educational activities carried out by the technicians, it is the action to promote the principles of the circular economy (reduce, reuse, recycle). Training workshops are offered on the restoration of parts, and services to builders, technicians, engineers, and architects to restore, reuse, or recycle building materials, prospecting for teaching about architectural, historical, cultural, and sustainability heritage.

12.6 Conclusion

The tile is a distinctive element of Portuguese cultural heritage. The protection and safeguarding of this heritage have gained importance in the country, however, the legislative framework is recent and lacks a more robust structure of resources and operational support. Therefore, municipalities need to play a leading role in the development of tools that allow for recovering and properly storing, both physically and digitally, the recovered elements. This study sought to map a process flowchart of the Porto Materials Bank to generate further discussions and reflections on the adopted process.

The flowchart clarifies the processes and related agents to allow the retrieval and storage of historical artifacts. Good process management depends on how well it is understood and documented. Standard operating procedures will help achieve consistency in operations and the pursuit of continual improvement in MB performance measures such as cost, quality, service, and scale. In addition, they provide the reduction of conflicts between those involved in the processes and the sharing of the process with other organizations.

The MB is a physical platform for safeguarding and protecting heritage, allowing the integration and coordination of different types of information (descriptions, location, areas, dates, photos, among others), different types of materials (tiles, tiles, stucco, iron, among others), and the integration of different purposes, such as the preservation of historical heritage, maintenance of Porto's urban image, diversity of interdisciplinary research, communication, and awareness of citizens, and tourism promotion.

Materials in buildings should sustain their value whereas buildings should function as banks of valuable materials and products. Efficient legislation and public policies that promote the reuse and recycling of construction materials and components are required. The joint action of the stakeholders with the government can further promote the Materials Bank development, strengthening the supervision and implementation of the secondary materials market.

The MB establishes opportunities for the reinsertion or reuse of historic artifacts, even on small residential facades. However, there was no monitoring of the volumes, typology, and quantities of materials that are reinserted in buildings, and that impact the reduction of carbon emissions from the built environment. Future studies should assess the environmental impact of the reinsertion of these secondary materials in the buildings.

Acknowledgements For the support of this research by the Municipal Directorate of Heritage and Culture the team of the Municipal Department of Cultural Heritage Management coordinates the actions of the Materials Bank of the city of Porto—Portugal and the technician Paula Cristina Alves Laje.

References

1. Circle Economy (2023) The circularity gap report 2023. Amsterdam
2. UNE (2019) Global resources outlook 2019: natural resources for the future we want
3. Krausmann F, Wiedenhofer D, Lauk C et al (2017) Global socioeconomic material stocks rise 23-fold over the 20th century and require half of annual resource use. Proc Natl Acad Sci 114:1880–1885
4. UNEP (2015) Global waste management outlook
5. Lanau M, Liu G (2020) Developing an urban resource Cadaster for circular economy: a case of Odense, Denmark. Environ Sci Technol 54:4675–4685
6. Stephan A, Athanassiadis A (2018) Towards a more circular construction sector: estimating and spatialising current and future non-structural material replacement flows to maintain urban building stocks. Resour Conserv Recycl 129:248–262
7. Arora M, Raspall F, Cheah L, Silva A (2020) Buildings and the circular economy: estimating urban mining, recovery and reuse potential of building components. Resour Conserv Recycl 154
8. Rose CM, Stegemann JA (2018) Characterising existing buildings as material banks (E-BAMB) to enable component reuse. Proc Inst Civ Eng Eng Sustain 172:129–140
9. Díaz-López C, Navarro-Galera A, Zamorano M, Buendía-Carrillo D (2021) Identifying public policies to promote sustainable building: a proposal for governmental drivers based on stakeholder perceptions. Sustainability 13

10. Miatto A, Schandl H, Forlin L et al (2019) A spatial analysis of material stock accumulation and demolition waste potential of buildings: a case study of Padua. Resour Conserv Recycl 142:245–256
11. Santos B, Gonçalves J, Martins AMT, Almeida PG (2020) Safeguarding portuguese traditional glazed tile cultural heritage with GIS. In: IOP conference series: materials science and engineering. IOP Publishing Ltd
12. Porto Vivo (2021) Centro histórico do Porto, Ponte Luiz I e Mosteiro da Serra do Pilar. Porto
13. Sá L (2018) United against illicit trafficking: SOS Azulejo. In: Kisić V, Tomka G (eds) Awareness raising & advocacy—learning kit for heritage civil society organisations, 1st edn. Europa Nostra, The Hague, p 65
14. Bertino G, Kisser J, Zeilinger J et al (2021) Fundamentals of building deconstruction as a circular economy strategy for the reuse of construction materials. Appl Sci (Switzerland) 11:1–31
15. Koutamanis A, van Reijn B, van Bueren E (2018) Urban mining and buildings: a review of possibilities and limitations. Resour Conserv Recycl 138:32–39
16. Munaro MR, Tavares SF, Bragança L (2022) The ecodesign methodologies to achieve buildings' deconstruction: a review and framework. Sustain Prod Consum 30:566–583
17. Arora M, Raspall F, Fearnley L, Silva A (2021) Urban mining in buildings for a circular economy: Planning, process and feasibility prospects. Resour Conserv Recycl 174
18. Dantata N, Touran A, Wang J (2005) An analysis of cost and duration for deconstruction and demolition of residential buildings in Massachusetts. Resour Conserv Recycl 44:1–15
19. Tatiya A, Zhao D, Syal M et al (2018) Cost prediction model for building deconstruction in urban areas. J Clean Prod 195:1572–1580
20. Heisel F, Rau-Oberhuber S (2020) Calculation and evaluation of circularity indicators for the built environment using the case studies of UMAR and Madaster. J Clean Prod 243:118482
21. Cai G, Waldmann D (2019) A material and component bank to facilitate material recycling and component reuse for a sustainable construction: concept and preliminary study. Clean Technol Environ Policy 21:2015–2032
22. Munaro MR, Tavares SF (2021) Materials passport's review: challenges and opportunities toward a circular economy building sector. Built Environ Proj Asset Manag 11(4):767–782
23. Porto (2019) Ordem Serviço No. I/194083/19/CMP (003). Câmara Municipal de Porto
24. Portugal (2017) Lei No.79/2017 de 18 de agosto. Diário da República, 1.ª série — No. 159, Portugal
25. Creswell JW (2014) Research design—qualitative, quantitative, and mixed methods approaches, 4th edn. SAGE Publications
26. Portugal (1999) Decreto-Lei No. 555/99, de 16 de dezembro. Diário da República
27. Ungan M (2006) Towards a better understanding of process documentation. TQM Mag 18:400–409

Part VI
Current Applications and Trends—Case Studies

Chapter 13
Focus on Skills for a Circular Built Environment in a New Curriculum Development

Matthias Haase, Isabelle Wrase, and Zifei Wang-Speiser

Abstract Despite the increased research regarding the sustainability transition towards the circular economy (CE) model, the existing literature on adopting and implementing the CE concept reinforcing educational approaches in secondary education seems limited. Considering the current challenges and the critical role of education to empower built environment management students to explore new paths of sustainable development and grow into active citizens, conscious producers, and consumers, this contribution investigates new ways of effective tools for teaching CE and sustainability concepts. The literature review has revealed a research gap regarding the formulation of educational approaches to effectively support CE concepts for higher education students, particularly in Swiss Facility Management education. The paper describes and critically discusses how an introduction of CE to master-level students to the circularity and sustainability perspective, prepare them to build prosperity, and act circularly in the future. A list of skills is presented which can be bundled into one holistic education. This provides valuable information for developing suitable programme in Higher Education which aims at the use and development of competitive methods and solutions for managing existing and new buildings that will contribute to lowering greenhouse gas (GHG) emissions related to the production, use, management, and demolition of architecture in a life-cycle perspective should be based on these skills.

Keywords Circular economy · Real estate · Facility management · Education programme

M. Haase (✉) · I. Wrase · Z. Wang-Speiser
Zurich University of Applied Sciences, 8820 Waedenswil, ZH, Switzerland
e-mail: matthias.haase@zhaw.ch

L. Bragança et al. (eds.), *Creating a Roadmap Towards Circularity in the Built Environment*, Springer Tracts in Civil Engineering,
https://doi.org/10.1007/978-3-031-45980-1_13

13.1 Introduction

Along with the European Union, Switzerland has committed itself to limit a global increase in temperature to 2 °C by 2050. On a global scale, the ambition is to reduce GHG emissions by 2050 to two tonnes CO_2-equivalents per capita and year if the concentration in the atmosphere is to be stabilized at 400–450 ppm CO_2-equivalents by the year 2100. The current Swiss average is 6 tonnes CO2-equivalents per capita per year [1].

The world population is constantly growing and is confronted with new needs for new buildings. This puts an enormous strain on our environment and our resources. The construction industry is responsible for approximately 33% of greenhouse gas emissions, 40% of waste generation and 40% of material consumption [2, 3]. Action plans for the Circular Economy (CE) include the gradual decoupling of economic activity from the consumption of finite resources and giving waste a second life. Recent efforts to take more concrete action include, above all, the development of new standardization activities such as CEN/TC 350/SC 1—"Circular Economy in the Construction Sector" and ISO/WD 59004—"Circular Economy" [4, 5].

The application of CE principles in real estate, building design and use (adaptability, durability, waste reduction and quality management) is mainly focused on new buildings where circularity can be embedded and enabled. Conversely, circularity has not been defined in the context of existing buildings [6].

CE must be viewed as a business strategy which adds value, not just a waste management or design strategy. Optimizing the use of buildings should also be brought into focus, rather than just considering them as potential material banks where components and materials can be recovered, reused, or recycled for new construction [6, 7].

IPCC (2007) and a study from Enkvist et al. (2007) report point to measures in the building sector as being the most economical, when compared to other important sectors [8, 9]. This reduction potential is "spread over hundreds of millions of individual buildings, each one presenting multiple and very diverse types of interventions" [10]. In addition, energy use in Swiss buildings contributes little to global GHG emissions due to extensive use of hydropower electricity [11]. However, real estate and the built environment exert a large influence on GHG emissions by means of transportation and land use planning, lifestyle and consumption, development of better materials and components, and life-cycle management of natural resources. The social dimension of sustainability is put more in focus. However, the development of a well-functioning society that provides a high quality of life for all citizens but only emits an average of two tons of CO2-equivalents per capita per year forms a huge challenge for all professions, not least in the building sector.

Despite these challenges there is need for a development of holistic built environment management skills for future workforce that can lead society to a more sustainable use of resources and adapt to future climate challenges. This linear model of resource consumption we are all used to has created an economy that largely depends on energy use and other resources, resulting in climate change, pollution and depletion of natural resources. Moving towards a CE requires changes in attitudes and behaviors, adopting a new way of thinking and developing new competencies and skills.

> How should educational institutions prepare their students for this reality? Which skills should we educate real estate and facility managers, architects, engineers and other building professionals to be able to contribute to this political goal in an optimal manner? Given the still ambiguous definitions and meanings of the concept, education for a CE plays a key role, and how we can teach and learn remains a challenge. New theories on how the CE concept can be applied to promote sustainability should be developed

13.2 Literature Review

The programmes that were analysed can be found on the homepage of European Real Estate Society (ERES): https://www.eres.org/index.php/activities/edueres [12]. Real estate programmes are being offered by a wide variety of universities across Europe. The list contains an overview of all Bachelor, Master, and postdoctoral programmes with a real estate focus. The list was downloaded in March 2022 and subsequently analysed. Details were recently published, and details of the analysis results can be found in Wrase et al. [13]. 374 programmes were analyzed in this study. To conclude whether there is a link for these sustainability-dimensions in the education programmes at all, a chi-square test of independence was performed [14]. The word "circular" was found only in two module designations: one made by the College of Estate Management in the M.Sc. in Real Estate and the other found in the description used for the M.Sc. in Real Estate and Facility Management by the Zurich University of Applied Sciences [13]. Real Estate Management is the most frequently mentioned discipline in the higher education programmes. However, the question of whether learners are taught interdisciplinary in all disciplines related to the real estate life cycle, only the ZHAW M.Sc. in Real Estate & Facility Management programmes and the College of Estate Management with its M.Sc. Real Estate score higher [13]. The word "Sustainability" and the dimensions "Processual" (Knowledge of real estate management, facilities management, architecture, and civil engineering as the base layer of the understanding of processes around the lifecycle in real estate), "Economical", "Ecological" and "Empirical" are significantly presented in the analysed higher education programmes in real estate [13]. However, there seem to be gaps between the content of the courses and the dimensions "Technical" and "Social". University programmes provide education not in all the described dimensions. The dimensions "Processual", "Economical", "Ecological" and "Empirical" are presented in the higher education programmes in real estate, whereas there is a strong correlation between these dimensions and the content of higher education programmes in real estate [13]. Remarkably, the dimensions "Social" and "Technical" are not presented in the higher education programmes in real estate.

13.3 Curriculum Development

According to this analysis, university curricula of European higher education programmes seem to focus to some extend on the required content to prepare graduates for real estate sustainability management, but rarely integrate all the required knowledge about the whole lifecycle of a building set in the megatrend sustainability towards a circular built environment. While there are studies that compare the content of courses [15], the focus of these studies is (supposedly) within one discipline in real estate [16]. In this study we applied the framing method proposed by Lahtinen and Yrjölä (2019) [17]. This study is in line with the logic of the dimension "social", "Economical, "Processual", "Technical", "Empirical" and "Ecological" similar to Babosa and de Oliveira (2020) for Managers [18]. In turn we examine the interdisciplinary share of architecture, engineering, facility management, and real estate management in European higher education programmes. Sustainable management requires an understanding of all the disciplines and related skills associated with the real estate lifecycle.

13.3.1 Future Workforce for Circular Economy

One could argue that being skilled in all disciplines or knowledge dimensions is not essential. However, a study analyzing 600 job advertisements in the real estate sector revealed that most of the dimensions listed are highly sought after [19]. Specifically, the "social" dimension appears to be in exceptionally high demand. In contrast, this study indicates that real estate programs tend to underrepresent social competencies. It is challenging to track the extent to which students are socially empowered.

Employability is an important criterion for testing training programmes for their raison d'être [20]. However, due to the macroeconomic situation, most real estate markets worldwide are still in a booming or at least preferable phase [13]. Workers in the field of real estate have been in demand for decades now. However, this positive development obscures the fact that companies are not only looking for real estate experts, but also those who are knowledgeable in areas related to the real estate lifecycle and can help manage and lead the sustainability transformation. Overall, Wrase et al. found an indication that there are still deficiencies in higher education programmes in real estate when it comes to whether to empower students in sustainability transformation [13].

13.3.2 Future Skills

Bearing in mind the findings of this analysis, the next important step towards introducing sustainability and circular economy management in real estate at higher education level would require a new set of skills. As stated above, the understanding of the real estate lifecycle and the inter- and transdisciplinary perspective of the different dimensions of sustainability taught in higher education can lead the future workforce to a holistic approach of sustainability and empower them to transition towards a circular economy of the built environment. The main scope of the programme must therefore relate to the general issue of sustainability including environmental as well as economic and social focus. The transformation of the existing real estate market is the most important step in developing an environment that fulfils the main key goals [21].

(1) Climate change mitigation, (2) climate change adaptation, (3) sustainable use and protection of water and marine resources, (4) transition to a circular economy, (5) pollution prevention and control, and (6) protection and restoration of biodiversity and ecosystems. Not mentioned are general skills related to "social, "processual" and "empirical". These will be assumed to be existing as well without detailing it here.

Considering the high demands of transdisciplinary skills to address the sustainable transformation and in particular the circular economy, the curriculum must thrive to educate built environment workforce on the:

- Sustainability consequences of transformation
- Quality of real estate concept with respect to circularity
- Quantity of real estate quality and circularity potential
- Quality of circularity concept
- Management consequences of circularity concepts.

Skills related to Sustainable Finance and Governance

The students need to be enabled to adopt a critical and reflective attitude towards today's sustainability practices throughout their studies. A well-founded introduction to the history of sustainability discourse and to the relevance of different sustainability concepts for the governance of real estate and facility management builds the basis. Then, skills of real estate-related financial and economic knowledge such as the time value of money, capital market theory and capital costs, financing and capital structure, company value, investments and risk management are needed. One focus is on the question of how factors from the areas of environment, social, and governance can be included. This enables students to are examine and evaluate most important players in the field of sustainable finance, green financing instruments, sustainability ratings and political influences.

The skills further base on knowledge of History-based overview of sustainability concepts and their different ethical implications (focus on sustainability and climate ethics). This provides insights to the relevance of different sustainability concepts for the governance of real estate and facility management as well as standard corporate finance concepts that also apply to real estate financing. It provides an understanding of the time value of money, knowledge of capital market theories and how financial markets work. This enables students to participate in discussions about capital budgeting considerations. Other skills include financial modelling and valuation and estimation of the cost of capital, definition/information on the origin of sustainability in finance and discussion of the relevance of sustainability for various actors in the financial sector. Finally, options for assessing and quantifying sustainability for the critical assessment of non-financial properties of projects, products and companies, followed by insights into how sustainability aspects can be integrated into common corporate finance instruments such as discounted cash flow. The skills include understanding of sustainability and financial analysis and ability to participate in discussions of the relationship between sustainability and profit-oriented factors, including sustainability aspects in various forms of financing and investment processes.

Skills related to Strategies in Sustainability

A comprehensive understanding of sustainability and its application to the area of real estate and facility management is conveyed. This includes ecological, economic and social aspects as well as consideration of the entire surrounding ecosystem. Skills further include presenting and analyzing common sustainability planning, certification, monitoring and reporting as well as methods for procurement and quality assurance of construction and FM services and name their advantages and disadvantages as well as convey the implementation in selected strategies.

The skills include areas for influencing the sustainability of buildings, districts and cities over the lifecycle with a focus on the transformation of existing structures. It enables the translation of abstract sustainability concepts into concrete sustainability and investment strategies and development of corresponding (quantifiable) goals and measures in a real estate-related context. Economic advantages and disadvantages as well as conflicting goals and synergies of different sustainability goals need to be presented and sustainability rating systems and criteria, sustainability planning, certification, implementation, monitoring and reporting concepts and tools introduced. The skills further include understanding of methods for the sustainable procurement and quality assurance of real estate services over the entire life cycle and offers opportunities and limitations of the tools, instruments and methods and to be discussed from the perspective of research.

Skills related to Circular Economy Management

The real estate industry is responsible for tying up many resources, sometimes for decades. To deal with the limited resources available in an economical and environmentally friendly manner, the change from linear to circular real estate management and sustainable building design, use and management is essential. The sustainability and circularity assessment is another element for assessing, (further) developing and applying the impact of the built environment on the climate and nature, also against the background of corporate, owner and user responsibility.

The necessary skills to handle these perspectives on converting the existing linear economic systems into circular economy systems include knowledge about circular economy strategies and design options as well as challenges for circular economy systems in the real estate context.

The understanding of building as a layered model—system separation in building construction and possible future Closed loop systems of the future (energy or material focus?) as well as principles of circular planning/building/use/operation are demanded skills. Further skills include methodological approaches and benefits of environmental assessment (also of buildings), impact on carbon accounting at portfolio level as well as cost accounting of waste avoidance and sustainably optimized dismantling in a quantitative manner. Further, basic understanding of enhanced models as "From waste to recyclables: recycling and recyclables management methods", "Urban Mining—Use today's inventory as a material reserve", "the Madaster vision: materials with identity" are needed and build the basis for developing own sustainability and circularity certificates.

Skills related to Built Environment Transformation

The stock of buildings and infrastructure outweighs the volume of new construction projects. The challenge lies in needs-based, multi-factorial, optimized modernization and/or conversion. In addition to the conventional developer aspects, sustainability, lifecycle and public concerns must be understood in a holistic solution. The transformation of the built environment is a complex task that indicates higher demands on real estate expert skills: The projects derived from a strategy and to be developed should enable the economic, ecological and social feasibility of sustainability at management level and be able to achieve implementation and conversion.

The necessary skills include the knowledge of classification of system boundaries of buildings, properties, areas, etc. It further includes peculiarities of

- re-development/transformation of existing buildings
- project management during re-development.

Then, it requires the appraisal of socio-cultural and institutional aspects in development, discussions of areas influencing sustainability in real estate and obstacles in practice; portfolio versus area perspective; Perspective space and infrastructure, but also people and organization.

The skills must provide a classification of digital (communication) options to be able to communicate and present complex issues in a way that is appropriate for stakeholders followed by discussion of multidisciplinary decision criteria in re-development. It also includes planning and construction accompanying facility management in re-development, risks in re-development, and the integration and implementation of economic, ecological and social sustainability using digital methods and tools. The students develop presentation skills to communicate their project work in a simulated management board meeting in the "Ethics, Leadership, Change" module.

13.3.3 Interdisciplinary Students and Staff

To co-ordinate professions, we need to co-ordinate education. Therefore, students need to be trained in interdisciplinary teams and by an interdisciplinary teaching staff. In this manner, the students get to know various distributions in responsibilities and tasks throughout the design process and learn in practice the integrated design strategies that can ensure usability and synergy of their design project with its surroundings and users.

The teaching staff ideally consists of a mixture of professionals from built environment practice and research to provide the students with professional perspectives from a wide variety of actors ranging from the latest scientific discoveries to actual management experiences. The fixed core of teachers, extended by invited guest teachers, present the students with a curriculum that aims at making maximum connections among the different institutions and professions.

One of the most important insights the students learn is the importance of building and managing interdisciplinary teams. Given the complexity of modern-day real estate tasks, it is an absolute necessity to complement one's own expertise with others into a constructive co-operation. While real estate and facility management students still occasionally consider interdisciplinary co-operation as unnecessarily time-consuming and frustratingly reducing their own chance at achieving a high grade, the best manner in which to counteract these prejudices is to guide the students towards a positive experience and give them time to recognize and respect the contribution the other profession's expertise can make to create a high-quality design project [22].

13.4 Discussion

For future built environment professionals to be able to translate the climate and resource challenges in society into a fitting form of the built environment, they need to not only understand the use and development of sustainable building methods and solutions, but, above all, to know how to integrate this knowledge into their everyday decision-making routines. They need to become creative, active professionals who can keep themselves updated on relevant theories, understand how they interact, and currently update their design routines according to this new knowledge. In this manner, the future workforce gets familiar with different ways in which to work professionally with the built environment and get a good foundation to mix and recombine the different elements into a meaningful whole.

All Real Estate and Facility Management professionals need to be able to handle challenges related to climate change and resource scarcity, and realise that this reality affects professional ethics, regulations and skills required in practice and research. Ever more strict building regulations require for even 'ordinary' transformation projects to stress low carbon emissions and high energy efficiency. As professionals, the students will be expected to not only work on projects with a set of measures and targets that put energy and greenhouse gas emissions at the heart of their performance, but also be able to negotiate the performance targets of the project with the building owner and real estate team while thinking about the long-term transformation of society and the development of innovative strategies.

In addition to management experience and theory, evaluation and reflection form an important part of the students' learning environment. To transform to a well-functioning real estate and built environment, the students need to build a thorough knowledge of a project's local physical and social site as well as its cultural history and recognize the consequences of the existing structures and dynamics for the design of their project. They need to know how to integrate sustainability issues with circularity models. The real estate industry is still in the beginning of recognising the long-term goals of CE.

It became obvious that a main task of future real estate and facility management research should focus on quantifying real estate qualities and qualifying facility management quantities with respect to circularity and sustainability. This must begin with the development of a common vision for real estate and facility management professional. The skills structure needed for these workforces support group work in different transformation stages that will enable participants to develop energy efficient transformation solutions and to effectively communicate them between the different domains (real estate and facility management).

The list of skills illustrates the importance of the questions a real estate and facility management professional should ask at the start of, during and after having finished a transformation project, to affect the building and environmental programme and the transformation's consequential performance. It also teaches the students how to assess which measures work best under given circumstances within a given budget, the different outcomes of various contracting frameworks and the negative consequences of types of actors not being engaged in the project from day zero.

13.5 Conclusions

We proposed a list of skills focusing on different aspects of the transition of the built environment towards circularity. This proposed list of skills for a Master course in Higher Education equips the students with extensive knowledge and experience that prepare them for a challenging, rapidly changing profession. There is a dire need for a real estate policy that encompasses the new technological opportunities at the same time as improving quality of life and circularity whilst reducing environmental impact. The development of a CE is pre-requisite for a zero-emission built environment. It creates a new physical framework for large parts of society, reflecting cultural values, existing structures, and the new layers of the future. In that sense the concept of CE challenges the current linear model of production and consumption, exploiting the planet's raw resources, manufacturing and using the product, and disposing of waste. That is a conceptual change in education from the traditional linear to the circular model of production and consumption. That involves educational curriculum improvements about resource reuse, recycling, sources prioritization, and their efficient use.

Theoretical challenges are plentiful when recognizing that the physical state of a building is the result of the complex interaction of a very large set of physical components. The integration of these interactions on behavioral simulation poses major modeling and computational challenges that must be discussed. Its ability to deal with the resulting complexity of scale and diversity of component interactions has gained building simulation a uniquely recognized role in the prediction, assessment, and verification of real estate performance. A real estate industry that does not look for definitive solutions and permanent projects, but for structures that can continue to adapt to the need of society and citizens: More built environment for less CO_2.

Green jobs and green entrepreneurship have a future in a modern form of economic activity that responds to business needs for profitability and growth, but at the same time considers the environmental limitations, treating circularity as an opportunity and not as an obstacle. We noticed that more and more real estate companies are seeking to help environmental protection and limit climate change by investing in the research and development of green technologies and practices. Therefore, to serve the new vision for society and the economy by respecting the principles and values of circularity and sustainable development, the need for research in education and the essential contribution of the educational system, of the formal, informal, and non-formal education, occurs. We hope we contributed to the current state-of-the-art on educational policies regarding the role of circular economy in sustainable development, to share a good and innovative practice of CE and sustainability teaching and learning in different countries, to open and encourage a critical discussion on the topic.

We recommend practical educational strategies hoping (i) to encourage teachers to adopt innovative teaching methods and share good practices of the CE teaching and (ii) to urge education policymakers to integrate the CE vision into school curricula.

References

1. Bundesamt für Umwelt (Bafu) (2022) Die Schweiz senkt die Treibhausgasemissionen, https://www.bafu.admin.ch/bafu/de/home/themen/klima/dossiers/klimakonferenz-cop21-von-paris--abkommen-ueber-die-international/die-schweiz-senkt-dietreibhausgasemissionen.html#-1491583269. (in German: Switzerland reduces its GHG emissions). Last Accessed 08 Feb 2023
2. Hossain MU, Ng ST (2018) Critical consideration of buildings' environmental impact assessment towards adoption of circular economy: an analytical review. J Clean Prod 205:763–780. https://doi.org/10.1016/j.jclepro.2018.09.120
3. Ness DA, Xing K (2017) Toward a resource-efficient built environment: a literature review and conceptual model. J Ind Ecol 21(3):572–592. https://doi.org/10.1111/jiec.12586
4. CEN/TC 350/SC 1—Circular economy in the construction sector. CEN
5. ISO/WD 59004—Circular economy. International Standardization Office. ISO

6. Kyrö RK (2020) Share, preserve, adapt, rethink–a focused framework for circular economy. IOP Conf Ser Earth Environ Sci 588:1–8. https://doi.org/10.1088/1755-1315/588/4/042034
7. Adams KT, Osmani M, Thorpe T, Thornback J (2017) Circular economy in construction: current awareness, challenges and enablers. Waste Resour Manag 170:10
8. IPCC (2007) Climate change 2007: contribution of working group III to the fourth assessment report of the intergovernmental panel on climate change. Cambridge University Press, Cambridge, pp 746–807
9. Enkvist P-A et al (2007) A cost curve for greenhouse gas reduction. McKinsey Q 1:35–45
10. Cheng C et al (2008) The Kyoto protocol, the clean development mechanism and the building and construction sector. In: A report for the UNEP sustainable buildings and construction initiative. United Nations Environment Programme, Paris, France, p 88
11. Jones P et al (eds) (2009) European carbon atlas. In : Low carbon urban built environment. The Welsh School of Architecture, Cardiff, p 220
12. ERES European Real Estate Society (2022) Real estate educational programmes, https://www.eres.org/index.php/activities/edueres. Last Accessed 13 Mar 2022
13. Wrase I, Haase M, Wang-Speiser Z (2023) Towards a circular built environment–focus on the new M.Sc. program in real estate and facility management. Front Sustain Sect Circ Econ 4. https://doi.org/10.3389/frsus.2023.1163394
14. Cohen J (1988) Statistical power analysis for the behavioral sciences. Taylor and Francis, New York
15. Keramitsoglou K, Litseselidis T, Kardimaki A (2023) Raising effective awareness for circular economy and sustainability concepts through students' involvement in a virtual enterprise. Front Sustain Sect Circ Econ 4. https://doi.org/10.3389/frsus.2023.1060860
16. Kans M (2021) A study of maintenance-related education in Swedish engineering programmes. Educ Sci 11(9):535. https://doi.org/10.3390/educsci11090535
17. Barbosa MW, de Oliveira VM (2021) The corporate social responsibility professional: a content analysis of job advertisements. J Clean Prod 279:123665. ISSN: 0959-6526. https://doi.org/10.1016/j.jclepro.2020.123665
18. Lahtinen S, Yrjölä M (2019) Managing sustainability transformations: a managerial framing approach. J Clean Prod 223:815–825. ISSN: 0959-6526. https://doi.org/10.1016/j.jclepro.2019.03.190
19. Wrase (2022) Real estate consideration in education. Submitted
20. Witt E et al (2013) Towards a frame-work for closer university-industry collaboration in educating built environment professionals. Int J Strateg Prop Manag 17(2):114–132
21. Envoria (2023) EU taxonomy overview, https://eu-taxonomy.info/info/eu-taxonomy-overview. Last Accessed 30 Mar 2023
22. Nicol D, Pilling S (2000) Architectural education and the profession. Preparing for the future. In: Nicol D, Pilling S (eds) Changing architectural education. Towards a new professionalism, Spon Press, London, pp 1–26

Chapter 14
Efficient Recovery of Valuable Resources from Construction and Demolition Waste Towards Circular Economy in Construction Industry—Sustainability Assessment and a Case Study

Emircan Ozcelikci, Alper Aldemir, and Mustafa Sahmaran

Abstract Considerably high amounts of CO_2 release to the atmosphere trigger global warming. Although there are several methods to reduce the level of CO_2 release, the extent is still very limited and recently-growing awareness of sustainability/global warming have been pushing the entire construction industry to seek alternative methods for rigorously lowering/eliminating the level of CO_2. The key strategic objective of the study is to develop eco-friendly/innovative, 100% CDW-based construction materials and demountable structural systems. This study aims to achieve higher levels of circularity in civil engineering materials/structures, contributing to the reduction/elimination of CO_2 emissions much more rigorously through the following key objectives: (i) Upgrading CDW recycling/reuse efficiency by capturing CO_2 from the atmosphere to improve properties of CDW-based constituents via accelerated mineralization/carbonation, (ii) Development of holistically-designed advanced material property improvement technologies to even enhance the greenness of 100% CDW-based materials/structures through efficient CO_2 binding/elimination capability, (iii) Validation of the ultimate products (materials/structures) with additional green perspective through a detailed large-scale field demonstration. Despite the abundance of studies in this area, there is currently very little work on demonstration activities on the real-time applicability of geopolymers development using industrial by-products and CDWs. Successful outputs of this study and their real-time demonstration will offer a fully sustainable construction system, including speed of construction/design flexibility/air purification/cost reduction/energy and material saving/avoidance of unwanted pollution-heavy demolition processes and make much larger audience to be influenced by the study's results.

Keywords Circular economy · Construction and demolition waste · Case study

E. Ozcelikci · A. Aldemir · M. Sahmaran (✉)
Civil Engineering Department, Hacettepe University, 06800 Ankara, Turkey
e-mail: sahmaran@hacettepe.edu.tr

L. Bragança et al. (eds.), *Creating a Roadmap Towards Circularity in the Built Environment*, Springer Tracts in Civil Engineering,
https://doi.org/10.1007/978-3-031-45980-1_14

14.1 Introduction

Portland cement (PC)-based traditional concrete is the second mostly used material globally, following water and cement industry alone is responsible for ~9% of total anthropogenic CO_2 emission [1]. Furthermore, current construction practice of materials' overproduction, insufficient longevity of concrete structures and accumulation of construction and demolition waste (CDW) are becoming increasingly. CDW industry is one of the largest global solid waste producers, accounting for 25–30% of total urban waste in Europe [2]. Such CDW production requires relatively high demand for proper handling not only to lower CDW going to clean landfills endangering health of people/environment, but to reduce concrete production that will be, otherwise, used to construct/renovate /repair/maintain new/existing infrastructure. It has been shown that, constituents from CDW, widely available and troublesome, can effectively be used singly/in combination to produce geopolymer binders, achieving major progress compared with most work utilizing common aluminosilicates as precursors (e.g. fly ash/slag/metakaolin) that are highly demanded by PC industry, expensive or even unavailable for certain regions for geopolymer production [3–5]. Additionally, aggregates conventionally constitute more than 70% of the volume of concrete mixtures and those diverted from recycling applications are mostly used in low-tech applications and/or are deposited to landfills, risking the health of individuals/poll ting environment. The properties of aggregates can be significantly improved by chemically binding CO_2 into their composition, achieving double benefits of better performance of CDW-based aggregates and deceleration of global warming [6].

With the successful production of CDW-based "green" concretes, the reliance and negative environmental impacts of PC, clean aggregates and traditional concrete whose mass production are major contributors of CO_2 emissions/global warming will believe to be significantly reduced. Moreover, incorporation of these "green" concretes into demountable structural components further reduces energy/material/workmanship needs and additional waste creation anticipated with conventional construction methods. Lowering these negative effects will have clear impacts on the quality of lives of people (environment-/health-/society-wise) worldwide considering the current issues of PC production/CDW generation.

To tackle the drawbacks of concrete production, achieve truly effective/easily appl cable/uncommon solutions for CDW problem and advance beyond the current state-of the-art, the main focus of this study is to the production of CDW-based PC-free "green" concretes to be incorporated in the development of lego-like/demountable structural elements that do not create additional waste, do maximize reductions in energy needs (>50%) and CDW upcycling and promote circularity in novel civil engineering materials/structures. Also, demonstration of study outcomes is crucial for wide-spread visibility/applicability and impact because almost never demonstration activity has been conducted in the current literature on geopolymer development according to the author's best knowledge. Therefore, final part of the study will be devoted to the design/construction of a real-time 1-storey residential building incorporated entirely with the study outcomes (i.e. fully demountable structural elements with 100% CDW-based green concretes having enhanced CO_2 capturing capability). The study will show easy-to-apply/effective/sustainable approach for a cleaner environment/energy efficiency/CDW upcycling/significantly cheaper/affordable housing targeting in-need people all around the world.

14.2 Laboratory Scale Experiments

The development of CDW-based green concretes, accelerated carbonation process of recycled aggregates, and design of fully-demountable structural elements such as column, beam and slab and their connection details were presented in this section.

14.2.1 Development of CDW-Based Geopolymer Concretes

In the development of CDW-based geopolymer concretes, unknown-origin CDW-based materials such as brick, tile, concrete and glass wastes were first collected separately from a demolition site. Thereafter, in order to achieve the suitable fineness for geopolymerization, these materials were put through a two-step crushing and grinding process. To enhance the mechanical and durability properties of CDW-based geopolymer concretes, 20% slag by weight was replaced with CDWs in the binder design. Three different alkali activators such as sodium hydroxide (NaOH), sodium silicate (Na_2SiO_3) and calcium hydroxide ($Ca[OH]_2$) were used to activate CDW-based materials. Fine recycled concrete aggregate (FRCA) and coarse recycled concrete aggregate (CRCA) obtained by crushing of concrete waste into different sizes were first subjected to an accelerated CO_2 capture process and then used as the aggregate phase of the mixtures. Figure 14.1 shows the visual images of materials used in the development of CDW-based geopolymer concrete.

In the development of CDW-based geopolymer concrete, several mixture designs were made in which various parameters such as precursor combinations, alkali activator concentrations, aggregate/binder and water/binder were investigated. According to the findings, the mixture contains 80% mixed CDW, 20% slag, 1:1 precursor/RCA ratio, 8 M NaOH, 1:2 wt% NaOH/Na_2SiO_3 ratio, $Ca(OH)_2$ at 5 wt% of the precursor and 0.33 water/precursor ratio were selected and used for the development of structural elements. In their unprocessed state, CDW-based materials are often inert and, when combined to produce geopolymers, they frequently form N-A-S-H type gel structures after geopolymerization reactions. In this study, CDW-based materials were supplemented with calcium sources such as slag and $Ca(OH)_2$ in order to reach structural strength. Thus, C-A-S-H type gel structures were also formed in the final matrix, which contributed to the strength along with the N-A-S-H gel structures. In addition, a certain amount of Na_2SiO_3 was added as an extra alkaline

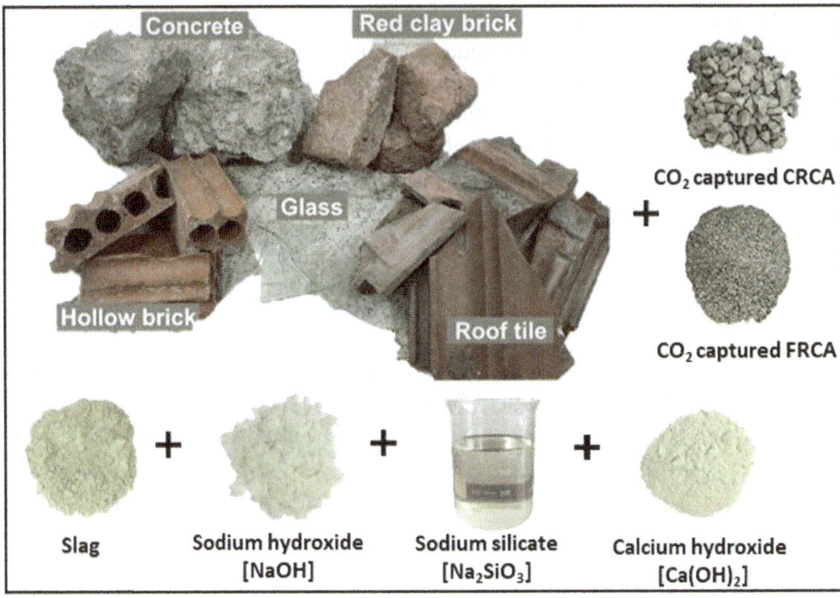

Fig. 14.1 Visual images of the materials using in geopolymer concrete production

activator to provide reactive Si ions to the system and to maintain the Si/Al balance of the matrix. To ensure worldwide reproducibility, the amount/concentrations of these inclusions used to boost geopolymer strength should be tailored based on CDW composition. For the CO_2 capture process of RCAs, a pilot-scale carbonation reactor with a rotational chamber (Fig. 14.2), which can provide to implement different ranges of operational parameters [e.g., temperature (25–100 °C), humidity (50–95%), CO_2 concentration (0–20%) and pressure (0–5 bar)] was developed. After carbonation, the water absorption value of RCAs decreased by 32.7%, while the flow value and compressive strength of the mortars containing carbonated RCA increased by 9.6 and 28%, respectively, compared to the reference mortars.

As a result of the experimental study, CDW-based geopolymer concretes reached a maximum compressive strength of 39.7 MPa and splitting-tensile strength of 2.93 MPa at the end of 28 days of ambient curing. Additionally, the durability performance of CDW-based geopolymer concretes were examined by conducting many tests such as drying shrinkage, water absorption, efflorescence, resistance against freeze-thawing cycles and sulfate solution. According to the results, the durability performance of CDW-based geopolymer concretes was found to be similar or better than that of PC-based structural concretes.

Fig. 14.2 Accelerated CO_2 capture process of RCAs

14.2.2 *Design of Fully-Demountable Structural Elements*

The structural design procedure of fully demountable buildings is composed of reinforcement design and the connection design (Fig. 14.3). For the case study, a 1-story residential building with a plan area of approximately 250 m^2 is designed. The structural design included both vertical load demands (i.e., dead load of structural elements, design snow load and live loads) and lateral load demands (i.e., earthquake loads and wind loads). The internal force demands under the effect of predefined loading patterns were obtained using a commercial structural analysis program (i.e., SAP2000 v21).

The computer model was formed using frame elements for beams and columns whereas the slab elements were modelled by utilizing shell elements. All the connections for structural elements (i.e., beams, columns and slabs) were assumed to be pin connection as all the proposed dry-joint connections developed in the scope of study permits rotations at the connections. After these demand calculations, the reinforcement design was performed using the fundamental rules in conventional concrete with a different material model for geopolymer concrete [7]. In addition, the design of connection was thoroughly based on scaled tests of each connections and their validated unscaled numerical models. The detailed design, material selection, testing and performance criteria of the structural elements developed within the scope of the demonstration activity were previously presented to the literature [8, 9].

Fig. 14.3 Details of connections of demountable structural elements

14.2.3 Sustainability Assessment for Circular Built Environment

In order to reveal the environmental advantages of CDW-based geopolymer concretes and demountable structural elements, life cycle assessment (LCA) was carried out. The LCA was applied followed by (ISO) 14,040 and 14,044 standards with the use of GaBi software [10]. In this analysis, cradle-to-gate approach with a functional unit of 1 m^3 CDW-based geopolymer concrete structural elements was determined (Fig. 14.4). According to the findings, demountable structural elements produced

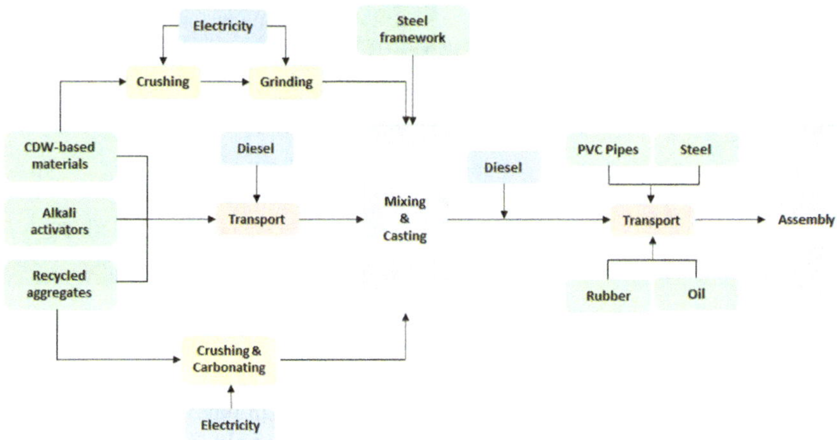

Fig. 14.4 Plan and system boundry of LCA study

with CDW-based geopolymer concrete exhibited 12.6% lesser global warming potential (GWP), 4.5% lesser eutrophication potential (EP), 11.8% lesser smog formation potential (SFP) and 2.4% lesser fossil fuel depletion (FFD) compared to demountable structural elements produced with conventional PC concrete. Additionally, considering to possible cyclic use of demountable structural elements several times in their whole lifecycle, it can be stated that demountable types of structural elements will be better in terms of environmental impacts overall.

14.3 Real-Time Field Demonstration

The construction of the first-ever real-time demonstration of a fully demountable 1-story building is one of the study's main goals in order to increase the credibility and trust of different stakeholders in dry/demountable reinforced concrete connections and CDW-based geopolymer concrete. At the initial stage of the field demonstration activity, the foundation of the building was built in the demonstration field. Firstly, a thin layer of gravel was laid on top of the soil in order to obtain a flat and workable environment (Fig. 14.5a). Accordingly, a mat foundation was constructed (Fig. 14.5b). The base plates that are used to anchor the demountable columns were installed inside the mat foundation before the concrete was cast.

The materials used for the production of prefabricated elements (CDW-based materials, slag, alkaline activators, chemicals, CO_2 captured recycled aggregates, steel profiles, reinforcement) were supplied and transferred to the prefabricated concrete manufacturer. Structural elements (columns, beams and slabs) were prefabricated to demonstrate the validity of the demountability of such elements when

Fig. 14.5 Construction of the mat foundation, **a** laying of gravel; **b** building of the foundation

dry connections are used (Fig. 14.6). At the end of the 28-day ambient curing, all structural elements were transferred to the demonstration field.

The construction of real-time 1-story building was started with the installation of vertical structural elements. All the columns were labeled and placed near the mat foundation with the help of a mobile crane. Afterward, the columns were lifted to their vertical position placed on base plants and bolted to maintain their aligned positions (Fig. 14.7). All demountable column elements were installed in less than three hours.

Fig. 14.6 Preparation of prefabricated geopolymer structural elements

Fig. 14.7 Installation of columns

Fig. 14.8 Installation of beams

Fig. 14.9 Installation of slabs

After the installation of column members, the beams were transferred to the field and placed between every two columns by simply sliding them through the beam-to-column connections (Fig. 14.8). All beam installation operations also took less than three hours. It should also be noted that the columns and beams were produced with relatively large sizes so it allows future uses in multi-story buildings.

The structural system of the prototype building was continued by placing the slab elements on top of the beams (Fig. 14.9). A membrane was placed between the slab and the beam to fill any manufacturing spaces. This operation was also easy to apply and the whole construction operation of the structural system was finished in the course of two days.

Besides the structural elements, the walls of the building were also made of precast panels, thus the building can be fully demountable. The panels were also mounted with the help of a mobile crane and each panel took a time around 15 min on average to be installed (Fig. 14.10). Additionally, other non-structural members such as windows, doors, etc. were installed on the prefabricated construction.

Finally, the first-ever real-time demonstration of a fully demountable 1-story building constructed with CDW-based geopolymer concrete has been completed as shown in Fig. 14.11. After the successful construction of the green building, installation of other work items such as electrical, plumbing and heating systems were carried out.

Fig. 14.10 Installation of non-structural elements

Fig. 14.11 1-story building field demonstration

14.4 Conclusion

In this study, it was aimed to develop fully-demountable structural elements produced with construction and demolition waste (CDW)-based green geopolymer concrete. With the achieving successful results from these elements, a demonstration of a real-time 1-story residential building was carried out as the next step. According to the findings, the following conclusions were drawn:

- CDW-based materials such as brick, tile, concrete and glass waste can be successfully utilized in geopolymer concrete production up to 40 MPa compressive strength after 28 d ambient curing.
- After the carbon capturing process on the CDW-based recycled concrete aggregates (RCA), the water absorption of RCAs decreased by 32.7% and the compressive strength of the mortars containing carbonated RCA increased by 28%.
- It was demonstrated that the modular demountable system design provides a fast, reliable and practical construction approach compared to the traditional construction system.
- Life cycle assessment revealed that CDW-based geopolymer concrete showed 12.6% lesser global warming potential, 4.5% lesser eutrophication potential, 11.8% lesser smog formation potential and 2.4% lesser fossil fuel depletion compared conventional PC concrete.

It is believed that real-time demonstration of the study's outcomes will help the impact of this study to last for years to come and will be used as a viable tool and advertisement element for anybody who is interested in making collaborations/ learning more about CO_2 capture/storage, innovative CDW recycling and design-for-demountability for future reuse. Owing to holistic solutions, highly-pure recycled building materials/higher traceability/market acceptance and thereby, higher levels of circularity will ensure widespread utilization of recycled materials in construction industry. The study will also contribute to higher circularity of the new building products, leading to even lower impacts in the subsequent lifecycles. Successful outcomes of the study can be reproduced through rigorous classification and characterization of country-specific CDWs, which is considered the main challenge in successful material development, selection of alkaline activators suitable for CDW composition, and integration of modular system design into the developed products.

Acknowledgements The authors gratefully acknowledge the financial assistance of the Scientific and Technical Research Council (TUBITAK) of Turkey and the British Council provided under projects: 218M102 and European Union's Horizon 2020 research and innovation programme under grant agreement No: 869336. This work was also supported by Newton Prize 2020.

References

1. Makul N (2020) Advanced smart concrete—a review of current progress, benefits and challenges. J Clean Prod 274:122899
2. Tan J, Cai J, Li X, Pan J, Li J (2020) Development of eco-friendly geopolymers with ground mixed recycled aggregates and slag. J Clean Prod 256:120369
3. Yıldırım G, Kul A, Özçelikci E, Şahmaran M, Aldemir A, Figueira D, Ashour A (2021) Development of alkali-activated binders from recycled mixed masonry-originated waste. J Build Eng 33:101690
4. Ilcan H, Sahin O, Kul A, Ozcelikci E, Sahmaran M (2023) Rheological property and extrudability performance assessment of construction and demolition waste-based geopolymer mortars with varied testing protocols. Cement Concr Compos 136:104891
5. Ozcelikci E, Kul A, Gunal MF, Ozel BF, Yildirim G, Ashour A, Sahmaran M (2023) A comprehensive study on the compressive strength, durability-related parameters and microstructure of geopolymer mortars based on mixed construction and demolition waste. J Clean Product 136522
6. Ruviaro AS, Silvestro L, Pelisser F, de Azevedo ARG, de Matos PR, Gastaldini ALG (2022) Long-term effect of recycled aggregate on microstructure, mechanical properties, and CO_2 sequestration of rendering mortars. Constr Build Mater 321:126357
7. Kocaer O, Aldemir A (2022) Compressive stress–strain model for the estimation of the flexural capacity of reinforced geopolymer concrete members. Struct Concr 1–20
8. Akduman Ş, Kocaer O, Aldemir A, Şahmaran M, Yıldırım G, Almahmood H, Ashour A (2021) Experimental investigations on the structural behaviour of reinforced geopolymer beams produced from recycled construction materials. J Build Eng 41:102776

9. Aldemir A, Akduman S, Kocaer O, Aktepe R, Sahmaran M, Yildirim G, Ashour A (2022) Shear behaviour of reinforced construction and demolition waste-based geopolymer concrete beams. J Build Eng 47 103861
10. Mir N, Khan SA, Kul A, Sahin O, Ozcelikci E, Sahmaran M, Koc M (2023) Construction and demolition waste-based self-healing geopolymer composites for the built environment: an environmental profile assessment and optimization. Constr Build Mater 369:130520

Chapter 15
Albanian Green Economy Trends

Llambi Prendi and Jonida Gashi

Abstract In the 1991s, Albania entered a great political and economic transition, passing from the communist to the capitalist system. All resources in common use were usurped. The environment and its conservation were the least concern for Albania. Even now its environment faces several problems. Some of the key issues are pollution, waste disposal, loss of biodiversity, deforestation, etc. Our paper uses data generated by OECD stat and abundant literature. We have selected OECD Green Growth indicators for monitoring progress towards green growth in Albania as environmental and resource productivity, the environmental dimension of quality of life, economic opportunities and policy responses, and the socio-economic context. The data values collected vary from 1980 to 2021. We rely on Albania's performance over time of many green growth indicators. This will allow us to observe whether Albania is moving toward a green economy and policies. From the data presented we see that Albania has made some advancements toward green growth.

Keywords Green growth · Policy responses · Quality of life

15.1 Introduction

During the 2008 financial crisis, the green economy comes to be a political approach to the solution of two problems: preventing environmental degradation and creating new mechanisms for economic growth and development.

Referring to Tamaki R. "*Green growth is about fostering growth and development while ensuring that natural assets continue to provide the resources and environmental services on which our well-being relies. Governments that pursue policies designed to promote green growth need to catalyze investment and innovation that underpin growth and give rise to new economic opportunities*" [1].

L. Prendi (✉) · J. Gashi
Aleksander Moisiu University, Durres, Albania
e-mail: llambiprendi@uamd.edu.al

L. Bragança et al. (eds.), *Creating a Roadmap Towards Circularity in the Built Environment*, Springer Tracts in Civil Engineering,
https://doi.org/10.1007/978-3-031-45980-1_15

In the course of the last twenty years in Albania, $PM_{2.5}$ quality improved. Air pollution is higher in cities as a result of the exponential growth of car use, the increase of productive activities that create pollution, and the reduction of green spaces. Massive deforestation has greatly increased carbon dioxide in the air. Water sources are contaminated by industrial effluent, human waste, and untreated sewage. Waste management companies don't use suitable technology to treat waste and protect the environment. Many species of flora and fauna are disappearing.

Many countries have developed green economy strategies, policies, and programs. They predict that in the future, humanity will face serious challenges relating to climate and resource scarcity. The countries have developed common policies and visions for a green economy that offers sustainable long-term growth. In this context, the European Union has established practices and strategies that focus on improving the well-being of future generations.

Governments of developed countries support and finance companies that apply the green economy. Certainly, individuals can contribute to all of this.

Our paper relies on Albania's performance over time of many green growth indicators. This will allow us to observe whether Albania is moving toward a green economy and policies.

The main research questions are: What is the trend of Carbon dioxide emissions? How is the performance of the Energy sector? What are the ways to improve the environmental quality of life? What is the trend of technology and innovation? And what are the recommendations for Albania to move toward a greener economy?

Our study can help policymakers to improve existing policies and implement better ones.

15.2 Literature Review

Several definitions of green have been suggested by various organizations and authors, some of the definitions will mention below:

"*A green economy is one in which policies and innovations enable society to use resources efficiently, enhancing human well-being in an inclusive manner, while maintaining the natural systems that sustain us.*" [2].

A green economy is defined as low-carbon, resource-efficient, and socially inclusive [3, 4]. A green economy is one that results in improved human well-being and social equity, while significantly reducing environmental risk and ecological scarcity [1].

According to Purvis et al. [5], there are three **pillars of sustainable development**:

(a) Economic sustainability: which aims to reduce extreme poverty by including job creation, profitability, and proper accounting of ecosystem services. (b) Environmental sustainability: limiting the impact of human activities on the environment and protecting nature. (c) Social sustainability: Social sustainability includes environmental justice.

United Nations Environment Programme in its report states that there are positive signs that trade-related practices are moving towards more environmental, social, and economic sustainability. These trends have to be encouraged as well as fully informed by the Rio+20 mandate to advance the green economy in the context of sustainable development and poverty eradication [3].

15.2.1 Green Development in Albania

There are a number of known environmental issues in Albania. The main problems include air and water pollution, poor waste management infrastructure, and deforestation. Air pollution in the biggest cities came as a result of an increase in car ownership and a decrease in urban greenery. Water pollution in Albania is caused by waste disposal and sewage discharge. The waste management system consists of weak collection systems. Recycling is done by private companies that employ poor people to collect plastic, metal, glass, and paper waste which is processed or packaged and then sold elsewhere. Deforestation is another issue of concern and illegal logging remains the main threat to Albanian forests [6]. Energy and air pollution are complex challenges and significant obstacles to future economic development and well-being. Albania's reliance on hydropower, and the absence of coal in its energy mix, meaning that the energy sector, climate, and environmental challenges it faces are very different from the rest of the region [7].

15.2.2 Green Recovery Path in Albania

Albania adopted a Law[1] on Climate Change in 2020 and it's National Energy and Climate Plan (NECP) in December 2021. Albania has had a Law on Energy Efficiency since 2015, to which ambitious amendments were recently introduced to include additional energy efficiency measures [9]. In 2017 Albania introduced renewable auctions for solar and wind energy.

15.3 Methodology

15.3.1 Data

Our paper uses data generated by OECD.Stat [10]. We have selected OECD Green Growth indicators for monitoring progress towards green growth in Albania:

[1] Law no 155/2020 on climate change [8].

- Environmental and resource productivity: indicate whether economic growth is becoming greener with more efficient use of natural capital and capture aspects of production that are rarely quantified in economic models and accounting frameworks, (OECD Green Growth database, 2022) [10];
- Environmental dimension of quality of life: indicate how environmental conditions affect the quality of life and well-being of people, (OECD Green Growth database, 2022) [10];
- Economic opportunities and policy responses, and the socio-economic context: indicate the effectiveness of policies in delivering green growth and describe the societal responses needed to secure business and employment opportunities, (OECD Green Growth database, 2022) [10].

The data values collected vary from 1990 to 2021. Some data is missing and this can distort the data evaluation, but this effect is extremely small for what we want to study.

15.3.2 Variables

Environmental and resource productivity includes three variables: CO_2 emission and energy productivity. For the environmental dimension of quality of life, we have elaborated on two variables: exposure to environmental risks with eleven components and access to drinking water and sewage treatment with two components. Economic opportunities and policy responses have in total nine components; four for the variable technology and innovation (patents) and five for environmental taxes and transfers. In the socio-economic context, we have presented two variables: economic context which has eleven components, and social context with nine components.

With the purpose of making the appropriate calculations, we have elaborated the specific weight formula below:

$$Z_t = Mean\left\{\sum_{1}^{n}\left(\frac{X_{n,t}}{\sum_{1}^{t} X_{n,t}}\right)\right\} * 100 \quad (15.1)$$

where: t—time, n—number of components, $X_{n,t}$—the value of the component n at time t and, Z_t—*the* specific weight of the specific variable at time t (in percentage).

15.3.3 Trend-Line Analysis

A common way to identify trends is using trendlines, which connect a series of highs or lows [11]. An uptrend situation is when variables move upward and keep increasing over the period and a downtrend situation is when variables move downward and keep decreasing over the period. A horizontal trend situation is when variables are not moving in any direction; they are moving sideways. We will create this trend line from the historical variables, and we can use this information to predict the future movement of the variables.

15.4 Results Analysis

Climate change is a major global issue that could have significant effects on green growth and sustainable development. Main concerns related to the effects of increasing gas concentrations on earth's climate, and the consequences for ecosystems. The main challenges are to limit the emissions of CO_2 and other GHG.

15.4.1 Carbon Dioxide Emissions

Carbon dioxide emissions are those stemming from the burning of fossil fuels and the manufacture of cement. They include carbon dioxide produced during the consumption of solid, liquid, and gas fuels and gas flaring [12]. In Fig. 15.1 we are able to see that the values of the CO_2 emissions, from 1980 to 2021, have had a negative trend. Albania's CO_2 emissions per capita were at the level of 1.56 tons of CO_2 per capita in 2021, down from 1.6 tons of CO_2 per capita previous year, this is a change of 2.53%.

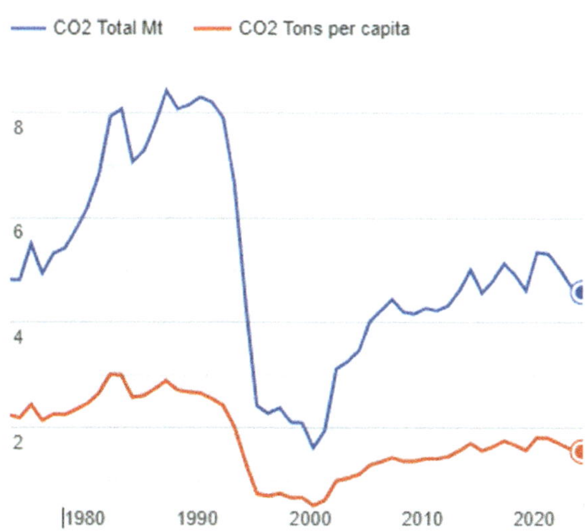

Fig. 15.1 CO_2 emissions in Albania [12]

15.4.2 Energy Sector

The country is predominantly mountainous, with eight major rivers crossing a basin with over 57% of its current administrative extension and a perennial flow of around 1300 m3/s.

As we see from Fig. 15.2, Albania relies almost entirely on hydropower for electricity generation, which accounts for 99.5% of domestically generated supply [13]. This gives it an advantage in decarbonizing its electricity sector but also makes it highly vulnerable to the changing climate. Albania is the only country in the Western Balkans to have completed new large hydropower plants in the last decade and as of the end of 2020, it had no fewer than 23 operational hydropower plants of more than 10 MW, as well as countless smaller ones. Nowadays the total installed capacity has reached 2400 MW. However, projects granted, but not yet developed, at 1485 MW (33%), and the hydropower potential studied still unexploited remains at around 615 MW (14%). Therefore, in a synthesis, considering the theoretical potential of 4500 MW, today, only 53% are exploited. As it is worth noting that the country can offer one of the lower costs of production (LCOE) of hydropower in the region starting from an average of 30 Euro/MWh [14].

Albania enjoys 220 sunshine days or 2,700 h of sunshine annually as one of the European nations with the highest number of sunshine hours per year, which makes it an excellent bet to host solar PV capacity. Albania has a potential of 3706 GWh of sun energy production annually. Together the solar and wind power potential of Albania is estimated at over 7 GW, of which by 2030, it can install 1,074 MW of solar PV capacity and 616 MW of wind energy can be deployed by 2030, with an annual generation potential of 1,697 GWh [15].

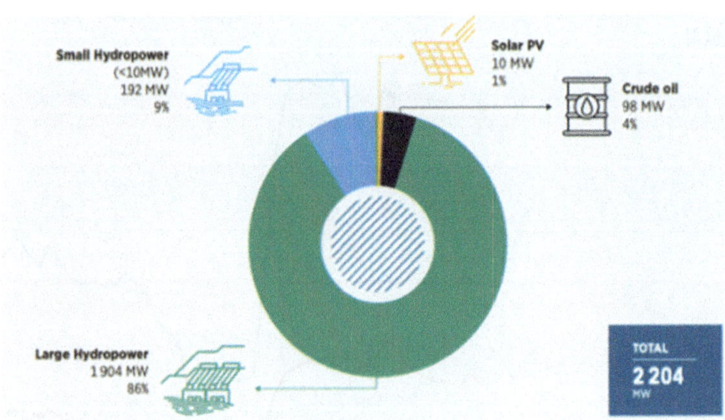

Fig. 15.2 Albania's electricity mix (*Source* IRENA, 2019)

15.4.3 The Environmental Dimension of Quality of Life

Albania has slightly improved its scores in the environment policy dimension, but air quality is still a predominant concern. PM 2.5 is the air pollutant posing the greatest risk to health, affecting more people than any other pollutant. Air quality has improved greatly in the course of the last ten years. Since 2005, emissions of sulfur oxides decreased by some 35%, and emissions of ammonia by around 10%, while emissions of NOx, NMVOC, and PM_{10} increased slightly [2].

Figure 15.3 reveals that there has been a slight fall in the percentage of Exposure to environmental risks since 1990 but a slight increase after 2015. What is interesting in this figure is the rapid decrease from 2000 to 2010. This compounded variable is expected to continue decreasing over time. The same figure shows that there has been a gradual increase in the percentage of access to drinking water. What is interesting in this figure is the rapid growth of the compound variable until 2015 and then stabilized until today. The rate of Access to Drinking Water and sewage treatment is likely to remain steady after 2021. According to OECD, the way to improve the environmental quality of life should follow this recommendation [16–18].

- Improve air quality by decreasing dependence on fossil fuels in the energy mix;
- Phase out coal subsidies and decarbonize the energy sector, and introduce incentives that support renewable integration;
- Include measures to prevent air emissions from industry;
- Promote sustainable transport options;
- Increase the number of wastewater treatment plants and reassess the fee structure so that fees cover the service costs;
- Introduce land-use management and soil protection legislative and policy frameworks.

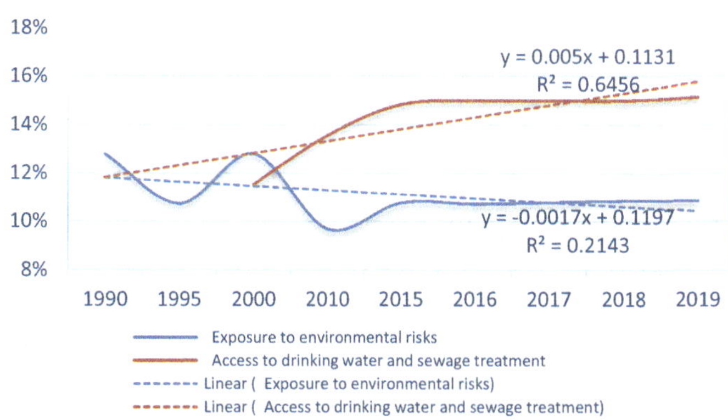

Fig. 15.3 Environmental dimension of quality of life

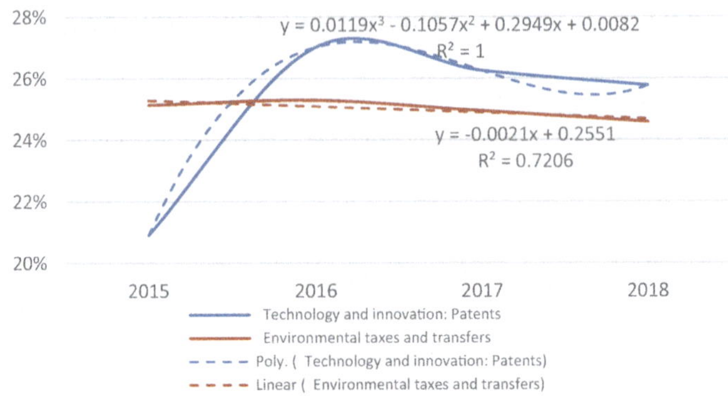

Fig. 15.4 Economic opportunities and policy responses

15.4.4 Economic Opportunities and Policy Responses

Albania's response toward the development of technology and innovation has not been in the proportion with the country's objective of being a member of the European Union. Production sector composition is heavily skewed towards traditional low-technology activities, based on labor costs rather than high value-added products or services, and competitiveness remains low generally [14].

In Fig. 15.4, it can be observed that the values for technology and innovation (patents) have increased quickly from 2015 to 2016, followed by a slight decrease in 2018. The trend of this variable had been the polynomial of order three. Recently this trend seems to be positive. The environmental policy in Albania is still underdeveloped, regardless of the fact that the legal framework is well-defined and according to EU guidelines. Taxes collected for environmental protection purposes are often not managed efficiently in terms of environmental protection. Conversely, the values of the variable environmental taxes and transfers (Fig. 15.4), have been stable throughout the 2015–2018 period, at around 25%.

15.4.5 Socio-economic Context

Albania is a transition economy but shows strong economic performance. About 80% of its exports are related to the countries of the European Union. The economic crisis caused by the Covid-19 pandemic has significantly reduced economic growth, which recovered, as GDP increased by around 5.3%, supported mainly by family consumption. High energy prices present negative risks for the future, also affecting the increase in the inflation rate. According to IMF data, public debt represented 81.5% of GDP in 2021, much higher than the pre-pandemic level of 67.8%.

As we see from the Fig. 15.5, what is curious about the values of this variable is that they have a rapid increase from 2000 to 2010. The values of the variable social

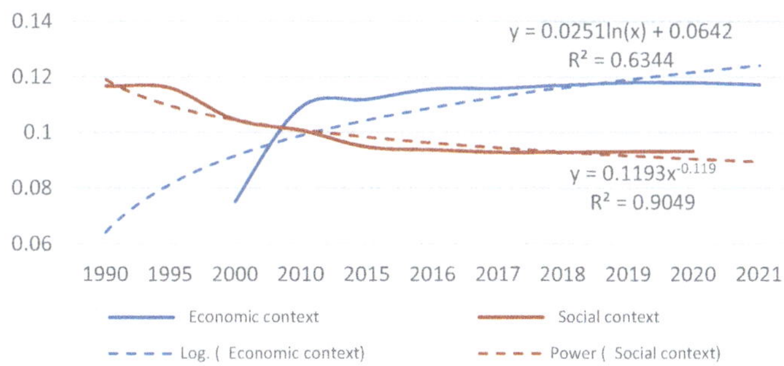

Fig. 15.5 Socio-economic context

context are monotonically decreasing. Strengthening measures and mechanisms of social development, at the national and local level, that reflect social inclusion should be urgent issues to be addressed.

15.5 Conclusions

Economic growth in Albania crucially depends on the availability and good quality natural and environmental resources, which are under increasing pressure. Albania has done some initiatives on renewable energy and energy efficiency.

CO_2 emissions are reduced in the period 1980–2021 and the government committed to reducing it by 12% compared with the baseline.

Substances such as $PM_{2.5}$, ambient ozone, lead, and residential radon, have negative effects on health, crops, ecosystems, and materials. Also, polluted water and untreated sewage lead to ill health and death.

The environmental policy in Albania is still underdeveloped, regardless the fact that the legal framework is well-defined and according to EU guidelines. And yet, with all the good intentions, promises, and commitments, it is visible that very little has been achieved.

In our study, the number of environment-related inventions is expressed as a percentage of all domestic inventions. The trend of this variable first results in an increase and then a decrease for the few years considered. Its values in recent years are almost unchanged. Investments in environmental management, water-related adaptation, and climate change mitigation technologies should be looked at.

In the economic context, we have included data such as GDP, real GDP, GDP per capita, etc. The trend of this variable in our paper turns out to be increasing, but its values from 2005 are almost unchanged. Albania needs investments in all sectors: industrial, service, and agricultural. There should be subsidies and tax breaks and also emphasis should be placed on those sectors where Albania can have comparative advantages.

15.6 Recommendations

Currently, the country remains highly dependent on weather hydropower technology due to which it has to also depend on energy imports. To deal with this, Albania's energy mix needs to diversify and become self-reliant with the help of solar and wind energy.

Among the recommendations we can give at this point is that Albania should apply specific policies that target domestic consumption and cleaner energy development to achieve environmental sustainability. It must direct its internal production towards the use of energy-saving technologies and clean and renewable energy.

Further reductions in $PM_{2.5}$ concentrations are required, and pollution controls need to be designed that simultaneously reduce $PM_{2.5}$ and O_3 concentrations.

The level of taxes for energy products for transport purposes and for stationary purposes, motor vehicles and transport, waste management, ozone-depleting substances, etc. remained the same for all the years considered, we suggest that Albania should apply new tax policies.

References

1. OECD (2017) Green Growth Indicators. OECD Publishing, Paris. http://www.oecd.org/greengrowth/green-growth-indicators-2017-9789264268586-en.htm. Last accessed 21 January 2023
2. European Commission: A European Green Deal. https://ec.europa.eu/info/strategy/priorities-2019-2024/european-green-deal_en. Last accessed 20 January 2023
3. UNEP: Green economy | UNEP—UN Environment Programme. Last accessed 1 February 2023
4. UNEP: Towards a green economy: pathways to sustainable development and poverty eradication. United Nations Environment Programme. http://www.unep.org/greeneconomy/greeneconomyreport/tabid/29846/default.aspx. Last accessed 19 January 2023
5. Purvis B et al (2019) Three pillars of sustainability: in search of conceptual origins. Sustain Sci 14:681–695 (2019). https://doi.org/10.1007/s11625-018-0627-5
6. CEE Institute (2021) Albania social briefing: green development in Albania—policies and actions, vol 42, no 3
7. OECD: Multi-dimensional review of the Western Balkans: assessing opportunities and constraints, OECD development pathways. OECD Publishing, Paris, https://doi.org/10.1787/4d5cbc2a-en
8. Kuvendi i Shqiperise: Ligji Nr. 155/2020. Per ndryshimet Klimaterike.
9. OECD (2018) Competitiveness in South East Europe: a policy outlook, competitiveness and private sector development. OECD Publishing, Paris. https://doi.org/10.1787/9789264298576-en
10. OECD (2023) Green growth indicators. OECD Environ Stat (database). https://doi.org/10.1787/data-00665-en,lastaccessed2023/1/19
11. Chen J (2022) A common way to identify trends is by using trendlines (2022)
12. Ritchie H, Roser M, Rosado P (2020) CO_2 and greenhouse gas emissions. https://ourworldindata.org/co2-and-greenhouse-gas-emissions. Last accessed 19 January 2023
13. Energypedia: Albania energy situation—energypedia. 5 January 2023
14. INSTAT, Energjia | Instat. Last accessed 5 January 2023
15. IRENA: Renewables Readiness Assessment: Albania (irena.org). Last accessed 5 January 2023

16. OECD: Toolkit for water policies and governance: converging towards the OECD council recommendation on water. OECD Publishing, Paris. https://doi.org/10.1787/ed1a7936-en
17. OECD: Exposure to air pollution. OECD Environ Stat (database). https://stats.oecd.org/Index.aspx?DataSetCode=EXP_PM2_5
18. OECD: Green budgeting and tax policy tools to support a green recovery, OECD policy responses to coronavirus (COVID-19). OECD Publishing, Paris. https://doi.org/10.1787/bd02ea23-en

Part VII
Barriers Against CE Implementation in Buildings

Chapter 16
PV and Thermal Solar Systems Application in Buildings. A State of Art in the Context of Circular Economy

Marilena De Simone

Abstract Solar energy is one of the most promising sources for low carbon energy production. In particular, PV panels and thermal solar collectors can be easily integrated into new and existing buildings to improve their energy efficiency and sustainability. On the other hand, solar-based technologies require extraction of natural resources and processing, thus materials conservation and recovery are vital to effectively contribute to the decarbonization of the construction sector. The paper is meant to be a brief state of the art that summarizes the relevant issues for achieving the goal of circular economy of buildings with the focus on solar energy application, with the novelty of considering and comparing two technologies, photovoltaic and thermal. Most of the scientific literature was dedicated to PV technologies due to the increasing importance of the electrification process and the usage of materials with reduced availability. Thermal solar collectors were mainly analysed developing LCA without a larger point of view embracing circularity concepts. Apart from the technological matters, the investigation highlights social, behavioral, and economic aspects that can be crucial to trace the route to circular economy.

Keywords Solar energy · Buildings · PV · Thermal solar collector · Circular economy

M. De Simone (✉)
Department of Environmental Engineering, University of Calabria, 87036 Rende, Italy
e-mail: marilena.desimone@unical.it

L. Bragança et al. (eds.), *Creating a Roadmap Towards Circularity in the Built Environment*, Springer Tracts in Civil Engineering,
https://doi.org/10.1007/978-3-031-45980-1_16

16.1 Introduction

16.1.1 Energy Efficiency of Buildings and Cities

In 2021, the European Commission proposed new legislation to promote the decarbonization of buildings and reduce methane emissions by 80% in 2030. Moreover, in 2022, new requirements to improve the sustainability of products and constructions were introduced with the aim to reduce primary energy consumption [1]. In the background of the energy transition, buildings are responsible for about 40% of the total end-use energy consumption. Generally, the residential sector consumes more energy than the commercial sector. Energy demands for space heating and hot water production predominate in the residential sector [2], while energy for HVAC systems and lighting are higher in commercial buildings [3]. In the future, due to the global warming, a significant increment in cooling demand of the residential sector is predicted [4]. The Nearly Zero Energy Buildings (NZEB) concept and diffusion could be a proper response to reduce energy consumption according to the Energy Performance of Buildings Directive (EPBD) [5].

The development of envelope materials is an important action to reach the NZEB target and apply eco-friendly and economical solutions in parallel with the installation of renewable energy technologies. The use of renewable energy systems can significantly contribute towards net positive energy balance of buildings [6] coupling smart electricity distribution grids and smart metering systems [7].

Buildings should become energy producers contributing to the opportunity to reach a balanced energy infrastructure. In the light of these changes, it is also important to underline the role of Micro-CHP, high performance HVAC systems, smart-grids and predictive Building Automation and Control Systems [8].

Moreover, the development of efficient building energy modelling (BEM) tools is valuable since there is a gap between measured and predicted energy demands. This gap depends on the specific building facility and on the building standards that generally simplify the definition of design parameters related to occupants' behaviour [9]. Occupants' behavior, habits, preferences, contribute to the mentioned mismatch between the designed (predicted) and real building energy performance. Occupant monitoring and the development of novel models, incorporated in building simulation tools, are still challenging research tasks [10].

The progressive integration of renewable energy in buildings and cities is an important step towards more circular material flows. But the materials needed to decarbonize electricity and mobility are supplied by natural resource extraction. Critical technologies to the clean energy transition, such as wind turbines, photovoltaic, batteries and vehicles, still primarily utilize feedstocks from natural resources instead from waste for processing and production. Moreover, robust and complete end of life (EoL) management strategies are not well developed for these materials. In the future, renewable energy manufacturing could be affected by depletion of metal reserves and consequent price volatility. Increasing metal recovery rates is a multilayered problem complicated by the inefficient collection and processing of metals at EoL and the still low-cost supply of certain metal reserves [11].

The aim of this paper is to synthesize current research themes, initiatives, and opportunities for the circular economy of solar technologies applied in buildings as pivotal to energy transition and decarbonization of the building sector.

The main novelty of the study is the consideration of both the systems, PV and thermal, following the leading concept that encourages energy mix and technologies integration.

The research questions that led to the study are the following:

(1) What are the application trends of PV and thermal systems and related implications in circular economy?
(2) What are the advantages and disadvantages of these two technologies in the light of the energy transition and circular economy?
(3) What are the main issues and potentialities in terms of materials used, EoL management, regulations, and social aspect in European countries?

To answer to these questions, documents were searched from the Scopus database and the websites of consortia, associations, and institutions that provide reports and information on renewable energy application at the European level.

16.2 Solar Energy Application in Buildings

Documents were searched from the Scopus database using the following key words: PV AND circular economy, solar thermal AND circular economy, PV AND LCA, solar thermal AND LCA. PV and thermal were also used in the same search. The documentation was enriched by reports and regulations collected on official websites. In total, 57 papers were found on photovoltaic panels, 15 papers on thermal systems (only considering LCA), 17 reviews on the solar systems application in buildings with related technical, environmental, and economic aspects. Only 4 research papers were collected on the comparison of the two technologies.

The documents listed in this paper, for the sake of brevity were limited to those that were informative to the research questions avoiding a pure description of a sole solar system. This section is dedicated to the two first research questions.

16.2.1 *Electricity and Heat Production*

Solar energy in buildings is used for electricity production from PV technologies and heat production from solar thermal systems. Integrated photovoltaic thermal PV/T systems are also available for production of thermal energy as well as electricity, but they are less market attractive due to relatively high investment cost [12]. The current problem with PV technologies is the high investment compared to their limited efficiency. PV market is mainly silicon based (about 95%) with a conversion efficiency usually ranging from 12 to 17%. The development of new PV technologies is oriented to organic and hybrid (organic–inorganic) Perovskite PVs, organic, or ones with the application of nano-technologies, such as nanopillar solar cells. The mentioned novel PV technologies are still not ready for an expected wide market application, due to present technical shortcomings and unfavorable economic viability [11].

The main types of solar thermal collectors can be classified in: glazed flat-plate solar collectors, the most versatile and widespread type that has good efficiency and guarantees the production of hot water with a variable temperature between 40 and 70 °C, and can be used all year round; unglazed solar collectors that are cheaper but also less efficient; evacuated and heat pipe solar collectors that have higher efficiency but present a more complex operation and technology.

The development of hybrid energy systems could be an effective solution to integrate in buildings renewable energy technologies based on the utilization of solar energy, and recent investigations, such as [13], evaluated different hybrid energy options for residential applications.

PV/T technology is particularly suitable in hybrid energy systems, but further development of PV/T plants is necessary to improve efficiency and reduce unit costs. Different PV/T configurations have been recently investigated and mainly focused on the technical aspects by simulation [14] or experimental approach [15].

16.2.2 Solar Systems and Circular Economy

The installation of silicon based solar modules continues worldwide. Waste management linked to this technology is a pressing problem considering that PV have 25–30 years of productive life. The predicted EoL PV modules could be 80 million tonnes globally by 2050 [16] and waste generated from solar panels is considered as one of the future challenging waste streams [17]. The materials that can be recovered from EoL photovoltaic waste include glass cullet, silicon wafers and granulates, silver, indium, tin, molybdenum, nickel, zinc, copper, aluminum, steel, tellurium, cadmium, selenium, gallium, ruthenium, and plastics. PV panels can also be a source of smelter flux by metals recovering. Plastics are in backsheets and encapsulants. The backsheet is usually made of polyvinyl fluoride that can be difficult to recycle due to toxic hydrogen fluoride halogen compounds [11]. Solar thermal (ST) installation recorded a market decline since 2010 [18]. Generally, this contraction is due to less attractive incentives, whereas the industrial sector and technology are well-experienced. This fact produced also the perception of more difficulties in this investment choice than PV one [19]. Overall, in the European countries, the ST market suffered the current economic crisis and can be found quite dependent on political support programmes. A comparison between the trends of diffusion of the two solar technologies is presented in Fig. 16.1a, b.

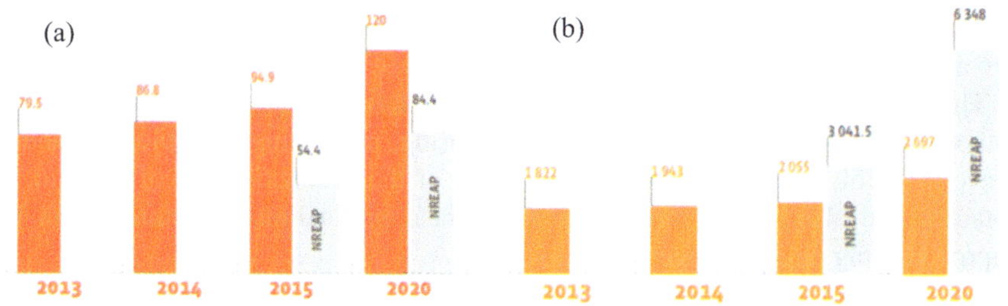

Fig. 16.1 **a** Comparison of the current trend of photovoltaic capacity installed (in GWp) and **b** of thermal solar systems (in ktoe) against the NREAP (National Renewable Energy Action Plans) roadmap. *Source* [20]

The variety of materials involved in ST systems is minor, but a large amount of metals and glass is implicated. As a consequence, recovery and recycling appear to be reasonable alternatives to final disposal to landfill or incineration [21].

The decline of ST application can also be the reason why a lot of the literature concerning circular economy development in the solar sector was focused on PV systems, neglecting the problems related to the ST already installed and the applicability of this technology in the future.

Both PV and ST systems can offer advantages and disadvantages from the environmental and economic point of view and potentialities in the light of the energy transition and circular economy.

Few studies are still available in the literature focused on the comparison of these two solar technologies. As an example, the authors of [22] compared the economic benefit related to the installation of Domestic Solar Water Heater (DSHW) and Building Integrated Photovoltaic (BIPV) systems for households and found that the convenience of these technologies is strongly related to the roof surface availability and incentives.

The authors of [21] posed the following question: "If a given surface of domestic roof is available, which is the best option in term of energy and environmental impacts for solar energy exploitation?". They considered the end of life of PV and ST technologies assuming that metals and glass can be recovered. In particular, LCAs were conducted to compare flat solar thermal collectors with four types of PV modules (mono-Si, multi-Si, CdTe and CIS) by considering a domestic usage. ST registered the highest number of favorable indicators: eight out of ten, in the case of no-recycling of materials after dismalting phase, and six out of ten in the case of recycling of materials after dismantling phase.

Also in [23] LCAs analyses were developed to compare the environmental performances of a PV system with mono crystalline cells and a ST system with evacuated glass tube collectors. The results demonstrated that the solar panels are responsible for the most impact in both the systems. ST system provided more than four times release to air than PV, and the outputs by a PV system to soil and solid waste were about one third that of a ST plant. The study also highlighted the importance of the selection of the system components (solar panels, battery, and heat storage), taking into account the problems related to toxicity.

The advantages of combining ST into PV panels were illustrated in [24]. The author promotes this integration in manufacturing, retrofitting of deployed panels, converting EoL panels. Overall, an extension of the productive life of PV can be obtained and also the performances of the heat production and storage can be improved using phase change materials.

16.3 Barriers, Challenges, and Opportunities

The implementation of circular economy embraces environmental, economic, and behavioral variables that can constitute positive drivers or obstacles.

This section provides information regarding the issues related to the usage of natural sources and materials, the problems still encountered in the EoL, the development of regulations at the European level, and some considerations on social aspects.

16.3.1 Materials and Manufacturing

Energy transitions and decarbonization will increase demand for certain materials and it is important to get product designers to think about recycling to reduce the impacts from mining. Elevate attention should be paid to metals, glass, and plastics.

Crystalline silicon PV panels require silver, quartz, copper, metallurgical silicon, aluminum, and tin. Thin-films PV made of CdTe require tellurium, cadmium, molybdenum, or tin. Thin-films made from copper indium gallium selenide (CIGS) require copper, indium, gallium, selenium, cadmium, tin, or zinc depending on the buffer layer used. Perovskites depend on lead and tin compounds.

The usage of plastics includes both recycling and degradation designs. Plastic degradation answers to the economic and environmental need for disposal at EoL. Different types of backsheets are available, and also batteries can include plastics that can be recovered or produced with recycled material. Bioplastics are alternative solutions. Bioplastics for batteries and PV panels must be stable during the operation phase and should provide end of life degradation under diverse environmental conditions [11].

Glass has an important role in energy transition. Flat glass is used both in PV (80–97% glass by weight) and thermal panels. Other kinds of glass can be found in batteries [25].

Solar technologies require glass with specific optical characteristics, generally low iron glass. Antimony and cerium are present in pre-2000 PV glass. The variety of glass composition and presence of impurities can complicate the recycling phase.

Overall, the percentages of recoverable materials that can be used in the production of modules are significant for glass and aluminum (97 and 100%, respectively), for copper and tellurium (about 80%), lower for rare metals such as indium and gallium (75 or 99%, respectively) that have a great value, despite are only 1% of the weight of the panel [26].

16.3.2 End of Life

The oldest generation of PV panels permits more than 80% recovery rates of components through the separation of plastics using a heating process and the manual recovery of the solar cells, glass, and metals. The process presents two critical issues: the necessity of manual activities, and the availability of many different silicon based solar cells that require specific treatments.

Automated processes for treating crystalline silicon modules need developments. The process employed by First Solar achieves elevate recycling rates for glass and semiconductor material (about 90%) [27].

Several studies are conducted in order to test processes applicable to emergent and new PV technologies (CIS, CIGS), for which it is still difficult to make predictions regarding the implementation of suitable EoL.

Moreover, a process of dematerialization in the production phase of the modules can have benefits in terms of costs and management of EoL. Recent and ample literature highlights as LCA is also a useful method to be applied to the solar technologies supply chain in order to ensure a "closure of the loop" [26].

16.3.3 Regulations and Standards

During the last decade, the EU enforced specific regulations concerning PV waste treatments. In particular, the extended producer responsibility principle (EPR) supports that producers are responsible for treating their EoL products [28].

The European Union's Waste and Electrical Equipment (WEEE) Directive included photovoltaics in waste regulations and established that producers have to prefinance anticipated reclamation and recycle costs. Moreover, Restriction of Hazardous Substances (RoHS) Directives prompted PV producers selling into the EU to plan take-back and collection on site [29].

Thus, basically, EPR organizations require compliance with EU standards (CENELEC or WEEELABEX) by contract operators, and EU prohibits collection of PV panels with construction and demolition waste, requires separate treatment of silicon and non-silicon PV panels; imposes depollution requirements on metals like cadmium, selenium, and lead. Moreover, thermal, chemical, and mechanical treatments of PV panels need the removal of dangerous substances [30].

The transition from linear to circular economy can provide benefits from diverse points of view: environmental protection, increasing security of the supply of energy and materials, stimulating innovation and competitiveness, and creating new job opportunities. With this aim, several countries in the world such as China, Japan, Canada, and also European nations like Germany, UK, The Netherlands, Sweden, and Finland promoted the CE concept [31].

16.3.4 Social Aspects

A recent and influential study based on the application of agent based model and machine learning techniques [16] highlights the importance of considering social factors in circular economy developments. The authors claim that a significant improvement of PV material circularity can be obtained by social interventions targeted to increase customer attitudes towards used PV modules. Also the authors suggested the question "Could it be possible to have a secondary market for used PV as strong as the market for used cars?".

In addition, as reported in [32], waste production is affected by recycling behavior that varies from the type of economy. This study underlines the difficulties encountered in the populations of the European countries to understand the relationship between waste reduction and resource efficiency.

The education of the population is a key action to increase people awareness and to understand the importance of the contribution of each of us to the waste management.

16.4 Conclusions

The paper presents an overview of potentialities and issues concerning the application of circular economy to solar systems for energy production in buildings. The study starts from two relevant challenges for the European Union: energy transition and sustainable protection of natural resources. In this framework, the buildings sector is recognized as a key target by the scientific community and governments.

The installation of solar systems in buildings can answer to the need for decarbonization and can reduce the dependence from fossil fuels. Solar thermal (ST) systems are a mature technology able to ensure low costs and adequate efficiency, but it is penalized by scarce incentives that decelerated a massive diffusion. This can be the reason why scarce scientific literature and governmental initiatives support the inclusion of ST in structured concept of circular economy, despite of the significant usage of materials such as glass and metals. Attention was devoted instead to PV systems that still have low efficiency and require the application of a higher variety of raw materials. LCA analysis was effective to demonstrate the clear environmental advantages that both these systems can provide, especially if recycling or reuse of components is prevised at EoL. Moreover, new systems can be developed in the future, such as the combined PV/T panels. Technical and environmental issues affect the treatments at EoL and regulations and standards were focused on resolve them.

Despite the availability of European directives, EU members considered the problem with different attention and efforts. Moreover, human and behavioral aspects related to the application of circular economy in the real society are neglected in the majority of initiatives.

The study demonstrates the complexity of the problem, in terms of the diverse nature of driver factors and contexts, that requires more cooperation between the scientific disciplines as well as more organized frameworks connecting public and private sectors of the economy. On the other hand, this state of art is limited to be a general overview, and deeper analyses could be drafted considering issues and initiatives for each European country.

References

1. United Nations Environment Programme. The Closing Window. Emissions Gap Report 2022: Climate Crisis Calls for Rapid Transformation of Societies. Nairobi. https://www.unep.org/emissions-gap-report-2022
2. De Simone M, Callea L, Fajilla G (2022) Surveys and inferential statistics to analyze contextual and personal factors influencing domestic hot water systems and usage profiles in residential buildings of southern Italy. Energy Build 255:111660. Elsevier B.V. https://doi.org/10.1016/j.enbuild.2021.111660
3. Building Codes Assistance Project. bcapcodes.org/topics/climate-change/
4. Jakubcionis M, Carlsson J (2017) Estimation of European union residential sector space cooling potential. Energy Policy 101(November 2016):225–235. Elsevier https://doi.org/10.1016/j.enpol.2016.11.047
5. https://epb.center/epb-standards/energy-performance-buildings-directive-epbd/. Accessed 26 January 2023
6. Nizetic S, Coko D, Marasovic I (2014) Experimental study on a hybrid energy system with small and medium scale applications for mild climates. Energy 75:379–389. Elsevier Ltd. https://doi.org/10.1016/j.energy.2014.07.087
7. Shukla A, Liu S, Gaterell M, Wood G, Day R, Iweka O, Hussain A, Van der Horst D, Petridis P (2019) Implementing an Integrated Meter and Sensor System (IMSS) in existing social housing stock. Energy Build 182:274–286. Elsevier B.V. https://doi.org/10.1016/j.enbuild.2018.10.012
8. Nižetić S, Djilali N, Papadopoulos A, Rodrigues JJ (2019) Smart technologies for promotion of energy efficiency, utilization of sustainable resources and waste management. J Clean Prod 231:565–591. https://doi.org/10.1016/j.jclepro.2019.04.397
9. O'Brien W, Tahmasebi F, Andersen RK, Azar E, Barthelmes V, Belafi ZD, Berger C et al (2020) An international review of occupant-related aspects of building energy codes and standards. Build Environ 179(May):106906. Elsevier Ltd. https://doi.org/10.1016/j.buildenv.2020.106906
10. Carlucci S, De Simone M, Firth SK, Kjærgaard MB, Markovic R, Rahaman MS, Annaqeeb MK et al (2020) Modeling occupant behavior in buildings. Build Environ 174(February):106768. Elsevier Ltd. https://doi.org/10.1016/j.buildenv.2020.106768
11. Mulvaney D, Richards RM, Bazilian MD, Hensley E, Clough G, Sridhar S (2021) Progress towards a circular economy in materials to decarbonize electricity and mobility. Renew Sustain Energy Rev 137(December 2020):110604. Elsevier Ltd. https://doi.org/10.1016/j.rser.2020.110604

12. Jia Y, Alva G, Fang G (2019) Development and applications of photovoltaic–thermal systems: a review. Renew Sustain Energy Rev 102(November 2018):249–265. Elsevier Ltd. https://doi.org/10.1016/j.rser.2018.12.030
13. Bartolucci L, Cordiner S, Mulone V, Rossi JL (2019) Hybrid renewable energy systems for household ancillary services. Int J Electr Power Energy Syst 107(November 2018):282–297. Elsevier. https://doi.org/10.1016/j.ijepes.2018.11.021
14. Alshibil AMA, Farkas I, Víg P (2022) Multi-aspect approach of electrical and thermal performance evaluation for hybrid photovoltaic/thermal solar collector using TRNSYS tool. Int J Thermofluids 16(September). https://doi.org/10.1016/j.ijft.2022.100222
15. Abdulmajeed OM, Jadallah AA, Bilal GA, Arıcı M (2022) Experimental investigation on the performance of an advanced bi-fluid photovoltaic thermal solar collector system. Sustain Energy Technol Assess 54(September):102865. Elsevier Ltd. https://doi.org/10.1016/j.seta.2022.102865
16. Walzberg J, Carpenter A, Heath GA (2021) Role of the social factors in success of solar photovoltaic reuse and recycle programmes. Nat Energy 6(9):913–924. Springer US. https://doi.org/10.1038/s41560-021-00888-5
17. Islam T, Iyer-Raniga U, Trewick S (2022) Recycling perspectives of circular business models: a review. Recycling 7(5). https://doi.org/10.3390/recy-cling7050079
18. EurObserv'ER consortium (2021) The state of renewable energies in Europe—Edition 2021
19. Statistics, Market, and ©ESTIF (2015) Solar thermal markets 2016(June):5. http://www.estif.org/fileadmin/estif/content/publications/downloads/Online_version
20. EurObserv'ER consortium (2016) The state of renewable energies in Europe—Edition 2016
21. Carnevale E, Lombardi L, Zanchi L (2014) Life cycle assessment of solar energy systems: comparison of photovoltaic and water thermal heater at domestic scale. Energy 77:434–446. Elsevier Ltd.https://doi.org/10.1016/j.energy.2014.09.028
22. Wei H, Liu J, Yang B (2014) Cost-benefit comparison between domestic solar water heater (DSHW) and building integrated photovoltaic (BIPV) systems for households in urban China. Appl Energy 126:47–55. Elsevier Ltd. https://doi.org/10.1016/j.apenergy.2014.04.003
23. Mahmud M, Parvez A, Huda N, Farjana SH, Lang C (2018) Environmental impacts of solar-photovoltaic and solar-thermal systems with life-cycle assessment. Energies 11(9). https://doi.org/10.3390/en11092346
24. Senthil R (2022) A technical note on integrating thermal energy systems into solar photovoltaic panels toward a circular economy. Renew Energy Focus 42:97–100. . Elsevier Ltd. https://doi.org/10.1016/j.ref.2022.06.002
25. Burrows K, Fthenakis V (2015) Glass needs for a growing photovoltaics industry. Solar Energy Mater Solar Cells 132:455–459. Elsevier.https://doi.org/10.1016/j.solmat.2014.09.028
26. Malandrino O, Sica D, Testa M, Supino S (2017) Policies and measures for sustainable management of solar panel end-of-life in Italy. Sustainability (Switzerland) 9(4):1–15. https://doi.org/10.3390/su9040481
27. Solar F (2017) The Recycling Advantage. Prog Photovolt Res Appl 19(5):614–626
28. https://www.europen-packaging.eu/policy-area/extended-producer-responsibility/-buildings-directive-epbd/. Accessed 18 February 2023
29. https://environment.ec.europa.eu/topics/waste-and-recycling/rohs-directive_en. Accessed 17 February 2023
30. Kastanaki E, Giannis A (2022) Energy decarbonisation in the European Union: assessment of photovoltaic waste recycling potential. Renew Energy 192:1–13. Elsevier Ltd. https://doi.org/10.1016/j.renene.2022.04.098
31. Erat S, Telli A (2020) Within the global circular economy: a special case of turkey towards energy transition. MRS Energy Sustain 7(1):1–10. https://doi.org/10.1557/mre.2020.26
32. Minelgaitė A, Liobikienė G (2019) Waste problem in European union and its influence on waste management behaviours. Sci Total Environ 667:86–93. https://doi.org/10.1016/j.scitotenv.2019.02.313

Chapter 17
Barriers and Opportunities in the Transition to a Circular Construction Sector in Portugal

Vanessa Tavares and Marco Frazão Pedroso

Abstract Transitioning from a linear to a circular economy (CE) is one of the main goals set by the European Union to achieve carbon neutrality by 2050. Portugal is currently revising the national CE action plan, and the construction sector has been identified as one of the key sectors. Barriers and opportunities in the transition to a CE were previously mapped in literature and national or sectorial CE action plans but still need to be identified for the Portuguese construction sector specificities. Over one-thousand stakeholders were interviewed during eleven working sessions to characterize the national construction sector, identifying and clustering barriers and opportunities in this transition. Barriers from the Political and regulatory pillar were critical, and an urgent need to adapt (and simplify) the regulatory framework to promote a circular construction sector was identified. In the Technology pillar, stakeholders acknowledged the need to support people and companies to make the transition. In the market pillar the lack of pilot projects applying CE principles and of a CDW market. Finally, in the Cultural pillar, the need to reskill workers and empower society with CE principles was highlighted. The actions to trigger the transition are a CE-prone framework, digitalization of construction, simplification of procedures, support research, and empowering the whole value chain. All these actions are in line with the just transition mechanism to ensure that "*no one is left behind*" and point the path towards a carbon–neutral construction sector.

Keywords Barriers and opportunities · Circular economy · Construction sector

---

V. Tavares (✉) · M. F. Pedroso
BUILT CoLAB, 4150-003 Porto, Portugal
e-mail: vanessa.tavares@builtcolab.pt

M. F. Pedroso
e-mail: marco.pedroso@builtcolab.pt

V. Tavares
CEAU, FAUP, Universidade Do Porto, Via Panorâmica, S/N 4150-564, Porto, Portugal

M. F. Pedroso
CERIS, IST, Universidade de Lisboa, Av. Rovisco Pais 1, 1049-001 Lisboa, Portugal

L. Bragança et al. (eds.), *Creating a Roadmap Towards Circularity in the Built Environment*, Springer Tracts in Civil Engineering,
https://doi.org/10.1007/978-3-031-45980-1_17

17.1 Introduction

Transitioning from a linear to a circular economy (CE) is one of the main goals set by the EU to tackle climate change and achieve carbon neutrality by 2050. In 2020 the European Commission reviewed its Circular Economy Action Plan (CEAP) [1], following the European Green Deal principles and "*making sure no one is left behind*" [2]. The new CEAP [1], has four main goals: (i) Reduce pressure on natural resources, (ii) Create sustainable growth and jobs, (iii) achieve the EU's 2050 climate neutrality target, and (iv) halt biodiversity loss. The Sustainable Development Goals (SDG) set by the United Nations (UN) in 2015 [3] have four aligned goals: (i) Responsible consumption and production (SDG12), (ii) Decent work and economic growth (SDG8), (iii) Climate action (SDG13), and (iv) Life on land (SDG15).

Following the CEAP, most country members have adopted National Circular Economy Action Plans. European and national action plans (also named roadmaps or strategic plans) for a circular economy (CE) identify people's knowledge and engagement as the key element in the transition from a linear towards a CE. Sectorial plans focusing on the architecture, engineering, and construction (AEC) sector underline the need for workers' reskilling, creating new professions and courses, and raising general awareness of CE principals as the main actions for the Construction sector transition.

Some barriers hinder this transition such as the current legal framework, lack of investment, and need for skilled labor, among many others. Some opportunities have been identified to overcome these barriers, such as including circularity in the regulatory framework and public procurement (through Green Public Procurement), financing the transition by supporting pilot case studies, and CE training courses, among many other opportunities.

Barriers and opportunities in CE adoption have been identified in the literature and described in current action plans in different EU countries. According to key players in Portugal what are the main barriers and opportunities in the transition to a circular construction sector? The Portuguese Construction sector is mainly composed of micro-, small- and medium-sized enterprises [4], most of which are in a fragile economic position. Moreover, the legal framework is complex and sometimes contradictory, making it difficult to change procedures into a CE. Key stakeholders in the Construction sector and respective value chains were consulted to characterize circularity. This work is based on the results achieved in 11 working sessions, with more than one-thousand actors. It is framed under the funded work to define and publish the Portuguese Action Plan for Circularity in the Construction Sector [5].

17.1.1 Literature Review

Barriers and opportunities in the transition from a linear to a CE model are presented in literature since 2018. Previous work based on systematic literature reviews and statistical analyses, allowed us to identify, and cluster, the most relevant barriers, and opportunities in the construction sector. One of the first and most cited studies about CE barriers and opportunities was conducted by Kirchherr et al. [5]. It presents the first large-N-study on generic CE barriers in the EU and identifies—with the help of 208 survey respondents and 47 expert interviews—15 main barriers, clustered into four categories: Cultural, Market, Regulatory, and Technological.

Later in 2022, the research published by Wuni [6] employed systematic literature review protocols to search, retrieve, evaluate, and extract relevant metadata from 53 studies. This allowed identifying several barriers, which when analyzed using a Pareto analysis, showed a prioritization of the most relevant barriers (57 barriers), with a similar number of opportunities, and distributed in clusters (11 clusters), directly related to the implementation of CE in the construction sector. At the same time, Oluleye et al. [7] conducted a systematic literature review with a special focus on building construction and demolition waste management. This study considered a PRISMA (Preferred Reporting Items for Systematic Reviews and Meta-analysis Approach) approach that identified 38 relevant papers, identifying 33 barriers, and a similar number of opportunities, divided into 7 clusters and by the actors involved. Focusing on the Australian context, Shooshtarian et al. [8] evaluated stakeholders' perceptions about CE, identifying the main barriers and opportunities in CE adoption, and focusing on responders' characterization. Ababio and Lu [9] found 45 barriers and 15 opportunities divided into five clusters: Social and cultural; Political and legislative; Financial and economic; Technological; Framework and theory related. More recently, Murano and Tavares [10] categorized barriers and opportunities by stakeholder level: internal, external, or mixed level. Table 17.1 presents clusters defined in the previously cited papers: [6, 7, 9, 10], and [11] The table shows that the four clusters first defined by Kirchherr [11] are similar to those proposed by the other three authors (in **bold**).

Table 17.2 presents a top-ten list of barriers and opportunities identified in the literature (unranked) for implementing a CE model in the construction sector with a focus on the European market [12–17]. The main barriers are related to the lack of knowledge and awareness of society and all stakeholders along the supply chain, the lack of a CE legal framework, followed by high costs, lack of financial support and technologies, infrastructure, and technical constraints. Identified opportunities are successful pilot projects and businesses applying CE best practices, the empowerment of workers and labor reskilled, investment and financial support, new technologies, and increased data sharing.

17.2 Materials and Methods

A five-step approach was developed to build an action plan for circular construction in Portugal: (i) **map** the regulatory framework; (ii) **identify** circular economy (CE) action plans in Europe; (iii) **characterize** the national Construction sector and related impacts; (iv) **survey** the key stakeholders to identify main barriers and drivers; and

Table 17.1 Clusters defined in the literature (in **bold** similar clusters)

Kirchherr et al. [11]	Wuni [6]	Oluleye et al. [7]	Ababio and Lu [9]	Munaro and Tavares [10]
Cultural	**Cultural**	**CE framework**	**Social & Cultural**	**Informational**
Market	**Market**	**Economic & Market**	**Financial & Economic**	**Economic**
Regulatory	**Regulatory**	**Institutional & Regulatory**	**Political & Legislative**	**Political**
Technological	**Technological**	**Technology & Information**	**Technological**	**Technological**
	Organizational	Organization	Framework & Theory	**Institutional**
	Management	Infrastructural & Process		
	Knowledge	Behaviour		
	Stakeholder			
	Supply chain			
	Technical			
	Financial			

Table 17.2 The top-ten barriers and opportunities in the most recent literature (unranked), based on [12–17]

Barriers	Opportunities
• Lack of knowledge, technical competencies, and expertise of CE CE practices • Limited awareness and knowledge of circular products and services of stakeholders along the supply chain network • Higher upfront investment costs • Unpredictable financial returns and economic savings • Lack of appropriate CE policies, legislation, and regulations, and financial support and tax incentives • Lack of technology and infrastructure readiness • Lack of enabling digital technologies and solutions • Lack of data transparency and information sharing	• Government-led large-scale CE demonstration projects • Establishment of CE training courses for practitioners and inclusion of CE in educational curriculum • Professional training of supply chain members in CE • A strong business case for circular construction • Benchmarking best practices for circular construction • Government financial support and tax incentives for CE • Innovation and technology to improve CE scaling in construction • Investment in CE technologies and infrastructure • Increased CE data and information sharing • Increasing technical training in circular construction operations

(v) **plan**, writing the action plan, measures, actions, and governance. This paper presents the fourth stage of that approach: a survey of the Portuguese key players in the sector to identify the main barriers and drivers in the transition of the Construction sector to a CE.

17.2.1 Research Approach

Key actors in the Portuguese Construction sector and construction value chain were consulted during eleven sessions, performed virtually due to pandemic restrictions (from January to March 2022) and using the Concept Board online platform and presential afterward (April to May 2022). The events first had an expository part that framed the subject under discussion, followed by an interactive part during which the various actors could participate in working groups using a methodology combining design thinking principles in a focus group environment. The analytical treatment of the interviews included the transcription of barriers and (after) opportunities, grouping, prioritization (through voting), and results interpretation, this last one including graphical representation and discussion. In addition to statistical data, the information collected from stakeholders' interviews supported Construction sector characterization. The main obstacles that hinder the transition to circular construction were identified and prioritized, and potential opportunities were pointed out. Table 17.3 presents the list of sessions held during the consultation stage. These sessions are available on YouTube.

17.2.2 Participants Characterization

The online sessions were public, but the application form allowed to cluster participants in eight categories, following a similar distribution of the actors as defined by the European Commission [18]: Academia & research centers; Contractors & builders; Demolition & CDW management teams; Construction products manufacturers; Government & regulators; Investors, developers & insurance providers; Designers; and Users & owners. Figure 17.1 presents the total number of participants per actor category, showing that Contractors & builders were the highest shares of participants (~30%), followed by Users & owners (~20%) and Government & regulators (~5%).

Figure 17.2 presents the participants' distribution per actor category in each one of the eleven sessions. From the online sessions, Session 3 (S03) "*Construction and demolition waste management: What has changed? What is new?*" had the highest number of participants (227), one-third of them being contractors and builders (88), followed by users and owners (47), and government and regulators (34). Both sessions, S02 "*How to move from theory to action in the incorporation of recycled materials in the Construction sector*" and S04 "*Innovation for a more circular*

Table 17.3 List of events

Session	Date	Title	Location	Participants
S01	01/2022	Circular economy: life cycle view (LCA) from design to the deconstruction of the work, the role of the building owners	Online	104
S02	01/2022	How to move from theory to action in the incorporation of recycled materials in the Construction sector	Online	135
S03	02/2022	Construction and demolition waste management: What has changed? What is new?	Online	227
S04	02/2022	Innovation for a more circular construction	Online	127
S05	02/2022	Good practices on transitioning to circularity in construction: The importance of life cycle assessment	Online	112
S06	02/2022	Good practices on transitioning to circularity in construction: The importance of green & innovative public procurement	Online	70
S07	03/2022	Good practices on transitioning to circularity in construction: The importance of the construction and demolition waste	Online	101
S08	03/2022	Construction and demolition waste management: What has changed? What is new?	Oporto	128
S09	03/2022	Construction and demolition waste management: what has changed? What is new?	Leiria	29
S10	04/2022	Living Labs as Technological Start-Up Incubators—A Marketplace for Circularity in Construction	Online	72
S11	05/2022	Construction and demolition waste management: what has changed? What is new?	Lisbon	137

construction" had around 130 participants, being the second and third most attended online working sessions. The number of participants in the face-to-face events was high, with 128 participants at the "*Construction and demolition waste management: What has changed? What is new?*" session held in Oporto (S08) and 137 held in Lisbon (S11). In Lisbon, most of the participants were Contractors & builders (90), while in Oporto they were Government & regulators (32), and contractors & builders (28). Attenders in Leiria were fewer (29) in line with population density and reduced Construction sector activity.

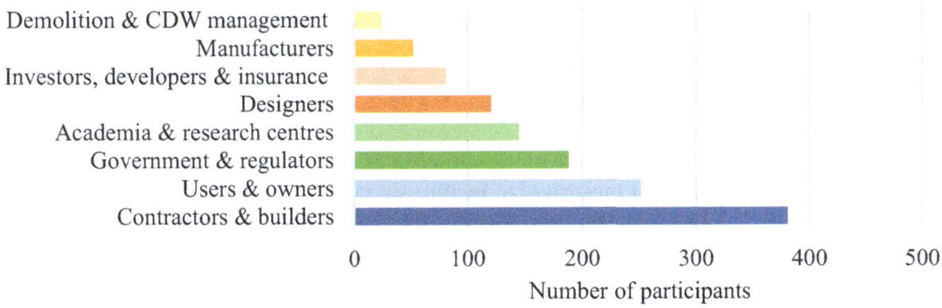

Fig. 17.1 Total number of participants per actor category

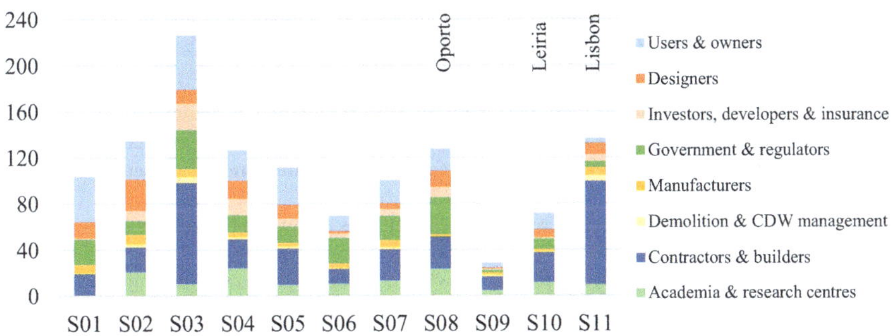

Fig. 17.2 Participants distribution per actor category in each one of the eleven sessions

17.3 Results and Discussion

The results and discussion section presents the main barriers identified by key stakeholders and possible opportunities in the circular economy (CE) within the AEC Portuguese sector and value chain.

17.3.1 Barriers Identified

The main barriers identified by key stakeholders were distributed along the four pillars following Kirchherr et al. [10]: (i) Political & regulatory; (ii) Technology; (iii) Market; and (iv) Cultural. In the Political & regulatory pillar, stakeholders pointed out the complex and sometimes contradictory legislation as the main barrier, followed by a lack of incentives and sustainable procurement to leverage the transition. In the Technological pillar, it was acknowledged that some building technologies are not fully developed (for disassembly and adaptability, modularity, and offsite manufacturing) and a lack of alternative CE materials or limited knowledge about them. The Market pillar identified the urge to have some pilot projects following CE principles,

Table 17.4 Main barriers to CE adoption identified by key players and divided into pillars

Pillar	Main barriers
Political & regulatory	**Complex legal framework with some legislation hindering circularity** **Tender specifications without CE requirements** **Lack of incentives to include sustainability and circularity** Environmental and social costs are excluded from Project Value Lack of knowledge about sustainable procurement Inexistence of scheduled adoption of Green Public Procurement
Technology	**Construction for deconstruction technology is not fully developed** **Lack of alternative and innovative materials** **Lack of information about LC performance and a national digital DB** Lack of repetition and modularity, with unique and unrepeatable works Insufficient or inappropriate computer skills Lack of awareness about CE certification
Market	**Lack of demonstrating projects following CE principles** **Lack of a CDW market and works incorporating recycled materials** **Product's short lifetime and hard maintenance, disassembly, and reuse** The high price of recycled materials compared to equivalent raw materials Manufacturers lack knowledge about recycled materials incorporation Public funding does not include environmental criteria
Cultural	**Lack of acceptance of recycled materials, preferring virgin materials** **Lack of CE training and shortage of new CE-qualified professionals** **Society's lack of knowledge on circularity and sustainability** Non-existent or little shared practical examples Owner´s lack of awareness of sustainable procurement Lack of understanding of differences between reuse and recycling

no market for secondary materials (from CDW), the short life span of building materials and components, and difficulties in maintenance. Finally, the Cultural pillar indicated the lack of qualified professionals and training, acceptance of recycled or reused materials, and the society's knowledge and awareness about the principles and benefits of circular construction. The main barriers identified by stakeholders are presented in Table 17.4, with the most important highlighted in **bold**. Barriers in the Political and regulatory pillar are further detailed in [19].

17.3.2 Opportunities Identified

The stakeholders also identified opportunities to overcome the main challenges presented by the barriers. Most of the opportunities are within the Political & regulatory pillar, including the incentives to support circular business models, simplification of processes, and adaptation of the regulatory framework to CE principles are the main opportunities identified. The Technology pillar has highlighted the support of training and research, and the development of new tools and technologies. In the Market pillar, the funding of circular products, the development of a marketplace with environmental data to support decisions, and public procurement, including environmental criteria. Finally, in the Cultural pillar, stakeholders identified the empowerment of workers and society with CE principles, based on demonstrators' projects, workshops, and dissemination actions, and reinforcing circular partnerships along the value chain. The main opportunities identified by stakeholders are presented in Table 17.5, with the most important ones highlighted in **bold**. Opportunities in the Political and regulatory pillar are further detailed in [19]

17.3.3 Uncertainty and Limitations

The main limitations that may impact the validity, reliability, or robustness of the findings can be due to sampling selection bias as the eleven sessions had an audience composed of professionals with no general public attendees. The sampling was vast (over one thousand people) and presential sessions were held in three different locations trying to overcome the geographical lack of representativeness that may occur: one session in the north region in Oporto, another in the center in Leiria, and finally one in Lisbon. The online events allowed a vaster coverage of the territory, although being held in Portuguese excluded a non-Portuguese-speaking audience. However, as the scope of the study was to listen to the Portuguese key stakeholders is not a limitation but reflects this study's boundaries.

17.4 Conclusions

Most of the barriers identified by Portuguese AEC stakeholders during workshops are similar to those stated in the literature, accentuated by the characteristics of the national construction sector: consisting of micro and SMEs, workers shortage, and low productivity. Stakeholders highlighted the need to redefine legislation to include CE principles, conciliate regulation, and simplify procedures. Moreover, the need to reskill labor and raise awareness of CE advantages is clear, as well as, to support enterprises and people in the transition, expressing the difficulties of the national Construction sector. Both barriers and opportunities should be considered by decision-makers, from regulators to business owners or designers, to better support decisions toward a more circular construction sector. To sum up, four opportunities (one per pillar) and underlying mottos are identified:

(1) Redefine and simplify the legal framework: *"better" legislation, not "more" legislation.*
(2) Digitalization will support *the twin transition: digital and green.*

Table 17.5 Main opportunities in CE adoption identified by key players

Pillar	Main opportunities
Political & regulatory	**Incentives for new circular business models** **De-bureaucratize projects, procedures and certification** **Redefining current legislation to support circularity** Motivate municipalities to define fee exemptions/reductions Waste Management Plan developed as a strategic project Promote environmental impact studies and pre-demolition audits Create a digital platform with legal information Define new construction guidelines to be included at the early design stage Improve material/waste separation, sorting, and classification EU taxonomy based on LC performance and including tax or financial incentives Encourage and support entities applying LCA and LCC methodologies The energy framework expanded to include a whole life-cycle perspective
Technology	**Digital literacy training with tools that include BIM and GIS** **Support research, development, and innovation** **New technologies for material recycling** Declare the material's performance during the whole life cycle Digital construction to support sustainability and circularity Technologies that allow the reuse and repair during use-phase Digital platform with recycled material content information
Market	**Digital platform with simplified access (marketplace and matchmaking)** **Voluntary adoption of a green public procurement framework** Framework for a demand/supply database Compel designers to incorporate recycled materials Financial support to sustainable and circular practices Design for adaptability, flexibility, disassembly, and deconstruction Extend use-phase through design (increase durability) Producers extended responsibility, including maintenance and EoL Professional recognition of Sustainability and Circularity practitioners Including environmental criteria in all contracts
Cultural	**Reskilling existing workforce and training actions (owners and contractors)** **Knowledge-sharing among partners (CE products, buildings, companies)** **Pilot case studies demonstrating selective (de)construction activities** Raise awareness and acceptance about secondary materials quality Co-responsibility of all stakeholders along the construction value chain Increased supervision with increased responsibility for the owner Case studies to reskill and empower new and existing labor Empowerment of public owners and regulators about green procurement Workshops and dissemination actions on sustainability and circularity Demand the environmental impacts (Environmental Product Declaration (EPD)) Eco-design is supported by green partnerships Description of CDW percentage incorporation and future recyclability

(3) Support research: *new ways of doing (and living).*
(4) Empower workers, companies, and society: *no one is left behind!*

Acknowledgements The authors gratefully acknowledge the COST Action CircularB for this opportunity. The authors would also like to acknowledge the project "Circular Agreement with the Construction Industry" funded by the Portuguese Environmental Fund and the contribution of the workgroup: AECOPS, AICCOPN, APA, CIP, CPCI, IMPIC, and PTPC. Vanessa Tavares would like to acknowledge the North Regional Operational Programme [NORTE-06-3559-FSE-000176], and Marco Frazão Pedroso the Lisbon Regional Operational Programme [LISBOA-05-3559-FSE-000014].

References

1. European Commission, Circular Economy Action Plan, European Commission (2020) 28. https://doi.org/10.2775/855540
2. European Commission: The European Green Deal (2019) https://ec.europa.eu/info/strategy/priorities-2019-2024/european-green-deal_en
3. United Nations, Sustainable Development Goals (2015)
4. IMPIC, Relatório do Setor da Construção em Portugal (2020)
5. Fonseca P, Pedroso M, Tavares V, Laranjeira L (2022) Plano de Ação para a Circularidade na Construção (PACCO)
6. Wuni I (2022) Mapping the barriers to circular economy adoption in the construction industry: A systematic review Pareto analysis, and mitigation strategy map. Build Environ 223:109453. https://doi.org/10.1016/j.buildenv.2022.109453
7. Oluleye B, Chan D, Olawumi T (2022) Barriers to circular economy adoption and concomitant implementation strategies in building construction and demolition waste management: a PRISMA and interpretive structural modeling approach. Habitat Int 126:102615. https://doi.org/10.1016/j.habitatint.2022.102615
8. Shooshtarian S, Hosseini M, Kocaturk T, Arnel T, Garofano T (2022) Circular economy in the Australian AEC industry : investigation of barriers and enablers. Build Res Inform. https://doi.org/10.1080/09613218.2022.2099788
9. Ababio B, Lu W (2022) Barriers and enablers of circular economy in construction : a multi-system perspective towards the development of a practical framework. Constr Manag Econ 0:1–19. https://doi.org/10.1080/01446193.2022.2135750
10. Munaro R, Tavares F (2023) A review on barriers, drivers, and stakeholders towards the circular economy: the construction sector perspective. Clean Respon Consump 8:100107. https://doi.org/10.1016/j.clrc.2023.100107
11. Kirchherr J, Piscicelli L, Bour R, Kostense-Smit E, Muller J, Huibrechtse-Truijens A, Hekkert M (2018) Barriers to the circular economy: evidence from the European Union (EU). Ecol Econ 150:264–272. https://doi.org/10.1016/j.ecolecon.2018.04.028
12. Bertozzi C (2022) How is the construction sector perceiving and integrating the circular economy paradigm? Insights from the Brussels experience. City, Cult Soc 29:100446. https://doi.org/10.1016/j.ccs.2022.100446
13. Giorgi S, Lavagna M, Wang K, Osmani M, Liu G, Campioli A (2022) Drivers and barriers towards circular economy in the building sector: stakeholder interviews and analysis of five European countries policies and practices. J Clean Prod 336:130395. https://doi.org/10.1016/j.jclepro.2022.130395
14. Klein N, Deutz P, Ramos B (2022) A survey of Circular Economy initiatives in Portuguese central public sector organisations: national outlook for implementation. J Environ Manage 314:114982. https://doi.org/10.1016/j.jenvman.2022.114982

15. Reich H, Vermeyen V, Alaerts L, Acker K (2023) How to measure a circular economy: a holistic method compiling policy monitors. Resour Conserv Recycl 188:106707. https://doi.org/10.1016/j.resconrec.2022.106707
16. Sáez-de-Guinoa A, Zambrana-Vasquez D, Fernández V, Bartolomé C (2022) Circular economy in the European construction sector: a review of strategies for implementation in building renovation. Energies (Basel) 15:4747. https://doi.org/10.3390/en15134747
17. Yu Y, Junjan V, Yazan M, Iacob M-E (2022) A systematic literature review on Circular Economy implementation in the construction industry: a policy-making perspective. Resour Conserv Recycl 183:106359. https://doi.org/10.1016/j.resconrec.2022.106359
18. European Commission, Circular Economy: Principles for buildings design (2020)
19. Pedroso M, Tavares V (2023) Circular economy supporting policies and regulations: the Portuguese case, in: CircularB 1st Workshop "Creating a Roadmap towards Circularity in the Built Environment," Skopje—Macedonia

Part VIII
Efficient Waste and Circular Resource Management

Chapter 18
Wood-Based Waste Management—Important Resources for Construction of the Built Environment

Jan Parobek and Hubert Paluš

Abstract The circular economy focuses on the utilisation of resources and the reutilisation of these resources and waste streams into value added products. Wood as a renewable resources represent one of the most important advantage of the forest based industry and all related industries. These sectors aim on the sustainable wood and different wood products production and utilisation, wood buildings including. Improved utilization of available industrial wood assortments and utilisation of wood waste to added value products generate profit for all actors in the supply chain. Analysis of wood flows take into account not only the uses of wood as a material, but also by-products and waste generated by the production to be used as inputs for further uses in construction, wood processing or energy sectors. This paper deals with the analysis of raw wood flows in Slovakia with a focus on wood-based waste management, utilisation of wood waste for long term wood products in the built environment. At the present time new approaches such as cascade use of woody biomass can be applied to ensure the sustainable utilisation of renewable resources. The material flow analysis (MFA)was used to identify relations between the resources and primary uses of wood. In particular the results show some particular examples of wood flows focused on possibilities of utilisation of wood waste from the wood processing industry (WPI) and recycled wood in the construction sector.

Keywords Construction · Wood-based Waste · Wood Flow Analysis · Cascading Factor · Environment · Slovakia

J. Parobek (✉) · H. Paluš
Technical University in Zvolen, T. G. Masaryka 24, 96001 Zvolen, Slovakia
e-mail: parobek@tuzvo.sk

L. Bragança et al. (eds.), *Creating a Roadmap Towards Circularity in the Built Environment*, Springer Tracts in Civil Engineering,
https://doi.org/10.1007/978-3-031-45980-1_18

18.1 Introduction

The path of raw wood material from its production to giving the final product to a consumer is relatively long, as it passes several stages of production and different types of markets until the final product fulfils the needs of the consumers. Before reaching the end-user, these stages include leaving the forest, primary wood processing, secondary wood processing, and subsequent wood-using industries. Within these stages, raw wood material is transformed into primary processed intermediate products (sawn wood, pulp), secondary processed products (furniture, construction, and joinery), and then has a role in the final production of different industries related to the use of wood (e.g., construction).

The economic growth of any national economy depends largely on the many different factors as, strength, competitiveness, effectiveness and investment attractiveness of the industry. In the last decades there are new challenges, such as transformation and adaptation to the new demands from the construction sector characterised by the process of globalisation, an extremely competitive business environment and new technologies and innovations. Therefore, wood consumption must be assumed and coordinated with the demand for traditional as well as innovative products. A lot of studies [1–3] assume increasing demand for wood in the European Union (EU) mainly due to an expected increase of energy consumption of the renewable resources. Thanks to new technologies and innovative products new opportunities of wood utilization are created in the building industry. It is necessary to take into account the fact that the wood processing industry (WPI) also produces significant amounts of wood residues (30–50% of the volume of processed wood), which can be used for industry purposes. Morover, the construction sector has been the primary consumer of sawn wood and other solid-wood products [4]. From this perspective, it is important to prefer outputs with higher added value, creating jobs and contributing to a better carbon balance (resource efficiency).

18.1.1 Wood in Construction

The construction industry is one of the most significant producers of waste in any given economy. In fact, construction is responsible for more waste globally than any other single area of economic activity. However, waste in construction has traditionally been considered as an inevitable by-product that is usually managed from a health and safety perspective rather than with recycling in mind [5]. According to a study in the United Kingdom, about 10 to 15 per cent of the wood used in new construction ends up in recycling. This statistic is a concern to policymakers, who observe that the recycling rate for construction and demolition (C&D) derived wood is considerably lower than for other C&D materials such as concrete (82 per cent) and structural steel (98 per cent) [6]. This suggests that there is considerable work to be done to make the construction sector more circular, including the use of waste wood as a part of a broader system loop [7, 8]. To facilitate more post-construction wood re-entering the supply chain, systemic developments are needed to enhance sorting, separation and recovery options through, for example, more efficient recycling during demolition so that wood waste can be cycled back at the end-of-life stage to other industrial processes. Moving away from the business-as-usual approach would furthermore require cross-cutting and networked systems with stronger collaboration between business ecosystems (e.g., municipalities, architects, designers, builders and inhabitants.

Another approach to improve the circularity of the construction sector relates to the design and detailing of mass timber buildings for greater durability, including measures to hold materials in place for longer, prolong the lifespan of wood to reduce the demand for new materials and standardize modular wood construction elements that could be re-used and recycled more easily. This requires that the wood's entire life cycle (from primary to tertiary processing) is taken into account when constructing new buildings to allow for more efficient usage of side products (e.g., recovered wood).

18.1.2 Circularity Concepts in Forest-Based Industries

A circular economy refers to a wide range of materials and processes, both in the technical and biological cycle of the economy. As an example, wood may enter the technical cycle when it is combined with technical materials in the construction of buildings.

The term "value chain" entails a series of manufacturing steps that link raw materials to final products through different sub-sectors of an industrial or economic sector. A value chain can vary in scale from being local to global and the range of activities along the value chain may be implemented by different actors, such as resource extractors, processors, traders, retailers and service providers. Each sub-sector (e.g., furniture manufacturing or construction) could be described as a distinct value chain, however, processing steps in different product groups downstream often have common sources upstream [9].

18.1.3 Analysis of Wood Flows—Theoretical Approach

Material flow analysis (MFA) was used to reveal and quantify relations between the resources and the primary uses of wood. EUROSTAT [10] distinguishes and explains the three basic dimensions of material flows: territorial dimension, product chain or life cycle dimension and the product dimension. Different approaches to material flow analysis and modelling have been used by e.g. [11–14]. The analysis of material flows can be also used as an analytical and modelling tool for different areas and sectors e.g., material balances of corporations and urban regions in industrialised countries [15], regional wood management [16], and the generation of waste in regional systems [17].

The analysis of wood flows enables one to determine a balance between the production and the use of wood in the country. The analysis results reveal relationships between the production, quality, and availability of data, the balance of foreign trade, and the importance of wood in domestic consumption. Wood flow analysis is focused on all uses of wood and takes into account by-products and waste generated by processing the material input for further use. Both sides of the balance, the resources and the use side, are specific, as they incorporate different markets and products; therefore, it is necessary to examine each side individually. The overall structure of the balance is not constant and may vary depending on the uses of wood and wood products. In most cases, the balance includes such uses of wood for which there are no official statistics available, and the total consumption therefore cannot be simply calculated. Consequently, the consumption of wood may be much higher than indicated by official statistics [18].

18.1.4 Concept of Cascading Use of Wood Products

The utilization of wood for building industry represent optimal solution from an economic, as well as, an environmental point of view. A different way of potential wood and wood residues utilization in the value chain is described in the concept of cascading use of wood products. The final and optimal use of wood, both from an economic point of view and from the point of view of binding carbon dioxide (CO_2), is precisely in the construction industry. A cascade use can be defined as multiple use of the wood from trees by using residues, recycling (utilization in production) resources or recovered (collected after consumption) resources [19, 20]. The more often by products and recycling products are used the higher the cascade factor gets. Cascade principle means to use wood in an effort to increase added value of wood raw material from forests. It also means that wood should be primary used in the construction, furniture or other products with a long life cycle and energy should primarily be generated from waste or recycled products. In this sense we consider energy uses of wood as the least preferred way of utilization. The concept of cascading use of biomass can be defined as cascading in function. It is actually co-production, which can be achieved by using bio-refinery. Co-production is the production of different functional streams (e.g. protein, oil and energy) from one biomass stream, maximizing total functional use. Of course, after cascading in function, cascading in value or time follows. The cascading in time meaning that the life span of biomass use is increased (e.g. paper recycling). Another approach can also be defined as 'cascading in value' meaning that the maximum value of the whole life cycle of biomass is gained through optimizing the use of biomass for multiple services [21].

The main objective of the study is the identification of raw wood flows in Slovakia with a focus on utilisation of wood waste for long term wood products in the built environment. To follow above mention aim the re-search applied the concept of cascading to optimise the use of wood in the chain of its processing. The approach was applied to identify relations between the re-sources and primary uses of wood focused on possibilities of utilisa-tion of wood waste from the WPI and recycled wood in the construction sector.

18.2 Methodological Approach

The domestic wood processing industry in the Slovak Republic is the major customer of the products of the forestry sector, and roundwood represents the main material input for this sector [22]. Similar links exist between the wood processing industry and other sectors that are dependent on wood products. MFA can be used for the quantification and modelling of wood flows. The analysis process includes the gathering of information and requires market experience and recognition of mutual relations in the "forest—wood—end-user" chain.

A single wood balance presents a global view of the resources and primary uses of roundwood in Slovakia. The main categories of resources are represented by the domestic roundwood production and imports, and the main uses by the domestic roundwood consumption and exports. The resource side is complemented by the recycled material and stock decrease, and the use side by the stocks increase. An increase in stocks causes a decrease in consumption, and vice-versa. The availability and consistency of data represent a limiting factor for the construction of the wood balance. Available data for 2021 from the FAOSTAT database [23] and the reports on forestry in Slovakia [24] were used. To achieve the state of wood balance, the resources should equal the uses. However, there were no data available on domestic consumption; therefore, it was deducted from the volumes of roundwood production and foreign trade.

Unlike the wood balance, which takes into account only uses of wood as a material, the wood resource balance is focused on different uses within the internal environment of the sector. The wood resource balance provides a detailed analysis of wood and wood products flows which has been done by survey. First of all, it takes into consideration by-products and waste generated by the production for use as inputs in wood processing or in the energy sector. The main categories of resources are (i) forest woody biomass, (ii) used material, (iii) other woody biomass, (iv) wood processing residues and (v) processed wood fuel. The main categories of uses consist of (i) wood processing industry (material stream) and (ii) energy use (waste stream). The analysis focuses on wood waste streams that can subsequently be used for the construction industry.

The quality of the final wood resource balance depends directly on the quality and availability of data on wood production and use in individual sectors. Generally, the availability of data on consumption is usually poor, and detailed data do not exist. Empirical research and expert estimations based on the available production data are commonly used to obtain the missing data. Under current conditions, wood resource balance data can be compiled as a mix of officially published and empirically collected data. Official statistics are available for highly concentrated sectors such as the pulp and paper industry. However, certain sectors of the wood processing industry and building sector, such as the sawmill industry, are poorly concentrated; thus, access to data is complicated. Therefore, to estimate the material flows the main streams of primary wood processing and utilisation were only considered, in particular sawmilling industry, veneer and plywood production, particleboard and fibre board production, processed wood fuel, pulp and paper industry, energy. To quantify flows and balances in a single measurement unit meter cubic (m^3), the UNECE/FAO [6] official input/output ratios for Slovakia were used. The flow of wood as raw material and flow of wood residues as a waste from the process of wood products production was identified separately. The flow of wood residues was complemented by recycled wood and paper (post-consumer material). Wood as an elementary input comes from domestic sources (forest) and from the import. The production of roundwood was analysed in the structure corresponding to the main groups of assortments in terms of their use and quality (logs, pulpwood and energy wood). The concept of cascading is focused on the domestic utilization and therefore it was necessary to estimate the domestic consumption (apparent consumption). The presented paper describe of primary wood processing flows, which generates a key by-products (waste) for cascade utilization of wood. Based on wood balance of the available resources and the description of wood flows it was possible to describe and quantify the cascade use of wood. Cascade use is defined as multiple use of wood from the forest with wood residues from the forest industry. The more times the industrial residues and by-products are produced during processing of wood the higher factor cascade gain. The sectors of WPI in Slovakia are interconnected. For example, the waste from sawmills is used for industrial purpose (use in the production of wood base panels and pulp), as well as a source of energy (the production of end products such as pellets or briquettes). It can be also used outside the WPI for the production of energy in other sectors, heating plants and households. In case that the inputs to the process is only roundwood without additional other sources, the cascade factor takes the value 1.00 [3]. The cascading factor (CF) is calculated using the following Eqs. (18.1, 18.2) taken from Mantau [6, 25]:

$$Cascade = RW + IR \tag{18.1}$$

where: RW—recycled wood and paper, IR—industrial residues.

$$CF = 1 + {}^{Cascade}/_{WR_{forest}} \tag{18.2}$$

where: WRforest is wood resources from forest.

18.3 Quantification of Wood Flows for Construction Sector in Slovakia

The volume of domestic consumption (9.59 mil. m^3) was deducted from the volumes of roundwood production and foreign trade. Wood balance presents a global view on the resources and primary uses of roundwood in Slovakia and is illustrated in following Table 18.1. Due to unavailability of data it does not consider the stock changes.

Wood balance is mainly oriented to estimate domestic consumption regardless the further use of wood and, unlike wood resource balance, it considers foreign trade in wood products. Logs are mainly processed by sawmills and only a smaller proportion is consumed by plywood or veneer producers. As a paradox, in spite of a higher composition of broadleaved forests in Slovakia coniferous logs are the main raw material used by sawmills. Non-coniferous pulp wood and other industrial roundwood is mainly used by pulp and paper industry. Results of the wood resource analysis show that the total domestic resources were 7.23 mil. m^3 roundwood (Table 18.1.). However, except of roundwood the resource side is supplemented by wood processing residues consisting mainly of sawmill residues and black liquor. The majority of residues was produced by the pulp and paper industry (black liquor 0.99 mil. m^3) and the rest was produced by sawmill industry—sawdust, chips and particles (0.89 mil. m^3). Total domestic supply includeing wood waste and black liquor were 9.5 mil. m^3 roundwood equivalents. The following Table 18.2 gives a summary of all domestic supplies and utilization.

Indirect wood flows can be expressed by a cascade factor, which considers the repeated use of wood originating on the use side and returning back to the resource side and vice versa. Considering the actual total consumption of wood and wood based inputs 9.5 mil. m^3 and the volume of wood supply from forest wood biomass 7.2 mil. m^3 (domestic wood resources) the value of cascade factor was 1.32.

Table 18.1 Domestic wood resources in Slovakia (m^3)

Wood resources		Use of wood	
Roundwood production	7 448 000		
Roundwood import	2 000 000	Roundwood export	2 290 000
Recycled paper	141 113		
		Domestic consumption	7 299 113
Total sources	**9 589 113**	**Total uses**	**9 589 113**

Sources Authors' own research, FAOSTAT 2023 [22] and Green Report 2022[23]

Table 18.2 Utilization of by-products derived from industrial wood processing in 2021

Wood based resources		m^3	Use	Use m^3	Production	
Forest woody biomass	Coniferous Logs	2 722 420	Coniferous Sawnwood	2 289 096	1 067 517 m^3	Wood processing Industry
			Non Coniferous Sawnwood	947 839	266 879 m^3	
	Non Coniferous Logs	951 839	Veneer, Plywood and other Large Boards Production	437 325	437 325 m^3	
	Pulp Wood and Other Industrial Wood	2 923 388	Particleboard and Fibre Board Production	1 138 644	882 487 m^3	
	Wood Fuel	560 352			3 299 t	
	Energy Wood and Logging Residues	47 125				
Recycled material	Post-Consumer Paper	141 113	Pulp and paper industry	2 310 188	1 238 430 t	
Industrial residues	Sawdust	254 585	Other Energy Wood	2 382 392	722 TJ	Energy Users
	Chips	472 543				
	Particles	166 950	Energy Use of Wood in Industry		12 536 TJ	
	Black liquor	988 467				
Processed wood fuel	Pellets, briquettes and others	276 702	Private Households		24 359 TJ	
Total		9 505 484	9 505 484			

* m^3-cubic meters, t-tons, TJ-Terajoule
Sources Authors' own research, FAOSTAT 2023 [22]

18.4 Discussion

Wood residues and by-products are produced during industrial processing of wood. The waste stream is represented by different types of waste generated during the logging operations (e.g., logging residues) as well as the waste generated during primary mechanical and chemical processing of wood (sawdust, chips, black liquor), which can be used either industrially or for the production of energy. The primary source of wood residues used for production of agglomerated wood-based panels, processed fuel wood, and energy generation is the sawmilling industry. Wood balance is primarily used to estimate domestic consumption, regardless of the further use of wood; unlike the wood resource balance, it considers foreign trade in wood products. Taking into account roundwood classification, the wood resource balance distinguishes wood flows for individual sectors according to the intended use of assortments. Logs are primarily processed by sawmills, and only a small portion is consumed by plywood or veneer producers. As a paradox, in spite of the large proportion of broadleaved forests in Slovakia, coniferous logs are the primary raw material used by sawmills. Non-coniferous pulp wood and other industrial roundwood is used by the pulp and paper industry for the production of pulp, or alternatively for the production of particleboard and fibre board. The importance of wood for energy production has been increasing recently. Wood fuel is used for energy production in either internal or external facilities. At the same time, it represents a significant source for heat energy in households. Wood, which was traditionally utilised as material for the production of wood products, is presently in demand for energy production. The increasing direct or derived demand for energy wood causes an increase in energy wood prices. Wood cascading considers complete wood using cycle and recognizing the differences in wood flows. The concept of cascading can help to optimize the use of wood in the whole chain of its processing and utilization in Slovakia. Results of the analysis can help in many innovations to increase the efficiency of the cascade of wood processing for construction industry.

18.5 Conclusion

The utilisation of wood is continually changing, and the demand for roundwood is changing depending on the technologies and on the demand of final wood products. On one side, wood and wood products production is subject to available resources and has been recently influenced by the high proportion of accidental felling. On the other side, wood production tries to adapt to rapidly changing on the market. The applied concept of cascading can describe the actual consumption of wood in various forms. The outcome of the analysis of wood material flows and cascading concept in Slovakia go out the balance between the resources and the primary uses of wood and wood residues. The analysis describes in detail the relationships between resources (wood and waste), basic production indicators, foreign trade relations, and the use of raw wood material and waste in the domestic conditions. Main conclusions can be considered as follow:

- The wood market in Slovakia is permanently developing and the demand for roundwood is changing depending on the possibilities of its use. There are many specifics influencing production and consumption patterns at the domestic market. On one hand, timber production is subject to available resources, which are the result of long-term forest management and long-term planning.
- Timber production has been recently influenced by the high proportion of accidental felling. However, it tries to adapt to rapidly changing market conditions and requirements of wood processing sector that vary over a relatively short period of time.

- The applied material flow analysis can transparently reveal the actual consumption of wood in its various forms. Such an approach also allowed pointing out the vulnerability of revealed relationships occurring due to the limited wood resources and exiting regulative as well as supporting measures.

This analysis is a key base for improving the use of circular economy principles in the above wood base sectors and construction. Better use and knowledge of wood material flows may ensure production of higher added value products as well as to lead to cost reduction. Despite the fact that the research is given as a model for Slovakia at a certain point in time, the methodology and approach is applicable to different types of industries, as well as, countries at different times for comparison possibilities. Concurrently, it may scale up the innovation process and increase investments to the wood processing sector, which may ensure the production of the long-life wood products such as wood buildings and consequently increase the volume of carbon stored in wood-based products and applications.

Acknowledgements The author would like to thank the European Cooperation in the field of Scientific and Technical Research—COST. This paper was elaborated within the frame of COST Action CA21103, Implementation of Circular Economy in the Built Environment and for the support of the Scientific Grant Agency of the Ministry of Education, Science, Research, and Sport of the Slovak Republic Grant No. 1/0494/22 Comparative Advantages of the Wood Based Sector under the Growing Influence of the Green Economy Principles and Grant No. 1/0495/22 Sustainability of Value Supply Chains and its Impact on the Competitiveness of Companies in the Forest and Forest-Based Sectors.

References

1. Buongiorno A, Raunikar R, Zhu, S (2011) Consequences of increasing bioenergy demand on wood and forests. An application of the Global Forest Products Model. J Forest Eco 17(2):214–229
2. Raunikar R, Buongiorno J, Turner JA, Zhu S (2010) Global outlook for wood and forests with the bioenergy demand implied by scenarios of the Intergovernmental Panel on Climate Change. Forest Policy Econ 12(1):48–56
3. Mantau U, Saal U, Prins K, Steierer F, Lindner M, Verkerk H, Eggers J, Leek N, Oldenburger J, Asikainen A, Anttila P (2010) EUwood real potential for changes in growth and use of EU forests. http://ec.europa.eu/energy/renewables/studies/doc/bioenergy/euwood_final_report.pdf, last accessed 2023/02/12
4. EUSTAFOR (2023) https://eustafor.eu/forest-europe-published-the-state-of-europes-forests-2020-report/, last accessed 2023/02/14
5. Osmani M, Villoria-Sáez P (2019) Current and emerging construction waste management status, trends and approaches. In: Letcher T, Vallero D (Eds), Waste: a handbook for management. Cambridge, pp 366–379. 2nd edn. Academic Press, Massachusetts
6. UNECE/FAO: Forest Product Conversion Factors for the UNECE Region, ECE/TIM/DP/49, UN, Geneva (2010)
7. Dangel U (2017) Turning point in timber construction: a new economy. Birkhäuser, Basel
8. LIGNATEC: Climate-Friendly and Energy-Efficient Construction with Wood: Basic Information and Implementation. Lignum, Zurich, Switzerland (2015)
9. UNECE: Circularity concepts in forest-based industries, United Nations, Geneva (2022)

10. EUROSTAT (2023). http://epp.eurostat.ec.europa.eu/cache/ITY_OFFPUB/KS-34-00-536/EN/KS-34-00-536-EN.PDF, last accessed 2023/02/18.
11. Binder CR, Hofer C, Wiek A, Scholz RW (2004) Transition towards improved regional wood flows by integrating material flux analysis and agent analysis: the case of Appenzell Ausserrhoden, Switzerland. Ecol Econ 49(1):1–17
12. Bringezu S (2003) Industrial ecology and material flow analysis. Basic concepts, policy relevance and some case studies. In: Perspectives on Industrial Ecology, D. Bourg and E. Suren (eds.), pp 24–30. Greenleaf Publishing, Sheffield
13. Fisher-Kowalski M (1998) Society's metabolism, the intellectual history of material flow analysis. J Ind Ecol 2(1):61–78
14. Zhou Z, Sun L (2008) Preliminary research on regional material flow analysis: a case study of Chengyang District in Qingdao. J Northeast Agric Univ 15(2):41–47
15. Baccini P, Bader HP (1996) Regionaler Stoffhaushalt. Erfassung, Bewertung und Steuerung, Spektrum, Heidelberg, Germany
16. Műller D (1996) Szenarien zur nachhaltigen regionalen Holzbewirtschaftung. Schweiz Z Forstwes 147(11):873–885
17. Schwaezenbach RC, Heitzer A, Stäubli B, Grossmann B, Scholz RW (1999) A regional perspective on contaminated site remediation—fate of materials and pollutants. Environ Sci Technol 33(14):2305–2310
18. Parobek J, Paluš H (2016) Material flows in primary wood processing in Slovakia. Acta Logistica 3(2):1–5
19. Hekkert MP, Joosten LAJ, Worrell E (2000) Analysis of the paper and wood flow in the Netherlands. Resour Conserv Recycl 30(1):29–48
20. Sokka L, Koponen K, Keränen JT (2015) Cascading use of wood in Finland– with comparison to selected EU countries. VTT, Helsinki, Finland
21. Piškur M, Krajnc N (2007) Roundwood flow analysis in slovenia, croatian. J Forest Eng 28(1):39–46
22. Parobek J, Paluš H (2008) Modelling of wood and wood products flow in the Slovak Republic. In: A European Wood Processing Strategy: Future Resources Matching Products and Innovations, Ghent University, Belgium, pp 93–99
23. FAO FAOSTAT – Forestry Homepage (2023) http://faostat.fao.org, last accessed 2023/02/21
24. MARD Green Report (2022). http://mpsr.sk/sk/download.php?fID=7840, last accessed 2023/02/24
25. Mantau U (2015) Wood flow analysis: quantification of resource potentials, cascades andcarbon effects. Biomass Bioenerg 79:28–38

Chapter 19
Agricultural Biomass Ash as a Circular Building Material: Connecting Agriculture and Construction Industry

Suzana Draganić, Slobodan Šupić, Mirjana Laban, Mirjana Malešev, Vlastimir Radonjanin, Vesna Bulatović, Ivan Lukić, and Olivera Bukvić

Abstract Previous studies have indicated that agricultural biomass ash is an important resource with great potential for the construction sector. To valorize agricultural waste (as a renewable energy source and as a supplementary cementitious material—SCM), the crucial steps in the integrated management system of the circular economy cycle are the establishment and maintenance of database on crop production, namely, on harvest residues amount and quantity and quality of available biomass ash. The purpose of the study was to establish a multi-level georeferenced interactive database (map) on the produced quantities of agricultural biomass ash and cement consumption in Vojvodina region (Republic of Serbia), based on the analysis of agricultural biomass ash stream through three sectors as potential actors of the supply chain: agriculture (biomass producers)—industry (biomass users)—construction industry (users of biomass ash). Conducted research indicates the annual potential of over 2.4 million tons of harvest residues from corn, wheat, soya and sunflower, available for energy purposes in Vojvodina region. The potentially available amount of ash that might be generated annually by harvest residues combustion is estimated at over 196 thousand tons. Identified available amount of biomass ash (4.2 thousand tons) indicates an extremely low utilization (~2%) of the biomass potential. On an annual basis, all current agricultural biomass ash production can be used for partial cement substitution up to 30% in six construction companies. However, the generated ash is mostly disposed of in municipal landfills, which represents the end of the waste stream.

S. Draganić (✉) · M. Laban · V. Bulatović · O. Bukvić
Faculty of Technical Sciences, University of Novi Sad, Trg Dositeja Obradovića 6, 21102 Novi Sad, Republic of Serbia
e-mail: suzanav@uns.ac.rs

S. Šupić · M. Malešev · V. Radonjanin · I. Lukić
The Society of Civil Engineers of Novi Sad, Maksima Gorkog 39, 21000 Novi Sad, Republic of Serbia

L. Bragança et al. (eds.), *Creating a Roadmap Towards Circularity in the Built Environment*, Springer Tracts in Civil Engineering,
https://doi.org/10.1007/978-3-031-45980-1_19

Keywords Agriculture · Industry · Construction industry · Biomass ash · Waste management · Circular building materials

19.1 Introduction

Agriculture and the construction industry are essential sectors for a country's economic growth [1]. However, both sectors are large consumers of natural resources and generate significant amounts of wastes and green-house gases, that negatively affect human health, ecosystems and the economy.

One of the main problems in agriculture is an inadequate management of waste, as agricultural activities generate huge volumes of biomass residues. It is estimated that globally 140 Gt of agricultural residues are generated each year [2]. In recent years agricultural biomass has gained a global value as an alternative fuel for energy production, being considered as a renewable and CO_2-neutral energy resource. Despite the increasing consumption of biomass as an energy source, enormous quantities remain unused in landfills or burned by households.

The processing and disposal of the ash resulting from biomass combustion has become another environmental and economic issue. Biomass ash is the solid residue produced during the incineration of biomass for heat and electricity production and represents ~2–20% of the input material [3]. The available literature reflects the current situation in which energy production from biomass is the primary intention and ash utilization is often neglected [4]. Biomass ash is a relatively consistent potential resource for manufactured products, but still recycling and recovery rates of these wastes remain low and much is landfilled, or "downcycled" and used as lower-value materials [5].

The largest emitter of greenhouse gases is the construction sector, reaching nearly 38% of global CO_2 emissions, mainly from the production of cement, that reached a 4.3Gt in 2021 [6, 7]. The sector needs to reduce these emissions by embracing the principles of a circular economy.

The inclusion of biomass ash into cement-based materials, valorised as a SCM, which has been investigated in various studies [8], is a promising circular economic strategy to reduce the CO_2 emissions and minimize waste through its utilization.

The Republic of Serbia belongs to the countries that have significant biomass potential, estimated at 3.448 Mtoe per year. In the total potential of renewable energy sources, biomass participates with 61%. The largest part of this potential is agricultural biomass potential (48%) and wood biomass potential (44%) [9]. Agricultural biomass refers to the harvest residues, pruning residues and livestock manure, while the most significant source of agricultural biomass is represented by harvest residues. Regardless of the amount of agricultural biomass that is generated on an annual basis, the utilization rate is not higher than 2% [9].

The research was conducted for the region of the Autonomous Province of Vojvodina where the majority of agricultural biomass available at the territory of the Republic of Serbia is located [9]. Vojvodina is situated in the northern part of the

Republic of Serbia, in the Pannonian Basin in Central Europe. It covers an area of 21.506 km^2 and is divided into 7 districts, with a population of 1.75 million people [10].

To valorize agricultural waste (as a renewable energy source and as a SCM), the crucial steps in the integrated management system of the circular economy cycle are the establishment and maintenance of database on crop production, namely, on harvest residues amount and quantity and quality of available biomass ash. The purpose of the study was to establish a multi-level georeferenced interactive database (map) on the produced quantities of agricultural biomass ash and cement consumption in Vojvodina region, based on the analysis of agricultural biomass ash stream through three sectors as potential actors of the supply chain: agriculture (biomass producers)—industry (biomass users)—construction industry (users of biomass ash). The area of interest covered 4 districts of Vojvodina: North Bačka, West Bačka, South Bačka and Srem.

19.2 Analysis of Agricultural Biomass Ash Stream in Vojvodina Region—Methods

The analysis of agricultural biomass ash stream was conducted through three phases that included:

1. assessment of agricultural biomass potential for ash production,
2. assessment of the availability of biomass ash and
3. assessment of the capacities for the production of cement composites based on agricultural biomass ash.

Collected geospatial data on the produced quantities of biomass ash and cement consumption were integrated into a common database, within the free open-source geographic information system program—QGIS, which enabled the establishment of multi-level georeferenced interactive map.

19.2.1 Agricultural Biomass Potential for Ash Production

The assessment of agricultural biomass potential for ash production included:

- collection of data on agricultural crop production;
- calculation of the amount of residues that remain on the fields after harvest;
- calculation of the amount of harvest residues available for energy use;
- calculation of potentially available amount of ash generated by biomass combustion;
- analysis of collected data and obtained results.

As a system for monitoring the agricultural biomass potential in the Republic of Serbia does not exist, the assessment is based on the production of major harvested crops, whose harvest residues can be used as an energy source in the industry, replacing traditionally used fossil fuels. Data on crop production for a period of 5 years (2017–2021) were collected from national statistical yearbooks. In order to calculate the quantity of residues that remains on the fields after harvest, the average value of the annual quantity of individual crops for the observed time period was determined and multiplied by the coefficients indicating the quantitative ratio of basic crop product and biomass.

Quantities of harvest residues available for energy use are calculated as 30% of the total amount of crop residues, in accordance with the estimation that 30% of the produced above-ground biomass can be collected from the field without reducing the humus content of the soil [11].

As the amount of ash generated by the combustion process is not usually measured, the potentially available amount of ash that could be generated by the combustion of corn stalks, corn cobs, wheat straw and sunflower stalks and heads is estimated at approximately 8% and in the case of sunflower husks at a maximum of 5% of the calculated amount of harvest residues available for energy use [12]. The assessment of the potential for ash production was carried out through the analysis of crop production stability, the amount of ash that theoretically can be generated by biomass combustion and usability in terms of biomass collection possibilities in the fields.

19.2.2 Availability of Agricultural Biomass Ash

The assessment of the availability of biomass ash included the analysis of available data on companies that use harvest residues for energy purposes [13]. Data on used biomass types, generated quantities of biomass ash per year and ash disposal were analyzed.

19.2.3 Capacity for the Production of Cement Composites Based on Agricultural Biomass Ash

The assessment of the capacities for the production of mortar and precast concrete elements, where cement partially can be substituted with agricultural biomass ash, is based on the analysis of available data on construction companies. The assessment included:

- collection of data on average annual concrete production;
- calculation of total annual consumption of cement for concrete production;
- calculation of potential average annual consumption of biomass ash for cement substitution (cement savings);

- comparative analysis of available biomass ash and its potential consumption.

Based on the average annual production of concrete, the total annual consumption of cement was determined. The assumed average amount of cement in 1 m^3 of concrete, for the production of prefabricated elements, is 350 kg [14]. Depending on the percentage of cement substitution (10–50%), potential average annual biomass ash consumption was calculated and compared with available biomass ash, identified in phase two.

19.3 Results and Discussion

19.3.1 Agricultural Biomass Potential for Ash Production

With 83% Vojvodina has a noticeably high share of arable land in relation to the total land area and a low share of forest area (7%) [15], which implies that the region in general has significant agricultural biomass potential. The greatest potential for energy utilization lies in harvest residues, which make up 58% of the total biomass [15].

Total quantities of major harvested crops whose harvest residues can be used as an energy source in the industry for the period 2017–2021 [16, 17] are presented in Fig. 19.1.

In the structure of analyzed agricultural crops, cereals have the largest share—corn 59.1% and wheat 24.2%, while the share of industrial crops is significantly lower—sunflower 8.5% and soya 8.2%. The lowest crop production was recorded in 2017. In 2018 crop production increased by 57% and this continuity of production was maintained during 2019 and 2020, with minor deviations. Compared to 2020,

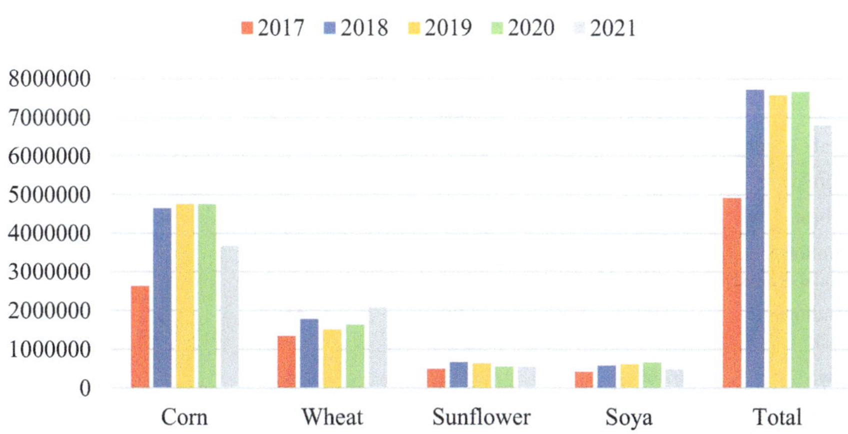

Fig. 19.1 Production of major crops (in tons) in Vojvodina for period 2017–2021

in 2021 production decreased by 12%. Minor oscillations were observed in wheat, sunflower and soya production. Larger oscillations are identified in the corn production. Compared to 2017, crop production increased by 76% in 2018, while in 2021 production decreased by 23%, compared to 2020. The corn production may vary significantly from year to year due to droughts that occasionally occur in the middle or in the end of summer and dramatically reduce production (in some cases by more than 50%) [18].

Table 19.1 presents the average annual agricultural biomass potential for ash production, in terms of amounts of ash that could be teoretically generated by combustion of analyzed agricultural biomass.

Presented data indicate the annual potential of over 2.4 million tons of agricultural biomass available for energy purposes in Vojvodina.

Industrial crops (sunflower and soya) have twice the volume of biomass than cereals (wheat and corn) in relation to the main product. However, as corn and wheat are the dominant agricultural crops, their residues account for 70.6% of the total analyzed biomass, with the ratio of corn biomass (stalk and cob) and wheat straw of 2.4.

The potentially available amount of ash that could be generated annually by combustion of analyzed biomass is estimated at over 196 thousand tons. In the context of determined amounts, corn stalk has the greatest theoretical potential for ash generation (40.1%), followed by wheat straw (20.5), sunflower stalks and head (14.5%), soy straw (13.8%) and corn cob (10%). Sunflower husk has the lowest theoretical potential (1.1%).

Table 19.1 The potentially available amounts of ash that could be generated annualy by combustion of analyzed agricultural biomass (for the 2017–2021 observation period)

Agricultural crop	The total product yield (t) [16, 17]	Product/ Biomass [19]	Biomass in the fields (t)	Biomass for combustion (t)	Biomass ash	
					(%)	(t)
	1	2	3 (1/2)	4 (3·30%)	5	6 (4·5/ 100)
Corn (stalk + cob)	4 106 000	1	4 106 000	1 231 800	8	98 544
Corn cob	–	5	821 200	246 360	8	19 709
Wheat	1 681 290	1	1 681 290	504 387	8	40 351
Sunflower (stalk + head)	592 738	0.5	1 185 476	355 643	8	28 452
Sunflower husk	–	4	147 185	44 456	5	2 223
Soya	566 460	0.5	1 132 920	339 876	8	27 191
Total	**6 946 489**	–	**8 253 871**	**2 476 162**		**196 759**

Although corn stalks have the largest share in the structure of the generated biomass, in addition to the stability of corn production, their potential is also questionable in terms of collection in the fields. Corn harvest is carried out in a period of frequent rainfall and low temperatures (October–November), which causes increased humidity, difficult drying and increased presence of dirt, in relation to other types of biomass.

Corn cobs have very good combustion characteristics, but their availability is limited due to the contemporary harvesting system, which involves harvesting corn in the grain, during which the cobs are broken and left in the fields. Harvesting corn on the cob is considered outdated and is less common in Serbia, compared to harvesting in the grain.

Sunflower seeds are usually harvested in September. After the harvest, the parts of heads and stalk mass comminuted with the combine are scattered over the field surface, and it is very difficult to collect such biomass [20]. At this stage, the sunflower stalk has high moisture (15–20%) and a very high ash content (10%) [18], which limits its use in energy production. Sunflower husks as production residues after oil extrusion have low moisture (9%), good and better thermal properties than stalk and heads and low ash content (ca. 2%) [21].

In the context of conducted research, wheat straw has been identified as agricultural biomass that has the greatest real potential for ash production, as agricultural biomass that is available in significant quantities and is most often used as a fuel, as well as it can be characterized by production stability. Soya straw is also identified by significant potential, due to the most optimal combustion properties, available quantities and production stability. Wheat is harvested in June–July, soya in September and collection of their residues can be carried out efficiently with appropriate mechanization.

19.3.2 Availability of Agricultural Biomass Ash

Twelve companies that use harvest residues as an energy source for obtaining thermal energy have been identified in the area of interest (Fig. 19.2). Data on used biomass types and generated quantities of biomass ash per year are presented in Table 19.2.

The available amount of biomass ash in the researched area of Vojvodina is over 4.2 thousand tons annually basis, which indicates an extremely low utilization (~2%) of the biomass potential determined in the first part of the research (196 thousand tons of ash).

Which type of ash will dominate the available ash structure depends on the climatic conditions and agricultural crop production. Most of the identified companies (7/12) use more than one type of biomass, which are usually mixed during combustion and which makes it difficult to determine the amount of ash by type. Consequently, a mixture of different types of ash dominates the available ash structure.

Five companies use one specific type of biomass—sunflower husk (two companies), soya straw (two companies) or corn cob (one company). Although sunflower

Fig. 19.2 Agricultural biomass users (potential biomass ash producers) and construction companies (potential users of biomass ash) identified in the Vojvodina region, with integrated data

Table 19.2 Identified quantities of biomass ash in Vojvodina [adapted from 22]

Company	Biomass type	Biomass ash (t/y)	Company	Biomass type	Biomass ash (t/y)
Sojaprotein	WS, SS, SW	1100	KNOTT Autoflex	WS, SS	60
ALMEX-IPOK	WS, SS, SH, CC	1100	PTK Panonija Mecker	WS, SS	60
Victoria oil	SH	720	Sava Kovačević	CC	30
Hipol	SS	700	Mitrosrem	WS, SS	15
Heating plant	SH	240	Victoria Starch	WS, SS	9
The veterinary institute	WS, SS	240	DTD Ribarstvo	SS	9
Total					**4 283**

WS—wheat straw, SS—soya straw, SH—sunflower husk, SW—silo waste (sunflower husk, soya husk and silo dust)

husk has the smallest share in the structure of available biomass, it is used in a significant amount—sunflower husk ash accounts for almost a quarter (22.4%) of the total available ash. Soya straw has a share of 16.5%. Only one identified company uses corn cob for energy production, which results in the smallest share of corn cob ash (0.7%) in the total amount of available ash. The small share of corn cob was expected based on the assessment of its potential, carried out in the first part of the research.

The use of corn stalks and sunflower stalks and heads is practically negligible in the identified companies.

Based on the analysis of biomass ash disposal, it was determined that generated ash is mostly deposited in containers for these purposes and transported to municipal landfills. Only a small amount of ash is deposited on the unregulated landfill within the company and is usually mixed with other waste materials.

19.3.3 Capacity for the Production of Cement Composites Based on Agricultural Biomass Ash

Eighteen active companies that produce precast concrete elements were identified in the area of interest (Fig. 19.2) [14], however the further analysis included only six companies (1/3), which were ready to provide the requested data.

In the analyzed companies, the average annual consumption of cement is 13,936 t.

Figure 19.3 compares the amounts of currently available agricultural biomass ash and the potential average annual savings of cement in case of partial replacement (10–50%) with ash. On an annual basis, total amount of currently available biomass ash produced in Vojvodina can be used for partial cement substitution up to 30% in six construction companies for production of precast concrete.

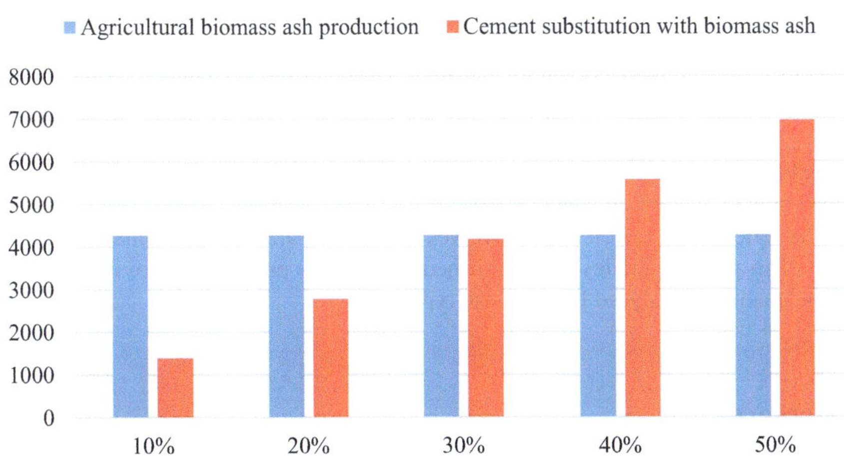

Fig. 19.3 The ratio of the total available amount of biomass ash and the required amount of ash for partial cement substitution (10–50%) in tons/year for 6 companies included in the analysis

19.4 Conclusion

Conducted research indicates the annual potential of over 2.4 million tons of harvest residues from corn, wheat, soya and sunflower, available for energy purposes in Vojvodina region. Nevertheless, enormous quantities remain unused in landfills or burned by households. By selling the biomass to industry, agriculturists would solve the problem of their waste disposal and also would make a profit. Biomass is cheaper than the commonly used fuel types and by its introduction as energy source, industry would benefit as well.

The potentially available amount of agricultural biomass ash that might be generated annually by combustion is estimated at over 196 thousand tons. Identified available amount of biomass ash (4.2 thousand tons) indicates an extremely low utilization (~2%) of the biomass potential. In addition, determination of the amount of ash by type can be challenging, as most of the companies usually mix different biomass types during combustion. On an annual basis, all current biomass ash production can be used for partial cement substitution up to 30% in six construction companies. However, the generated ash is mostly disposed of in municipal landfills, which represents the end of the waste stream. Networking of industry and construction companies would introduce a possibility of biomass ash use in civil engineering structures as an integral part of building construction. Acknowledging the possibility of ash application in civil engineering would give a new value to this type of waste and thus solve the problem of its disposal.

Identifying waste streams is an essential component of waste management plan development. The integration of all collected data and the results of the conducted research into a common database within the QGIS program enabled the establishment of multi-level georeferenced interactive map. The incorporated database provides the review, editing, analysis and update of geospatial data on the produced quantities of biomass ash and cement consumption in Vojvodina. The database can be connected to the Internet and become a publicly available database, which would enable an easy and simple review for all stakeholders.

The following research phase involves the characterization of biomass ashes as SCM. This phase upgrades the established database by providing more information on the quality of analysed biomass ash types.

References

1. Duque-Acevedo M, Lancellotti I, Andreola F et al (2022) Management of agricultural waste biomass as raw material for the construction sector: an analysis of sustainable and circular alternatives. Environ Sci Eur 34:70
2. Hills CD, Tripathi N, Singh RS et al (2020) Valorisation of agricultural biomass-ash with CO_2. Sci Rep 10:13801
3. Salvo M, Rizzo S, Caldirola M et al (2015) Biomass ash as supplementary cementitious material (SCM). Adv Appl Ceram 114(sup1):S3–S10

4. Tosti L, van Zomeren A, Pels JR et al (2021) Evaluating biomass Ash properties as influenced by feedstock and thermal conversion technology towards cement clinker production with a lower carbon footprint. Waste Biomass Valoriz 12:4703–4719
5. OECD: Towards a more resource-efficient and circular economy, The role of the G20, A background report prepared for the 2021 G20 Presidency of Italy, https://www.oecd.org/environment/waste/OECD-G20-Towards-a-more-Resource-Efficient-and-Circular-Economy.pdf, last accessed 2023/02/26
6. Ryłko-Polak I, Komala W, Białowiec A (2022) The reuse of biomass and industrial waste in biocomposite construction materials for decreasing natural resource use and mitigating the environmental impact of the construction industry: a review. Materials 15(12):4078
7. Chen C, Xu R, Tong D et al. (2022) A striking growth of CO_2 emissions from global cement industry driven by new facilities in emerging countries. Environ Res Lett 17
8. Skevi L, Baki VA, Feng Y et al (2022) Biomass bottom ash as supplementary cementitious material: the effect of mechanochemical pre-treatment and mineral carbonation. Materials 15(23):8357
9. Energy Sector Development Strategy of the Republic of Serbia for the period by 2025 with projections by 2030, Republic of Serbia Ministry of Mining and Energy, Belgrade (2016)
10. The first results of the 2022 Census of Population, Households and Dwellings (in Serbian), https://publikacije.stat.gov.rs/G2022/Html/G20221350.html, last accessed 2023/02/26
11. Establishing a biomass exchange in AP Vojvodina—Elaborate (in Serbian), Faculty of Technical Sciences Novi Sad and Provincial Center for Energy Efficiency (2012)
12. Zekić V (2018) Study—Expertise on distributions, types and quantity of available biomass ash (in Serbian)
13. Agricultural Waste—Challenges and Business Opportunities (Eco Build) project, https://www.interreg-croatia-serbia.eu/project/eco-build/, last accessed 2023/02/26
14. Plazina Pevač D (2018) Study—Expertise on distributions and capacities of precast concrete elements services (in Serbian)
15. World bioenergy association: Bioenergy action plan Vojvodina, (2015).
16. Statistical Office of the Republic of Serbia: Statistical Yearbook of the Republic of Serbia 2020, ISSN 0354–4206, Belgrade (2020)
17. Statistical Office of the Republic of Serbia: Statistical Yearbook of the Republic of Serbia 2022, ISSN 0354–4206, Belgrade (2022)
18. Weisar H, Milijić V (2017) Availability of agricultural biomass in Serbia (in Serbian), DKTI (GIZ) Program "Development of sustainable bioenergy market in Serbia"
19. Kovačević V (2018) Utilization of agricultural biomass for energy needs in Serbia (in Serbian), Belgrade, UNDP Serbia
20. Geletukha G, Drahniev S, Zheliezna T et al. (2020) Prospects of sunflower residues use for energy, Position paper, Bioenergy Association of Ukraine
21. Maj G, Krzaczek P, Kuranc A et al (2017) Energy properties of sunflower seed husk as industrial extrusion residue. Agric Eng 21(1):77–84
22. Šupić S, Malešev M, Radonjanin V (2021) Harvest residues ash as a pozzolanic additive for engineering applications: review and catalogue. Build Mater Struct 64(1):1–18

Part IX
Circular Buildings Lifecycle Management and Decision-Making

Chapter 20
Potential Synergy Between Agile Management and the Mindset of Circular Economy in Construction Projects

Flávia Luisa Pires Enembreck, Maria do Carmo Duarte Freitas, Luís Bragança, and Sergio Fernando Tavares

Abstract The focus of project management on processes and activities in the construction sector presents high variability and uncertainty. This article suggests the adoption of agile methodologies with an emphasis on people as an alternative that encourages reflection and anticipation of building maintenance and demolition problems. The objective is to identify the theoretical proximity between the main characteristics of agile construction project management and the circular economy mindset. The methodology used was a systematic bibliographical review and state-of-the-art content analysis on the application of agile in construction projects, with the discovery of the most cited characteristics and, subsequently, the establishment of a correlation with the elements of the circular economy mindset from inferences regarding the similarity of concepts and perspectives between both. From this analysis emerges the synergy between the circular and agile mindset. The conclusive trend is that architecture and construction benefit from the transitions from traditional management methodology to agile management and from the linear economy to the circular economy, given its limited technological stage in Brazil, in addition to the possibility of reducing waste generation and maintenance of the added value of buildings.

Keywords Circular economy · Project management · Agile · Construction industry

---

F. L. P. Enembreck (✉) · M. do Carmo Duarte Freitas · S. F. Tavares
Federal University of Parana, Curitiba, Brazil
e-mail: flavia.enembreck@ufpr.br

L. Bragança
University of Minho, Guimaraes, Portugal

L. Bragança et al. (eds.), *Creating a Roadmap Towards Circularity in the Built Environment*, Springer Tracts in Civil Engineering,
https://doi.org/10.1007/978-3-031-45980-1_20

20.1 Introduction

The emergence of new technologies, the internet and mobile phones, provides and generates a large amount of content that is stored and that revolutionizes the way organizations manage. These facts are transforming the world of work, the focus now dedicated to the method and the tools with their excessive documentation is reoriented to give importance to people. The agile approach emerged to promote the iterative and incremental development of projects with high uncertainty and variability—frequent cases in the construction sector—encouraging people to anticipate impediments before their realization. Decision making is oriented towards the quality of projects and the elimination of maintenance and instability problems in the future of the delivered product.

The effectiveness of the traditional management methodology, predominant in construction projects, has been questioned by researchers in recent years due to the intense appearance of uncertainties and changes in this industry [1]. Professionals in this sector are looking for alternatives for project management, in a way that preserves the profits and value of the projects, in addition to achieving more efficient and effective projects [2]. Traditional project management is characterized by advance planning of all sequential stages, in which estimates are made at the beginning of the project, and resistance to changes prevails among professionals in this area. One of the problems generated in this type of management is noncompliance with the schedule due to divergences between planning and reality, resulting in customer dissatisfaction. Another characteristic of the traditional method of managing projects is the hierarchy in project design, with the main scope being defined by the contractor and the other stages following with other stakeholders, negatively impacting communication, whose deficiency brings gaps in the scope and schedule, increments on budget and risks to the quality and success of the project [3]

In this context, the agile approach is revealed as an alternative to mitigate the problems that traditional management faces [4] and promote a satisfactory management of changes, such as high costs, late schedule and productivity and quality problems [5]. In particular, due to the need to bring discussions about the transition from a linear economy to a circular economy as a way to minimize negative impacts on the environment by the construction industry [6]. There is a growing movement of policies, academic research and initiatives aimed at promoting the circular economy in construction [7]. It is understood that there is a need to migrate to a circular economy in the built environment, but there is a lack of standardization of methods and practices to guide its implementation in construction projects [8].

Thus, in view of this scenario, the research aims to answer the following question: What are the main theoretical characteristics of agile project management in the construction industry associated with the circular economy? For this, the state of the art on the application of agile in the construction industry was investigated, identifying the mentioned fundamentals and establishing a correlation between such concepts and the principles of the circular economy.

20.1.1 Circular Economy

The circular economy (CE) is an evolution of the linear economy whose product flow ends with disposal. Aiming to promote principles such as increasing resource efficiency, closing material flows and extending the life cycle of products through reuse and recycling, CE has inserted itself in the midst of organizations in the global market. This insertion occurred with approaches such as industrial symbiosis, closed supply chain cycle and circular product design [9].

The transition from linear economy to circular economy depends on a change of mindset. In practice, the aim is to broaden the vision beyond the production processes, also seeking to benefit the ecosystems to which the product is associated. The premise for the circular mentality goes beyond the efficiency assumed by the linear economy, being necessary to consider the effectiveness of a certain activity, taking into account its effects and consequences. The value proposition changes from the search for lower costs (linear economy premise) to the generation of value from resources throughout the product's life cycle, maintaining the added value for as long as possible. The focus is redirected to the circular economy, to innovation and the addition of new values. As for the personas, the EC emphasizes the importance of the participation of all interested parties of the business (stakeholders) in the project and not just the shareholders (shareholders). Ethics in a circular mindset prioritize collaboration over competition. And as for the role, it focuses on the user experience, strengthening the relationship with the customer and improving their satisfaction. From the transition from the linear economy to the circular economy, it is possible to bring resilience and sustainability to production processes [10].

For CE to be incorporated, it is necessary to understand a building as dynamic structures adaptable to current needs, which are composed of layers of different materials and products. The understanding of the construction system in its entirety collaborates in the adoption of circular techniques such as dismountable buildings, modular parts, use of bio-based and/or renewable materials and the incorporation of reused materials. There is a challenge in relation to managerial aspects, due to the insertion of circularity in traditional project management. Three points are seen as paramount in the transition to a circular economy in construction projects, they are: knowledge of CE, integration of the circular principle in project preparation and management [11].

20.1.2 Agile Project Management

Since the launch of the Agile Manifesto in 2001, the application of agile values and principles has spread to sectors beyond software development [12]. The 15th State of Agile Report [13] showed an expansion, due to the distribution of the global workforce, of the adoption of the agile approach in areas such as IT, but also finance, human resources, marketing. This dissemination has evolved more and more since the first report published in 2006. It was also mentioned that, even after two decades of the Agile Manifesto, agile practices are relevant to the current market, mainly due to the growing pressure to meet the needs of customers and stakeholders quickly. The period of the Covid-19 pandemic, which started in 2020, brought uncertainty to companies and reinforced the importance of having project management prepared for scenarios of sudden and sudden changes. The adoption of agile by industries of various modalities came as an alternative to achieve greater flexibility, dynamism and practicality in management processes [14].

The use of the agile approach is recommended for projects with a high degree of uncertainty due to the variety of changes, high complexity and identification of risks. As the environment becomes more and more uncertain, the efficiency of traditional management drops, as the traditional approach is conditioned to defining the full scope early and reacting late to changes. Management in short cycles and adaptability through evaluation and feedback are agile objectives that add to the conduction of projects full of uncertainties. Through incremental delivery, the team is able to anticipate problems and change the subsequent increment, reducing rework and improving understanding of requirements with the customer, since each delivery requires validation [12].

20.2 Research Method

The initial methodological approach used was based on a bibliographic review and content analysis. A search was carried out for peer-reviewed articles from 2018 to 2022 in three different search bases: Scopus, Science Direct and Web of Science. Inclusion criteria were peer-reviewed articles that bring contributions on agile methodology in the construction industry. The exclusion criteria were duplicate articles and evasion of the investigated theme. With the application of these criteria, a sample of 26 articles was qualitatively analyzed to identify the main characteristics of agile management directed to construction projects. At the end of the bibliographical review of agile management in the construction industry, a systematic correlation is made [15] through inferences to establish theoretical approximations between the agile mindset and the circular economy mindset. The research protocol with the sequence of methodological referral processes is found in Table 20.1.

20.3 Results and Discussions

Based on the systematic review carried out, the eight agile characteristics (Table 20.2) most recurrent among the authors when using agile management in construction projects were raised.

The improvement in communication between stakeholders stems from continuous contact between team members and other stakeholders, with the practice of the Daily

Table 20.1 Research protocol

Step	Description
Defining the Problem	Which agile characteristics bring improvements to project management in the construction industry and what are the theoretical approaches with the circular economy mindset?
Determining the search parameters	Through the adhesion test, the following keywords found were: Project management, Agile and construction or buildings
Choose the scope of the research	Scopus, Science Direct and Web of Science
Delimitation of search criteria	Keywords were combined with the Boolean operator "AND" and quotation marks were used for the compound term "project management" Search for publications from the last 5 years
Filtering of studies	Scanning titles, abstract and keywords Selection of articles available to access to the full text on agile methodology and the construction industry obtained from the databases The exclusion criteria included repeated articles that were not related to the investigated topic
Data collection and analysis	Reading of full articles and selection of articles relevant to the theme Qualitative data analysis with the categorization of bibliographic research Content analysis correlating the raised codification of agile characteristics in civil construction and the circular economy mindset defined by the National Confederation of Industry in the bibliographic review [10]
Quality assessment	The search sources are secure and the articles considered were peer-reviewed
Drafting the summary report	Elaboration of a table with the main characteristics of Agile in construction projects Summary of theoretical approaches between Agile and the circular economy

Meeting being one of the main agile premises [40]. Regarding change management, there was great potential for adopting agile when incorporating change as a strategy to add value to the final delivery [41]. In addition, agile management is prepared for scenarios of sudden changes with the concept of adaptability [14]. Agile collaboration concerns project team management with active participation in decision-making [42] and collaboration with the client and/or user [43]. Flexibility is one of the main factors needed in view of the needs brought by the user, even if they were not mapped at the beginning of the project. It is directly related to change management [31]. The iteration is the division into periods of time with the objective of finishing a deliverable content of the projects, also called increment. Iterations and increments are a way to achieve continuous improvement and innovative solutions. From this installment of deliveries, the team is able to anticipate problems and change the subsequent increment, reducing rework and improving understanding of requirements with the customer, since each delivery demands validation [41]. A multidisciplinary and self-organized team is a key foundation in the agile organizational structure, so that all members gather the necessary knowledge for the progress of the project [27].

Table 20.2 Main characteristics of agile applied in the construction industry

Description	Authors
Improved communication between project stakeholders	[4]; [16]; [17]; [18]; [19]; [20]; [21]; [22]; [23]; [24]; [25]; [26]; [27]; [28]; [29]; [30]; [31]; [32]; [33]
Change management	[4]; [5]; [16]; [17]; [18]; [21]; [24]; [25]; [26]; [27]; [28]; [29]; [30]; [31]; [32]; [33]; [34]; [35]; [36]
Collaboration between the managing team and the customer	[4]; [16]; [18]; [22]; [23]; [24]; [25]; [26]; [27]; [28]; [30]; [31]; [32]; [33]; [34]; [35]; [36]; [37]
Flexibility	[4]; [16]; [17]; [20]; [21]; [24]; [25]; [26]; [27]; [28]; [30]; [31]; [33]; [35]; [36]; [37]; [38]; [39]
Iteration	[4]; [5]; [16]; [17]; [18]; [19]; [20]; [21]; [26]; [27]; [28]; [31]; [33]; [34]; [35]; [36]; [37]; [39]
Multidisciplinary team	[16]; [17]; [20]; [23]; [26]; [27]; [28]; [30]; [31]; [33]; [37]
Lessons learned	[20]; [19]; [27]; [17[; [36]; [31]; [33]; [25]; [28]
Cyclical process	[20]; [37]; [4]; [35]; [27]; [5]

The mapping of lessons learned during the project is a way of promoting continuous improvement with constant sharing of knowledge among team members and exchange of perceptions [40]. The cyclical process is guided by management in short cycles and adaptability through evaluation and feedback are agile objectives that add to the conduction of projects full of uncertainties [12].

From the agile elements brought to civil construction, correlations with the circular mindset were established, based on the principles of the Circular Economy advocated by the Ellen Macarthur Foundation [10]. Table 20.3 presents these correlations, with the first column bringing the elements of the mindset of a circular economy described by the National Confederation of Industry [10] and the second column bringing the agile characteristics (Table 20.2) associated with each circular element.

When correlating the configuration of the circular economy mindset with the elements found in agile, it was noted the similarity of concept and perspective between both. It was possible to infer the theoretical approximation of the eight agile characteristics with the seven aspects of the circular economy. A multidisciplinary team manages to gather a repertoire of information about the associated ecosystems and assess the impacts at each stage of the project. The circular premise is the effectiveness, which is an expected result in flexible and iterative management. Creating value through cycles is synergistically associated with the cyclical process of agile, characterized by closing cycles and delivering maximum value. One of the main objectives of change management is innovation, with the vision that each modification is an opportunity to improve and include advantageous novelties in the project's life cycle. As premeditated by the circular economy and agile, it is of paramount importance to involve all stakeholders in the project and promote a collaborative spirit. As a way to bring improvements to the user, the practice of

Table 20.3 Theoretical approaches to the characteristics of agile project management in the construction industry associated with the circular economy

Mindset circular	Related agile feature
The scope considers ecosystems	Multidisciplinary team
The premise is the effectiveness	Flexibility e Cyclical process
The value proposition considers the generation of value	Iteration
Focus on innovation and new values	Change management
Personas include all stakeholders	Improved communication between project stakeholders
Ethics based on collaboration	Collaboration between the managing team and the customer
Role is user-oriented	Lessons learned

listing the lessons learned during the project cycles is essential to identify points to be corrected.

20.4 Conclusion

Based on the results obtained in the systematic bibliographic review, a promising trend in the construction sector is evident in incorporating agile practices and principles into the management method, replacing the traditional methodology. Despite the great research potential on agile in the construction industry, the bibliography presents a limited number of publications. Thus, there are still gaps for further research on this topic, with regard to greater theoretical and practical deepening on adaptations in the implementation of construction projects, especially in the execution phase, and to investigate conflicting concepts between the software industries (origin of agile) and construction.

In the same way that problems are identified in project management, there are also environmental issues that need to be solved in the civil construction sector. Currently, the main issue is the exacerbated generation of waste arising from the unbalanced consumption of raw materials. As a strategy to mitigate negative impacts on the environment, the circular economy emerges. The transition to the circular mindset theoretically resembles agile characteristics in project management in the construction industry. This work allows the identification of insights that connect both concepts. The foundation in closed cycles and the delivery of maximum value, even if for different contexts, are similarities to be deepened. Principles of stakeholder communication, collaboration, and user focus are found in both agile and a circular economy mindset.

It is concluded that there is a synergy between the circular and agile mindset due to the inferred theoretical approaches. It is also complemented that construction projects tend to benefit from the transitions from traditional management methodology to agile management and from the linear economy to the circular economy. In this way, it appears that there is also a potential synergy in the practical incorporation of both changes together.

From this research, it is concluded that understanding the synergy between the agile framework and the circular economy will bring relevant contributions to the Architecture, Engineering and Construction (AEC) sector, increasing future researches. Agile management is considered as a potential facilitator in the implementation of circular economy practices in civil construction projects, promoting the continuous monitoring of the circulation of materials in the project phases. Based on frameworks that unite agile and the circular economy, it is expected to optimize the management of the environmental impacts of the construction industry and, consequently, contribute to a more sustainable society.

References

1. Akel RP et al. (2019) Estudo comparativo entre a metodologia tradicional e ágil de gerenciamento de projetos. XXXIX ENEGEP, https://www.emerald.com/insight/content/doi/https://doi.org/10.1108/REGE-01-2019-0021/full/html, last accessed 2022/12/21
2. Durante FK et al. (2015) Avaliação de aspectos fundamentais para a gestão integrada do processo de projeto e planejamento com uso do BIM. In: Proc., VII Encontro de Tecnologia de Informação e Comunicação na Construção—Edificações, Infra-estrutura e Cidade: Do BIM ao CIM, Editora Edgard Blücher.
3. Ekanayake et al (2019) Traditional project management approach, industry challenges and key attributes: a case study of construction project management in Sri Lanka. Asia Pac Inst Adv Res (APIAR) 5(1):52–64
4. Arefazar Y et al (2022) Prioritizing Agile project management strategies as a change management tool in construction projects. Int J Constr Manag 22(4):678–689
5. Mohammed KN, Karri SC (2020) An analytical approach in usage of agile methodologies in construction industries—A case study. Mater Today: Proc 33:475–447
6. Wuni IY (2022) Mapping the barriers to circular economy adoption in the construction industry: A systematic review, pareto analysis, and mitigation strategy map. Build Environ 223:109453
7. Guerra BC et al (2021) Circular economy applications in the construction industry: A global scan of trends and opportunities. J Clean Prod 324:129125
8. Benachio GLF, Freitas MCD, Tavares SF (2020) Circular economy in the construction industry: A systematic literature review. J Clean Prod 260:121046
9. Mies A, Gold S (2021) Mapping the social dimension of the circular economy. J Clean Prod 321:128960
10. Confederação Nacional da Indústria (CNI) (2018) Economia circular: Oportunidades e desafios para a indústria brasileira. ISBN: 978-85-7957-166-4, https://www.researchgate.net/publication/324729539_Economia_Circular_Oportunidades_e_desafios_para_a_industria_brasileira, last accessed 2022/09/28
11. Többen J, Opdenakker R (2022) Developing a framework to integrate circularity into construction projects. Sustainability 14(9):5136
12. Project Management Institute (PMI) (2017) Agile practice Guide. Newtown Square
13. Digital.Ai. (2021) 15th State of Agile Report. Digital.ai, https://Digital.ai/resource-center/analyst-reports/state-of-Agile-report, last accessed 2022/09/08
14. Marnada P et al (2022) Agile project management challenge in handling scope and change: A systematic literature review. Procedia Comput Sci 197:290–300

15. Bardiin L (1977) Análise de Conteúdo. Lisboa, Portugal, Edições 70
16. Ahmed MN, Mohammed SR (2019) Developing a risk management framework in construction project based on agile management approach. Civ Eng J 5(3):608–615
17. Albuquerque F, Torres AS, Berssaneti FT (2020) Lean Product development and Agile project management in the construction industry. Revista de Gestão 27(2):135–151
18. Almarar MS (2019) EPC strategies for a successful project execution. SPE Gas & Oil Technology Showcase and Conference
19. Campbell DTS (2019) Construction information management systems: Conceptual model to improve residential construction project resilience and Productivity in New Zealand and Saint Vincent and the Grenadines. International Conference on Sustainable Infrastructure, pp. 205–215
20. Carlos R, Amaral DC, Caetano M (2018) Framework for continuous agile technology roadmap updating. Innov & Manag Rev 15(3):321–336
21. Chumpitaz B et al. (2020) Application of the Scrum framework to optimize time in construction projects. Congreso Internacional de Innovación y Tendencias en Ingeniería (CONIITI), IEEE: 1–6
22. Hamerski DC et al. (2019) Bringing lean and agile project management in a multi-project environment: Case study in a retail company. In: Pasquire C, Hamzeh FR (eds.) Proc., 27th annual conference of the international. Group for Lean Construction (IGLC), pp. 239–250
23. Jethva SS, Mirosław JS (2022) Agile project management for design-build construction projects: A case study. Int J Appl Sci Eng 19(1):1–11
24. Lalmi A, Fernandes G, Boudemagh S (2022) Synergy between traditional, agile and lean management approaches in construction projects: Bibliometric analysis. Procedia Comput Sci 196:732–739
25. Lalmi A, Fernandes G, Souad BS (2021) A conceptual hybrid project management model for construction projects. Procedia Comput Sci 181:921–930
26. Malla V, Prasad KV (2022) Critical success factors and perceptions of stake holders to Lean-Agile adoption in construction projects (LACP). AIP Conf Proc 2393:020017
27. Mohamed B, Moselhi OA (2019) Framework for utilization of Agile management in construction management. Can Soc Civ Eng Annu Conf, CSCE: CON21 1-10
28. Ozorhon B, Cardak F. Caglayan S (2022) Investigating the agile hybrid approach in construction. J Manag Eng 38(4): 04022022 1–13
29. Shah A et al. (2022) Influence of agile leadership on project success; a moderated mediation study on construction firms in Nepal. Eng Lett 30(2)
30. Sljivar I, Gunasekaran A (2018) Agile-scrum for facility design project management. SPE West Reg Meet Proc, SPE: 1–11
31. Sohi AJ, Bosch-Rekveldt M, Hertogh M (2020) Four stages of making project management flexible: Insight, importance, implementation and improvement. Organ, Technol Manag Constr: Int J 12(1):2117–2136
32. Vaz-Serra P, Hui F, Aye L (2021) Construction project managers graduate agile competencies required to meet industry needs. ICSECM 2019, Springer, Singapore 94:601–607
33. Zender YO, de Soto BG (2021) Use of Scrum in the rehabilitation of a commercial building in Peru. Constr Innov 21(2):145–163
34. Besenyői Z, Krämer M, Husain F (2018) Building information modelling in agile environments—an example of event management at the Airport of Tempelhof. MATEC Web Conf 251:03064
35. Hussien A et al. (2019) Optimizing project delivery through augmented reality and Agile methodologie. In: Proc., 12th International Conference on Developments in eSystems Engineering (DeSE), IEEE, pp. 1006–1013
36. Lamacchia D, Chowdhury K, Sharif O (2020) A novel way of project management to ensure engagement for successful digital transformation. Offshore Technology Conference, OTC
37. Mnqonywa S, Von Solms S, Marnewick A (2018) A systematic literature review of the Agile methodology applied during construction project design. In: Proceedings of the International Conference on Industrial Engineering and Operations Management (Oct), pp. 1430–1442

38. Burmistrov A, Siniavina M, Iliashenko O (2018) Project management life cycle models to improve management in high-rise construction. E3S Web Conf 33:03005
39. Ingle A (2019) Agile project management: Feasible methodology in construction industry. Int J Eng Adv Technol 9(1):5210–5213
40. Radhakrishnan A et al. (2021) The impact of project team characteristics and client collaboration on project agility and project success: An empirical study. Eur Manag J
41. Owen R et al. (2006) Is Agile project management applicable to construction?. In: Proc., 14th Annual conference of the international group for lean construction, Ponteficia Universidad Católica de Chile
42. Franzolin CTS (2019) Barreiras na implementação de Metodologias Ágeis: estudo de caso em uma empresa de meios de pagamento. Dissertation (Master of Science). USP, São Paulo
43. Beck K et al. (2001) Manifesto for agile software development. http://Agilemanifesto.org/, last accessed 2022/08/15

Part X
Stakeholders Relationships and New Potential Actors in Circular Management Models

Chapter 21
How Does 'Locality' Matter in Enabling a Circular Built Environment?: A Focus on Space, Knowledge, and Cities

Mustafa Selçuk Çidik, Georg Schiller, Ning Zhang, Agatino Rizzo, Tatjana Tambovceva, Diana Bajare, and Mennatullah Hendawy

Abstract There is a growing interest in understanding and using local knowledge, resources, and stakeholders to achieve tailored and effective circular solutions in the built environment. Although the importance of clear centralised guidance and regulations are emphasised in the existing literature, there is also an emerging acknowledgement that understanding the 'local context' will be key to achieving tailored solutions that can effectively work in practice. However, there is a lack of discussion around the meaning and significance of 'locality' in terms of circularity solutions in the built environment. This discussion paper introduces space (both physical and social) and knowledge as two key aspects of 'locality' for enabling effective circular solutions in the built environment. Further, it argues that the cities can be seen as the locus of circular economy because of their role in localising space and knowledge. Thus, the paper enables a starting point to structure research towards an improved understanding of (i) the role of space and knowledge co-production for a circular built environment, (ii) the relevant local stakeholders, as well as (iii) city-level governance of locality in supporting a circular built environment.

Keywords Circular Economy · Context · Knowledge · Locality · Space · Urban

M. S. Çidik (✉)
University College London, London WC1E 7HB, UK
e-mail: s.cidik@ucl.ac.uk

G. Schiller · N. Zhang
Leibniz Institute of Ecological Urban and Regional Development, 01217 Dresden, Germany

A. Rizzo
Luleå University of Technology, 97187 Luleå, Sweden

T. Tambovceva · D. Bajare
Riga Technical University, Riga 1048, Latvia

M. Hendawy
Ain Shams University, Cairo 11517, Egypt

Impact Circles E.V, 12435 Berlin, Germany

L. Bragança et al. (eds.), *Creating a Roadmap Towards Circularity in the Built Environment*, Springer Tracts in Civil Engineering,
https://doi.org/10.1007/978-3-031-45980-1_21

21.1 Introduction

Circular economy (CE) is a concept that has attracted significant attention from academia, policy and practice. Although there are multiple ongoing discussions around CE, the CE discourse tends to adopt a deterritorialised approach, which focuses on innovation and growth [1]. Similarly, most research on circular built environment emphasise centralised or generalisable solutions and approaches such as digital systems, product/building design principles, life cycle or material flow assessment tools, as well as business models [2]. Debates around CE tend to be deterritorialised, and focused on sectors, because their focus is on economic system and production principles, which operate beyond local scales. However, such a perspective falls short in articulating the practical implementation and operation, and hides the full scale of politics of CE, thus creating a serious implementation gap which must consider local scales [3, 4].

On the other hand, the wider research on sustainability and socio-technical transitions widely recognises the importance of 'locality' to develop effective solutions in practice [5]. Here, 'locality' broadly refers to the specific set of socio-material relationships and practices associated with a place (e.g., region, city, neighborhood etc.), which affects the outcomes of sustainability efforts in practice for that place. Indeed, there is also an emerging acknowledgement in circular built environment research that understanding and working with the 'local context' (i.e., local stakeholders, knowledge and resources) is essential for effective circular initiatives [6].

However, currently there is a lack of clarity and discussion about how 'locality' matters in enabling a circular built environment [7]. Particularly, there has not been any discussions on what constitutes the 'local context', and what needs to be harnessed, or worked with, in that context, to enable effective circular solutions in practice. Based on a critical literature review, this conceptual discussion paper reflects on the meaning and significance of 'locality' in terms of enabling circular solutions in the built environment. It discusses space (both physically and socially) and knowledge as two peculiars of 'locality' in the context of circularity in the built environment. This discussion is then extended to the role of cities in enabling a circular built environment. It is argued that it is through the localising of space and knowledge that cities become effective in governing circular actions and strategies. The conclusions highlight the implications of the presented discussions on conceptualising local context, local stakeholders, and city-level governance in circular built environment research and practice.

21.2 Locality of Space and Circular Economy

21.2.1 Locality of Physical Space

Circular economy (CE) is often seen as a means to achieve sustainability. A 'value-retention process' can be envisioned in a CE system by intentionally closing, slowing and/or narrowing the cycle of resources [8]. The built environment is the expression of a material bank, which prompts us to consider the transition of the built environment to a CE to improve resource efficiency. The built environment consists of immovable buildings and infrastructure whose main consumption is bulk mineral materials [9]. The unit and added value of these bulk construction materials is low compared to other products, making their flow and supply usually limited to a local (e.g., regional scale). Thus, achieving circularity of materials in the built environment requires understanding and utilizing the local physical space, which means considering the physical characteristics of the space [10].

Much of the literature related to materiality of built environment in cities and regions has implicitly and sporadically demonstrated that the physical characteristics of space, such as density, distance, and material availability, can influence the circularity of building materials [9]. These characteristics of physical space can be reflected in areas of different scales from building to urban, and even to a national scale. Different types of spatial features can influence the circularity of materials in the built environment by giving relative advantage to slowing (extending the lifespan of the structure), closing (ending the loops between use and production) and narrowing (reducing the consumption of materials) the cycle. In other words, different spatial features such as terrain, land availability, availability of natural resources etc. [11] can affect the material circularity. Their role is direct or indirect in the CE due to the correlation between them. For example, regions with flat terrain are more habitable, where populations congregate to create a high-density built environment. High-density construction shrinks available land resources, reducing the landfills for waste materials on the one [12], and reducing available building plots on the other hand; thus, shifting construction activity to renovating existing settlements in order to improve durability [13].

Considering such influence of physical space on potentials of circularity, it is worth establishing the spatial characteristics at a local level to design locally adapted circular strategies. Particularly, the decrease in spatial scale can make the impact of physical space on material circularity more evident by reducing the geographic access to natural and anthropogenic materials. Therefore, the local spatial characteristics are crucial in determining the most appropriate strategies and actions for achieving circularity.

For example, high building density usually implies a concentration of population and scaled-up cities where efficient recycling activities are constrained by construction space and where newly built activities are limited by land resources. Therefore, in such localities slowing strategies can be considered a priority because the goal of such CE activities is to improve the lifespan of the structure and avoid new construction and demolition activities. Another example is the issue of supply and recycling of materials in construction. A CE perspective suggests that the smaller the transportation distance, the better it is, to minimise the resources associated with transportation, which is a non-value adding activity [14]. Therefore, generally, for recycling, on-site and proximity disposal are considered as the best options, while for supply, locally available materials are preferred. However, on the one hand, construction material recycling facilities are located either in localities with frequent construction and demolition activity or near highways [15]. On the other hand, in high building density areas, the construction space on site is compressed, resulting in a lower possibility of on-site recycling, and a shift to off-site recycling, which increases the transportation distance. Overall, an adequate understanding of the local physical space is important for the effectiveness of CE strategies and actions in the context of the built environment.

21.2.2 Locality of Social Space/Place

An understanding of space as a material/physical entity has long been seen as limited from the perspective of social sciences. From this perspective, space is explained as social phenomena resulting from situated and located social activities. Gotham [16] defines social space as a social relation and social construction that participates in the production and reproduction of social structures, social actions, and relations of power. In the context of a circular built environment, a social perspective on space becomes crucial for at least two reasons. First, a good understanding of the affected (local) socio-spatial contexts is key to developing and implementing circular initiatives that can deliver the anticipated benefits in practice. Second, such an understanding is also crucial for enabling a socio-ecological transition through CE initiatives instead of establishing an exploitative development logic that benefits only few actors [3, 17].

For the practicability of circular initiatives in the built environment, understanding locality from a socio-spatial perspective becomes critical, because in practice, it is through the socio-spatial relationships that various actors' resources, interests, and visions are configured; and thus, aligned or misaligned. It is this alignment/misalignment that eventually determines whether the new circular way of doing things could be established in practice as anticipated. Thus, there is a growing amount of empirical work showing that the social aspect of locality is key for enabling effective implementation and practicability of circular initiatives.

For example, the literature on Urban Living Labs (ULL) where real-life sustainable transition experiments take place, vividly show how such alignment happens in practice. Cuomo [18] demonstrates that, by bringing various local stakeholders together on a real-life experimentation, ULLs enable cooperative governance, which fosters creativity and positive cooperation that develop and implement practicable CE solutions. Further, acknowledging the effectiveness of ULLs in understanding and addressing social aspects of locality, Gundlach et al. [19] argue that the concept of ULL can be extended to cooperative urban development, where the built assets themselves become ULLs towards a circular built environment. On a parallel note, a socio-spatial perspective is also critical for ensuing the business viability of circularity initiatives at a local level. Howard et al. [20] adopts a placed-based view on SMEs to study SMEs role and resilience in CE. They find that SMEs become more resilient when their business value creation also help societal and environmental value creation. This means that "SMEs who recognize the role of placed-based societal identities and ecosystems not only become more resilient, but their considerations for community welfare and labour are intertwined with geographic-specific natural capital and the circular economy". This is a particularly important argument for circularity in construction, where majority of the companies are SMEs that are operating locally. Finally, Oyinyola et al. [6] demonstrate how understanding and working with local social groups reinforced a circular experiment of housing through various mechanisms including incentivisation for engagement, job creation, upskilling, user acceptance, and replication of new circular solutions.

It is also important to see 'locality' from a socio-spatial perspective to ensure a just sustainable transition through circular initiatives, and prevent an exploitative development logic [3]. The field of urban geography has long studied the relationship between the dominant political and economic models, and the ways in which the built environment is planned, developed, and operated to varying benefits and costs of different social groups in the society [21]. Some concerns have been raised in CE literature that the deterritorialised, technically- and growth-focused discourse of CE, can work as a disguise of the real politics involved in shifting the existing economic models (and so the politics) in an unjust way. Keblowski [22] argue that in this case, CE can just be used as "an urban sustainability fix by selectively incorporating ecological goals in urban governance strategies". However, because CE is as political and economic as it is technical and operational, when 'locality' is understood and incorporated from a socio-spatial perspective, it could provide a chance for people to (re)gain agency to enable a just socio-ecological transitions [3, 22].

21.3 Locality of Knowledge and Circular Economy

Throughout the past years there has been a paradigm shift in the way knowledge and science, and their roles in society are perceived. One of these shifts is the turn towards knowledge co-production, which calls for democratizing knowledge and science considering the hierarchical connection between knowledge and power that confers hegemony on science [23]. Knowledge co-production aims at achieving not only scientific but also societal impact through the collaboration between scientific and non-scientific actors as well as collective dialogue among actors with different expertise [23, 24]. As such, this transition in science towards knowledge co-production had focused on linking science with practice and linking science with policy. These calls aimed at making science more interdisciplinary and transdisciplinary by bringing in knowledge from diverse perspectives [6].

Norström [24] demonstrate the effectiveness of knowledge co-production for sustainability defining co-production in this context as "iterative and collaborative processes involving diverse types of expertise, knowledge and actors to produce context-specific knowledge and pathways towards a sustainable future". In the context of sustainability, knowledge co-production becomes essential to address the complex challenges of urban sustainability [25].

In a similar plea, and with regards to circular built environment, more recently, D'amato [25] demonstrate that "knowledge co-production entails the integration of different knowledge types and collaboration across multiple societal actors with potentially conflicting viewpoints and agendas". In this regard, they suggest that there are three main kinds of knowledge that are recognized by all the actors: lay knowledge, expert knowledge, and scientific knowledge. However, these kinds of knowledge are "dealt with, to different extents, according to the roles played by different actors in the process of knowledge generation". Similarly, Fratini et al. [27] argue that there is a need to bridge between scientific and practice-based knowledge to develop and stabilize CE concept because these are inevitably influenced by geographically and culturally contextualised 'socio-technical imaginaries'—i.e. "collectively held, institutionally stabilised and publicly performed visions of desirable futures, animated by shared understanding of forms of social life and social order attainable through, and supportive of, advances in science and technology".

With the above views, knowledge can be perceived as a key aspect of 'locality' for enabling effective tailored circular solutions in the built environment. As Durose et al. (2018) cited in Broto et al. [28] state "knowledge co-production recognizes that citizens hold knowledge, particularly in urban development planning action, requiring experiential learning to deliver sustainable outcomes". In this light, they claim that "community accountability is crucial in tracking the implication of multiple knowledges and intentions in co-production processes". As such, it is important to consider that although "co-production strategies may address epistemic injustices" they may also "generate new ones" [27].

Therefore, it is critically important to consider those situated knowledge processes as co-producing governance concepts, such as 'circular economy', which have major impacts on the people inhabiting the affected localities. Literature on sustainability transitions in cities [29] demonstrate how "knowledge and power are inevitably interlinked in the governance of urban transformations" [27, p. 4], Jassanoff [23] shows how "scientific knowledge both embeds and is embedded in social identities, institutions, representations and discourses". Addressing the challenge of epistemic injustice for local people in knowledge generation requires broadening the source of knowledge by respecting and incorporating multiple ways of knowing and engaging with diversified actor groups; not only elite actors with high power and influence (e.g., not only government bodies, technocrats, scientists, large NGOs, but also local and indigenous communities, small businesses and NGOs).

21.4 Localising Role of Cities and Circular Economy

Understanding circularity in terms of locality speaks for the need to focus on people's living environment. Projections show that by 2030 more than half of the world population will live in cities, and so, it is natural to 'localise' circular thinking and circularity in cities. This makes particularly sense when the ambition is to embrace an understanding of sustainability and circularity that is open to, and benefits, large masses, and not only a limited number of actors [30].

Cities are well placed to support a circular eonomy due to high concentrations of resources, capital, and talent [31]. Cities can provide that optimal space for interaction to explore the nexus between the resources (e.g., energy, materials) needed for building and urbanization, where this latter term is understood as a socio-economic-cultural process that embraces a social view of space and knowledge creation. Thinking of circularity as the nexus between resource flows and socio-economic-cultural processes of urbanization allows to centralize the role of urban planning and design as a method to both optimize physical space and empower local actors for knowledge co-production and active participation [32].

From a land-use/physical space point of view, cities represent an adequate scale for the governance and facilitation of circular economy due to their existing governance and infrastructures that are already in place for regulating economic processes and providing essential services (e.g., energy and waste management). As highlighted by Williams [10], urban planning and design can localise looping of resources through the co-location of producers and consumers of various value streams. Williams [33] gives examples of this: The London Plan [34] which allocated land for a variety of 'low value' circular activities (e.g., urban farming, storage and logistics facilities, waste management and green space), as well as Stockholm's plan [35] which encouraged urban form (high-density, mixed-use development linked by public transport and district heating networks) that supports the operation and expansion of a circular system (ecocycles) across the city.

It follows from this that land-use planning in cities has an important role also in enabling circularity in the built environment. There are already studies which report findings that support this argument. Tambovceva [36] evaluated the level of awareness and attitude towards construction and demolition waste among Latvian construction companies. Participants of the study mentioned the lack of space on the construction site as an obstacle to waste sorting and later recycling. Arora [37] used a bottom-up stock analysis approach to estimate the material and component stock of public housing developments in Singapore, where material stock estimations and potential outflows is crucial from matching demands and achieving resource efficiency. Furthermore, to facilitate finding the location of publicly available sorting containers and areas in Latvia, ZAAO Ltd. has developed an interactive map for residents and contractors to easily find the nearest sorting site around.

From a socio-economic-cultural perspective, it is a major challenge to identify the right circularity priorities, strategies and actions that holistically address multiple interests in urban contexts. It is this challenge that requires a good consideration of social aspects of local space and knowledge co-production with local actors. For example, on the one hand, changing consumer behavior is the most important and challenging issue for local policymakers [38]. On the other hand, there are several successful examples of implementing circular economy at city level, such as, creation of networks of reuse points for discarded items, such as household appliances and furniture, collection of household food waste to turn it into biogas, and implementation of circular urban farming principles.

Ultimately, these suggest the need for developing holistic urban planning and design frameworks that consider physical space, socio-spatial relationships as well as effective ways of knowledge co-production with local actors. For example, the city of Lappeenranta in Finland has focused its efforts in six areas: public procurement, land-use planning, construction, recycling and waste management, nutrient recycling, sharing economy and smart services. It is emphasized that communication, as well as ensuring the commitment of stakeholders to the task, is especially important for achieving the goals [39]. In a similar way, Williams [33] argues that 'circular development' (not circular economy) should be the focus of urban planning. She puts forward a framework for circular development that involves not only land-use considerations but also complementary socially- and knowledge-driven issues, such as market capacity building, partnership building and new regulations. Although there is a paucity of research on exactly how a circular built environment can be best supported through such holistic urban planning and design approaches, there seems to be an agreement in the literature about the need for urban governance to consider local physical space, knowledge and socio-spatial relationships to achieve any circular transition [40, 41].

21.5 Conclusions

Circular approaches are needed in the built environment due to its enormous impact on resource depletion and climate change. However, the dominant generalized and deterritorialised approaches have limited effect in enabling the desired circular transformation in the built environment. Therefore, locally-informed circularity strategies and activities are required to fundamentally change the existing linear system of building towards a stronger circularity. The paper argues that 'locality' matters in enabling a circular built environment because local materiality, stakeholder interests, social relationships, and knowledge are crucial in achieving the desired circular transformation. It also argues that cities are key to understanding and addressing these aspects of locality due to (i) high concentrations of built assets, resources, capital, and talent in cities, and (ii) the prospects of organising these for circularity through urban governance.

Although many circular economy initiatives are underway across industries, some of the barriers of circular transformation in the built environment are unique. A major challenge for the built environment is to promote integrative initiatives that bring together a large variety of stakeholders, while another one is to bundle diverse local initiatives and align them along overarching goals. These challenges critically require a joint consideration of local physical space, stakeholder knowledge and socio-spatial relationships. In this context, cities seem to present the ideal scale to govern the circular transformation by integrating local and general through a clearer picture of the relationship between use and production, back and forth effects, and cross-cutting relationships and politics between sectors and actors.

However, knowledge-related issues and socio-spatial relationships have so far been formulated too abstractly (ie., theoretically) in the context of circular built environment. They need to be aligned more clearly with the practical concerns and characteristics of the socio-technical system of the built environment. Hence, it may be possible to use the large international pool of empirical examples of the CircularB community to explore how locality plays out in practicing circularity in the built environment, and bridge the current gap between the theory and practice. This would help developing systematic knowledge about how to approach, and work with, local actors and spaces, as well as revealing how such local aspects could be aligned along overarching circularity ambitions. Although limited to a critical discussion of the literature, this paper provides the basis for such an effort, and calls for further empirical work in this area.

References

1. Wuyts W, Marin J (2022) "Nobody" matters in circular landscapes. Local Environ 27(10–11):1254–1271
2. Munaro MR, Tavares SF, Bragança L (2020) Towards circular and more sustainable buildings: a systematic literature review on the circular economy in the built environment. J Clean Prod 260:121134
3. Bassens D, Keblowski W, Lambert D (2020) Placing cities in the circular economy: neoliberal urbanism or spaces of socio-ecological transition? Urban Geogr 1–5
4. Hubmann, G (2022) The socio-spatial effects of Circular Urban Systems. In: IOP conference series: earth and environmental science, vol 1078, no 1. IOP Publishing, p 012110
5. Hodson M, Marvin S (2010) Can cities shape socio-technical transitions and how would we know if they were? Res Policy 39(4):477–485
6. Oyinlola M, Whitehead T, Abuzeinab A, Ade A, Akinola Y, Anafi F, Farukh F, Jegede O, Kandan K, Kim B, Mosugu E (2018) Bottle house: a case study of transdisciplinary research for tackling global challenges. Habitat Int 79:18–29
7. Rahla KM, Mateus R, Bragança L (2021) Implementing circular economy strategies in buildings—from theory to practice. Appl Syst Innov 4(2):26
8. Stahel WR (2016) The circular economy. Nature 531(7595):435–438
9. Schiller G, Gruhler K, Ortlepp R (2017) Continuous material flow analysis approach for bulk nonmetallic mineral building materials applied to the German building sector. J Ind Ecol 21(3):673–688
10. Williams J (2020) The role of spatial planning in transitioning to circular urban development. Urban Geogr 41(6):915–919
11. Schiller G, Bimesmeier T, Pham AT (2020) Method for quantifying supply and demand of construction minerals in urban regions—a case study of Hanoi and its hinterland. Sustainability 12(11):4358
12. Casas-Arredondo M, Croxford B, Domenech T (2018) Material and decision flows in non-domestic building fit-outs. J Clean Prod 204:916–925

13. Keßler H, Knappe F (2013) Anthropogenic stock as a source of raw materials: optimized utilization of recycled building materials to conserve resources. In: Factor X. Eco-efficiency in industry and science, vol 30. Springer, Dordrecht, pp 187–202
14. Enviterra (2020) Improved Management of Construction and Demolition Waste (CDW) in Greece—Final Report [Online]. https://www.giz.de/en/downloads/Final%20Report%20CDW%20management%20EN.pdf
15. Robinson GR Jr, Menzie WD, Hyun H (2004) Recycling of construction debris as aggregate in the Mid-Atlantic Region, USA. Resour Conserv Recycl 42(3):275–294
16. Gotham KF (2003) Toward an understanding of the spatiality of urban poverty: the urban poor as spatial actors. Int J Urban Reg Res 27(3):723–737
17. Monstadt J, Coutard O (2019) Cities in an era of interfacing infrastructures: politics and spatialities of the urban nexus. Urban Stud 56(11):2191–2206
18. Cuomo F (2022) Urban living lab: an experimental co-production tool to foster the circular economy. Soc Sci 11(6):260
19. Gundlach K, Marlow F, Peters N, Wall R (2022) Model projects as tools for cooperative urban development: the case of Haus der Statistik in Berlin. In: IOP conference series: earth and environmental science, vol 1078, no 1. IOP Publishing, p 012110
20. Howard M, Böhm S, Eatherley D (2022) Systems resilience and SME multilevel challenges: a place-based conceptualization of the circular economy. J Bus Res 145:757–768
21. Harvey D (1978) The urban process under capitalism: a framework for analysis. Int J Urban Reg Res 2(1):101–131
22. Keblowski W, Lambert D, Bassens D (2020) Circular economy and the city: an urban political economy agenda. Cult Organ 26(2):142–158
23. Jassanoff S (ed) (2004) States of knowledge: the co-production of science and social order. Routledge, Abingdon
24. Norström AV, Cvitanovic C, Löf MF, West S, Wyborn C, Balvanera P, Bednarek AT, Bennett EM, Biggs R, de Bremond A, Campbell BM (2020) Principles for knowledge co-production in sustainability research. Nat Sustain 3(3):182–190
25. Polk M (ed) (2015) Co-producing knowledge for sustainable urban development: joining forces for change. Routledge, London
26. D'amato D, Korhonen-Kurki K, Lyytikainen V, Matthies BD, Horcea-Milcu AI (2022) Circular bioeconomy: actors and dynamics of knowledge co-production in Finland. For Policy Econ 144:102820
27. Fratini CF, Georg S, Jørgensen MS (2019) Exploring circular economy imaginaries in European cities: a research agenda for the governance of urban sustainability transitions. J Clean Prod 228:974–989
28. Broto VC, Ortiz C, Lipietz B, Osuteye E, Johnson C, Kombe W, Mtwangi-Limbumba T, Macías JC, Desmaison B, Hadny A, Kisembo T (2022) Co-production outcomes for urban equality: learning from different trajectories of citizens' involvement in urban change. Curr Res Environ Sustain 4:100179
29. Fratini CF, Jensen JS (2017) The role of place-specific dynamics in the destabilization of the danish water regime: an actor–network view on urban sustainability transitions. In: Urban sustainability transitions. Routledge, pp 86–105
30. Rizzo A, Ekelund B, Bergström, J, Ek K (2020). Participatory design as a tool to create resourceful communities in Sweden. In: Smaniotto Costa C, Maciulienc M, Menezes M, Marušic B. G (eds.) Co-Creation of Public Open Places, pp 95–107 C3Places, Lisboa
31. Ellen Macarthur Foundation (2017) Cities in the circular economy: an initial exploration. [Online]. https://www.ellenmacarthurfoundation.org/assets/downloads/publications/Cities-in-the-CE_An-Initial-Exploration.pdf
32. Rizzo A, Sordi J (2020) Resources and urbanization in the global periphery: perspectives from urban and landscape studies. Cities 100:102647
33. Williams J (2023) Circular cities: planning for circular development in European cities. Eur Plan Stud 31(1):14–35
34. Greater London Authority (2020) London Plan. City Hall, London

35. Stockholm County Council (2015) RUFS 2050—Programme for a New Regional Development Plan for the County of Stockholm
36. Tambovceva T, Bajare D, Tereshina MV, Titko J, Shvetsova I (2021) Awareness and attitude of Latvian construction companies towards sustainability and waste recycling. J Sib Fed Univ Humanit Soc Sci 14(7):942–955
37. Arora M, Raspall F, Cheah L, Silva A (2019) Residential building material stocks and component-level circularity: the case of Singapore. J Clean Prod 216:239–248
38. Dawkins E, André K, Axelsson K, Benoist L, Swartling AG, Persson Å (2019) Advancing sustainable consumption at the local government level: a literature review. J Clean Prod 231(10):1450–1462
39. Urban Innovative Actions: Lappeenranta circular economy. [Online] https://uia-initiative.eu/en/uia-cities/lappeenranta. Accessed 03 Apr 2023
40. Bolger K, Doyon A (2019) Circular cities: exploring local government strategies to facilitate a circular economy. Eur Plan Stud 27(11):2184–2205
41. Turcu C, Gillie H (2020). Governing the circular economy in the city: local planning practice in London. Plan Pract Res 35(1):62–85

Chapter 22
The Rehabilitation of Buildings from the Perspective of Circular Economy Principles

M. C. D. Freitas, S. F. Tavares, L. Bragança, and S. Barbosa

Abstract The Material Bank (MB) of the city of Porto (Portugal) is an emblematic case that operates in the design phase of the rehabilitation of buildings from the perspective of the circular economy, acting as intermediation of information as a strategic resource in favor of the creation of a network of partners for the preservation of the old quarter with its historical and architectural heritage of the city. In this context, the study analyzes the actors, expectations and values with an emphasis on the circular economy and mitigation of construction and demolition waste. It consists of an exploratory-descriptive research, from qualitative nature, which uses bibliographical and documentary research to describe and analyze the Constructive Materials Bank of Porto City Council as a public policy capable of acting in the transition of the circular economy, noting that information is the key element in the mechanisms of reputation and trust, in the implementation of building rehabilitation projects and in the relationship between citizens and the city. The MB is an innovative model of public policy in managing and providing building materials of historical and architectural value. The idea is to align the heritage and cultural value of buildings with the principles of the circular economy.

Keywords Collaborators network · Circular Economy · Architectural · Material bank

M. C. D. Freitas (✉) · S. F. Tavares
Universidade Federal Do Paraná, Curitiba, Brazil
e-mail: mcf@ufpr.br; lncs@springer.com

L. Bragança · S. Barbosa
Universidade Do Minho, Guimarães, Portugal

L. Bragança et al. (eds.), *Creating a Roadmap Towards Circularity in the Built Environment*, Springer Tracts in Civil Engineering,
https://doi.org/10.1007/978-3-031-45980-1_22

22.1 Introduction

Concerns about the future of the planet involve changing people's mindsets, the processes that stimulate these transformations and the way products and services are produced and used in their daily lives. Circular thinking applied to building materials will result in a more prosperous, livable and sustainable society, but it will require a radical rethink of all aspects of the production of the built environment. Circularity changes the very environment in which we live, work and invest [1].

Cities have a massive stock of existing buildings that are considered real banks of material resources and stimulating their rehabilitation can stimulate the creation of banks of building materials, acting to reduce the generation of construction demolition waste. The adaptations of existing buildings help to preserve natural resources and extend the life cycle of buildings [2]. Data from the European Union show that 85% of the existing building stock (more than 220 million) has its construction dated before the year 2001, and the majority (between 85 and 95%) of the buildings standing will still exist in the year 2050 [3].

Circular economy actions can deliver up to 60% reductions in material-related greenhouse gas emissions over the material-related greenhouse gas emissions over the life cycle of buildings [3].

In Circular Economy models, end-of-life building materials should be reused and their components and parts deconstructed, to act as material banks for new buildings, keeping the components and materials in a closed circuit [4]. Furthermore, buildings produce a quarter of the world's waste [4].

This research is part of a network project called "Circular Economy as a Strategy for a More Sustainable Construction Industry—EcoEico" (portuguese acronym) which, among other goals, seeks the "development of strategies for the implementation of the Circular Economy, aiming to encourage the development of an industry of more sustainable construction, especially in Ibero-American countries".

City councils in many cities provide information services to citizens with a view to introducing the principles of the Circular Economy. The old quarter of the city of Porto in Portugal is on the list of UNESCO World Heritage cities [5]. This fact makes it an intermediary of information for the preservation of the old quarter with its historical and architectural heritage. A good part of these activities are delegated to the attributions and competences of the Municipal Board of Culture and Heritage through a more specific body, called the Material Bank.

This research is motivated by understanding the Material Bank as a unit that operates in the design phase of rehabilitation of buildings from the perspective of the circular economy. In view of this, the idea of identifying with whom, about "what" and "how" the Material Bank engages with partners to implement its actions to disseminate CE principles. The discussion through the case study intends to prospect these partners, the stakeholders, and the strategy for creating and/or encouraging a "culture" in the construction industry, based on the principles of circular economy and sustainability. As limitations encountered during the preparation of the article, we found that there is no quantitative data in the case study (MB) to allow justify the volume of materials that no longer go/went to replacement.

22.2 Background

22.2.1 Circularity of Materials in Building Rehabilitation

The verticalization of city centers to maximize and take advantage of the built area promoted a culture among investors and builders of demolishing to rebuild new buildings. Discussions related to reuse are joined by concerns about sustainability and the scarcity of raw materials, which raise the concern to stimulate the reuse of construction and demolition waste [6] on a large scale, reusing it in industrial areas, housing complexes, roads and shoots among others. Such actions, with an emphasis on the Circular Economy, bring together the different actors in the construction sector, such as: governments, builders, engineers and customers.

Materials in buildings should act as material banks as they are valuable products. This can be done using smart design and circular value chains, which is crucial for an industry to reduce its waste and the amount of virgin resources used. Therefore, a circular economy describes an economic system based on new business models which replace the perception of "end of life" concept by reducing, reusing, recycling and recovering materials in production/distribution and consumption processes with the aim to accomplish sustainable development, which implies creating environmental quality, economic prosperity and social equity, to the benefit of current and future generations [7]. The transition to circular construction involves changes in value chains, from product design to new markets, from new models of consumer behavior to new ways of turning waste into resources. Knowledge and experience about reuse and reuse options for construction products and materials is very limited. The reuse of materials may require minimal processing before their reapplication in a similar context [8]. It should be added that the constructive environment lacks information regarding the type and quantity of materials used in buildings with a view to potentially increasing circularity at the end of the life cycle, which culminates in the proposals for material passports [9].

In the case of cities and historic buildings, digital inventories emerge as a heritage management strategy, expanding with identification data; research and analysis of historical value; asset impact assessment; planning research activities, conservation and handling of materials used; and provide information to the public on architectural and historical value, awareness actions aimed at authorities and decision makers [10]. These inventories are essential informational tools to make decisions, apply laws and policies related to heritage [10]. Investigations on the subject point out: absence of circular economy policies at building level—technical and social, need for more coordinated actions and policies to encourage circularity—laws and regulations, lack of political actions promoting recycling, reuse and new models of public and private businesses—fiscal and economic incentives [11], lack of assessment of the environmental sustainability of circularity strategies—environmental and cultural and of enabling tools to promote circular supply chains and new production models—technology and economy [12–14].

22.2.2 *Strategy and Projects for Circular Economy Within Cities*

The importance of the circular economy (CE) leads to the stimulation of countless action plans applied in the creation of informational resources to manage and minimize the generation of waste in cities, increasing the efficiency and environmental performance of buildings. The transition from the linear economy to the circular economy within cities and institutions that provide services to citizens requires information and government policies for regulation and reduction in the taxation of materials or for the rehabilitation of existing buildings. For example, in the UK, taxes on new buildings are increased by up to 20%. On the other hand, the retrofit of the building is exempt from Value Added Tax (VAT). In Sweden, VAT rates are lower for renovations aimed at preserving unity. The Dutch government encourages the formation of partnerships to promote the design of new products inspired by circular business, through the program Realization of Acceleration of a Circular Economy—RACE [1].

The Building as Material Banks (BAMB) consortium has gathered some actions that provide information on "buildings as material banks" on its website, the challenges and perceptions regarding the circular economy in the built environment. In Denmark, a step-by-step tool has been developed that describes how to make the transition to CE and explains it in a local case study [15].

Another initiative is Circular Flanders as a Flemish government service responsible for waste, material and soil decontamination policies. This action aims to provide a hub and inspiration for the circular economy in Flanders. It is built from a partnership of governments, companies, civil society and knowledge institutes that act together. To this end, it promotes public tenders for city projects, businesses and circular purchases, launched the Green Deal 'Circular Construction' which brings together organizations that support circular economy actions in the built environment and Living Labs on 'Circular Construction' to support circular transition policies sectors either by experimentation or applied research [16]. The CityLoops project focuses on waste streams in Europe: Construction and Demolition Waste and Biowaste, and brings together seven small and medium-sized European cities—Apeldoorn, Bodø, Mikkeli, Porto, Seville, Høje-Taastrup and Roskilde. The idea of making cities circular is to eliminate all types of waste. The project proposes a digital material passport platform that aims to ensure that materials are recovered and traceable from demolition to their new uses. Passports store data such as a material's history, quantity, quality, estimated value, location, required maintenance, reuse or recycling potential, and instructions for disassembly and storage [17].

Another European Commission initiative gathered 71 practices in the Culture for Cities and Regions Project document, one of which is the Material Bank: capitalizing on architectural heritage [18]. The BAMB project stimulated the increase/sustaining of value in buildings through six tools: Material Passport, Reverse Building Design, Circular Building Assessment, Circular Business Models, Policies and Standards, and Pilot Case Studies [19]. These tools were underpinned by information and variety of stakeholders within the construction value network. It should be noted that the quality of this information results from the task of collecting, manipulating and exchanging digital data among stakeholders. But after all, which actors dialogue with the construction sector to promote the circular economy?

22.2.3 Collaboration Network in Favor of the Circular Economy

The aforementioned initiatives lead to the conclusion that a built environment, circular and reversible, can only be supported by a network of interconnected actors, that is, when relevant public and private companies in the construction industry work together, it becomes possible to create strategies to promote of the economy. Actions to reduce, reuse and recycle building materials are a temporary way to materialize investments and expand opportunities for exploring circular business models, in which economic and environmental value is conserved. It is noteworthy that the government is an active actor that helps with tax incentives that require low resource input and that facilitate the circulation of materials in the sector and accelerate the socialization and education of citizens and other actors in the adoption of the circular economy.

Perception/awareness about the CE concept, practices and transition, as well as managing this process, requires the creation of a network of collaborators that includes researchers, academics, policy makers, companies, workers, employees, consumers and others [20]. Organizational actors and society, in general, agree with the implementation of institutional and/or governmental guidelines as instruments to encourage CE in the construction, renovation and use of housing [21].

In the opinion of survey respondents [21], the most valuable actors to drive sustainability in housing are funders, followed by inspectors and government interventions. This reveals the opportunity for governments to exercise their legal powers to implement measures aimed at implementing these three types of instruments (financial, fiscal and government interventions) and their interest in doing so. These policies involve decisions at different levels of government—local and/or central—, which must act in a coordinated way in their measures to promote sustainable construction. On the other hand, in terms of government intervention, public policies should be based on priority, on technical support mechanisms, on the implementation of support tools for housing projects, on facilitating access to databases and on granting subsidies for finance public services [21].

22.3 Methodology

Research is exploratory-descriptive, of a qualitative nature, in which bibliographical and documentary recovery was used as a technique to describe and analyze the Materials Bank of the city of Porto. In this way, we sought to identify elements concerning the strategic function of the information that describes and analyzes the Bank of Constructive Materials as a public policy that acts in the promotion of CE and identifies the interested parties. The investigation explores partner and collaborator engagement practices that facilitate the dissemination of CE principles [22].

22.4 Material Bank of the City of Porto

The World Heritage List by the United Nations Educational, Scientific and Cultural Organization (UNESCO) brings together cities and the set of properties of Outstanding Universal Value (OUV) that it highlights as a cultural and/or natural value that transcends borders with a focus on current and future generations of all mankind. Climate changes caused by the intensive use of raw materials call for public policies to value processed products and reduce the scale of materials [5].

Historic cities such as Porto—being a UNESCO heritage site since 2001—need policies for the rehabilitation of their buildings, so that future generations can understand their origins and learn to value the achievements that are part of the strategic objectives of the Committee of the World Heritage: Credibility of the World Heritage List; Effective conservation of World Heritage properties; Effective competences in the States Parties; public awareness, participation and support for World Heritage through Communication; and Communities in the implementation of the World Heritage Convention [5] (Fig. 22.1).

The actions focused on the preservation of the OUV building act on the reduction of construction demolition waste. At the moment, it brings together professionals from architecture, engineering, history, archeology and restoration, in projects of national and international interest (organizational, institutional collaborators and society). UNESCO guidelines and the attributes present in non-monumental buildings—residential housing—also contribute to the OUV of Architectural Good: constructive systems—of buildings, circumscribed by walls, based on foundations and walls of local stone, with floors in wooden beams, in some cases resorting to structures made with a lattice on wooden beams; and materials and techniques: and landscaping of the set using materials and construction techniques, with emphasis on the worked stonework, the sloping, coated and ceramic roofs, wooden window

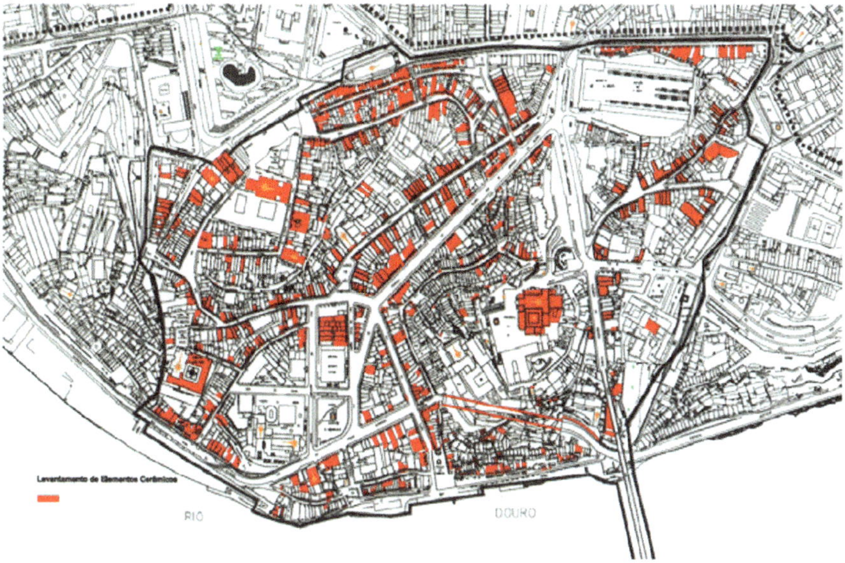

Fig. 22.1 Distribution map of ceramic elements in CHP/PM–2007 [−5]

frames, wrought iron elements, and the tile heritage (in the cladding facades and inside), skylights, fresh air and ceilings.

The Material Bank is a unit subordinated to the Municipal Board of Culture and Heritage and its principle is to preserve Porto's buildings, with their historic tiles—Hispano-Arabic—from patterns from the 17th to the twentieth centuries, such as: the panels in cobalt blue from the nineteenth century, artifacts of stonework, iron, wood and toponymic plaques collected on the public road. It should be added that it acts in the custody, preservation, restoration and return of building materials to the city in the rehabilitation of facades with the same topology regulated on December 16, 1999, by Decree-Law 555/99, as a public policy and "mechanism of protection of the tile heritage of the city of Porto".

The MB, by playing the role of preserving the architectural and historical heritage, participates as an important factor in intermediating the rehabilitation of buildings that favors project management and the relationship with employees. Citizen service is an activity that requires interactions with the actors in the environment and all the groups that are around them, who are implicitly or explicitly interested in their results, or tasks (technicians, architects, engineers, builders, among others). As a public institution, the unit offers services to the citizens and, therefore, any and all actors or groups that have an interest and influence in the processes of rehabilitation of the historical, architectural and cultural heritage of the city of Porto.

MB employees are actively involved in mapping the architectural and cultural heritage of historic buildings and works that are part of the city's memory and identity, recording, organizing and archiving the constructive elements that tell the city's history, providing guidance to professionals in the architecture and construction engineering such as lectures and a workshop to support the restoration project of historic buildings, recovery of decorative objects characteristic of Porto architecture (tiles, iron and stucco, etc.), management and physical and documentary storage of construction elements, information for tourists, young people, children and citizens on the heritage and architectural value of the materials that characterize the city's image, management in the donation of recovered elements free of charge to residents, among others. The information and building materials on the storage and management of the unit make it possible to generate indicators for the reduction of construction demolition waste, with an emphasis on reuse, reduction and recycling as one of the first fruits of the circular economy. Furthermore, this information makes it possible to create value in the buildings where the materials are used.

This study will consider the MB as an organizational actor—even though it is an institutional actor—that brings together a network of partners and/or collaborators in a mediation relationship in all services. Reasons that motivate the evaluation of organizational management, while seeking to identify the specific needs or benefits of these actors, respecting the actors (Porto City Council), other government bodies, society—students, professionals, universities, researchers, tourists, etc.—, and competing companies.

Table 22.1 presents an inventory of who are the main collaborators or potential partners, whether they are individual actors or groups, with a positive or negative opinion due to their particular interests. The public managers of the city of Porto, as employees of the material bank, take care of the materials that beautify the city. Investments must be raised with institutions for the management of documentation in the generation of a passport of matters of buildings in historic cities, allowing in the future the monitoring and exchange of building materials between citizens and perhaps cities. It appears that the team encourages design for disassembly and recyclability of materials removed from housing, whether reintegrated into the same building or repositioned at other points.

22.5 Conclusion

The MB works with an emphasis on preserving the historical and cultural heritage of the city of Porto, but the focus of this investigation is how these actions approach the citizen's awareness of the principles of the circular economy. It is important to highlight the alignment of policies, strategies and projects with the search for circularity in construction materials. The MB faces barriers with internal and external actors involved in the construction value chain. The network of MB collaborators has the support of public managers, inspectors, entrepreneurs, builders and citizens who work to expand actions that favor the circularity of buildings and prolong the life cycle of construction materials.

Table 22.1 shows who is part of the network of MB collaborators and the perception of this investigation regarding the expectation and the desired value when involving these partners to implement the actions of the EC principles associated

Table 22.1 The actors, expectations and desired value in relation to the circular economy

Actors		Expectation	Desired value
Organizational Material Bank	Public Managers Collaborators Internal partners	– Preservation of the historic and architectural heritage of the city of Porto – Encouragement to reduce, reuse and recycle constructive materials – Planning to comply with public policies related to the unit	– Beautification of the city and generation of income with rehabilitation of buildings – Attenuation of construction demolition waste – Competitive intelligence on good circular economy practices
Institutional Porto City Council	Government Partners Financiers	– Support for rehabilitation projects that encourage reduction, reuse or recycling, increasing the life cycle and adding value to building materials – Undertakings for Historical and Architectural Inspiration of cities – Structuring based on digital technology	– Promotion and generation of citizen, sociocultural and circular economy knowledge – Good knowledge management practices to guide decision-making by project – Shared management—speeding up retrieval of information and indicators

(continued)

Table 22.1 (continued)

Actors		Expectation	Desired value
Society	Universities, Environmentalists and Partners Builders and Real Estate Engineers and Architects Tourists—Historians and Students Municipalities	– Technical information that guides the suitability of the building with a focus on the 3Rs – Dissemination and encouragement of historical and architectural culture – Dissemination with professionals and academics in university training – Improving assets by meeting tax obligations	– Storage of technical knowledge and experiences – Educational and training practices focused on rethinking the use of construction materials – Assistance and guidance on good practices with a focus on the 9Rs (Refuse, Rethink, Reduce, Reuse, Repair, Refurbish, Remanufacture, Repurpose, Recycle and Recover) – Increase public and private participation
Contestants	Competitors and Detractors	– Brand Management in contrast to companies with similar nomenclature	– Highlight for competence, ethics and social responsibility

with the historical and cultural heritage value. The study leads to the assimilation that the MB has a lot of information about materials and systematized projects of interests of the Municipality of Porto (Skills and attributions). It notes that information is the key element in the mechanisms of reputation and trust (Credibility), in the implementation of building rehabilitation projects (Conservation) and in the relationship between citizens and the city (Community). Although, the discussion on the heritage and cultural value of building rehabilitation promotes the principles of the circular economy (Communication—public awareness). MB action also furthers the World Heritage Committee's strategic objectives. All of this, even in the face of the challenges faced by the lack of awareness on the part of builders.

It should also be noted that the MB team offers training workshops for professionals, students and citizens with a focus on tile heritage and on the identity building elements of historic buildings, which also encourages CE. It is concluded that the MB indirectly acts in the transition from linear to circular economy, involving citizens and beautifying and preserving the city, establishing partnerships with other internal actors of the City Council and the citizens. The MB is an innovative model of public policy by managing and providing construction materials of historical and architectural value, which modifies citizens' behavior regarding the importance of rehabilitation and the application of the principles of reducing, reusing, recycling as a decision strategy in the adoption of construction methods with secondary materials arises based on the perspective that material resources are finite.

The expectation is to adapt these actions to local contexts related to the Circular Economy in the construction sector for Latin American countries, considering their social, economic and cultural differences in different realities.

For future articles, we propose research on two axes of analysis: 1—analyze the existing laws and regulations for the treatment of solid waste (construction and demolition waste) and the entities responsible for verifying their compliance; 2—Evaluate, through LCAs or ecological footprints, the evaluation of the environmental impacts mitigated by the actions of the Material Bank.

Acknowledgements For supporting this research by the Municipal Board of Culture and Heritage, in particular Paula Cristina Alves Lage—from the Materials Bank of Porto City Council—Portugal.

References

1. ARUP (2016) The circular economy in the built environment. London. https://www.arup.com/publications/research/section/circular-economy-in-the-built-environment. Accessed 17 Jan 2023
2. Shahi S, Esfahani ME, Bachmann C, Haas C (2020) A definition framework for building adaptation projects. Sustain Cities Soc 63. https://doi.org/10.1016/j.scs.2020.102345
3. Comissão Europeia (2020) Impulsionar uma Vaga de Renovação na Europa para tornar os edifícios mais ecológicos, criar emprego e melhorar as condições de vida, Bruxelas
4. Benachio GLF, Freitas MCD, Tavares SF (2020) Circular economy in the construction industry: a systematic literature review, 260:121046
5. Porto City Council (2021) Historic Centre of Porto, Luiz I Bridge and Monastery of Serra do Pilar. Porto, 274p
6. Tam VWY, Lu W (2016) Construction waste management profiles, practices, and performance: a cross-jurisdictional analysis in four countries. Sustainability 8(2):190–206
7. Kirchherr J, Reike D, Hekkert M (2017) Conceptualizing the circular economy: an analysis of 114 definitions. Resour Conserv Recyc 127:221–232
8. Ghaffar, Hamidreza S, Burman M, Braimah N (2020) Pathways to circular construction: an integrated management of construction and demolition waste for resource recovery. J Clean Prod 244:118710
9. Munaro MR, Tavares SF (2021) Materials passport's review: challenges and opportunities toward a circular economy building sector. Built Environ Proj Asset Manag 11(4):767–782
10. Santos B, Gonçalves J, Martins AMT, Almeida PG (2020) Safeguarding Portuguese traditional glazed tile cultural heritage with GIS. In: IOP conference series: materials science and engineering. IOP Publishing Ltd.
11. Bianchini A, Rossi A (2021) Design, implementation and assessment of a more sustainable model to manage plastic waste at sport events J Clean Prod 281
12. Giorgi S, Lavagna M, Wang K, Osmani M, Liu G, Campioli A (2022) Drivers and barriers towards circular economy in the building sector: stakeholder interviews and analysis of five European countries policies and practices. J Clean Prod 336. https://doi.org/10.1016/j.jclepro.2022.130395.J
13. Garcia-Quevedo J, Jové-Llopis E, Martinez-Ros E (2020) Barriers to the circular economy in European small and medium-sized firms. Bus Strat Environ 29(3)
14. Bressanelli G, Perona M, Saccani N (2019) Challenges in supply chain redesign for the circular economy: a literature review and a multiple case study Int. J Prod Res 57(23):7395–7422. https://doi.org/10.1080/00207543.2018.1542176

15. BAMB, Building as Material Banks. About BAMB. 2018. https://www.bamb2020.eu/about-bamb/. Accessed 2 Feb 2023
16. Chang RD, Zuo J, Zhao ZY, Zillante G, Gan XL, Soebarto V (2017) Evolving theories of sustainability and firms: history, future directions and implications for renewable energy research. Renew Sustain Energy Rev 72:48–56
17. ECESP—European Circular Economy Stakeholder Platform. Circular buildings and infrastructure: state of play report ECESP leadership groupon buildings and infrastructure, 2021. https://circulareconomy.europa.eu/platform/sites/default/files/circular_buildings_and_infrastructure_brochure.pdf. Accessed 18 Jan 2023
18. Eurocities (2015) The culture for cities and regions team. Case Study. https://epale.ec.europa.eu/sites/default/files/catalogue_03122015.pdf. Accessed 18 Jan 2023
19. Sharp J, Hobbs G, Henrotay C, Steinlage M, Debacker W, Regel S, Sjögren S (2023) Framework for policies, regulation and standards. Build Material Banks—BAMB (2019). https://www.bamb2020.eu/. Accessed 2 Feb 2023
20. van Langen SK, Vassillo C, Ghisellini P, Restaino D, Passaro R, Ulgiati S (2021) Promoting circular economy transition: a study about perceptions and awareness by different stakeholders' groups. J Clean Prod 316 (2021). https://doi.org/10.1016/j.jclepro.2021.128166.
21. Díaz-López C, Carpio M, Martín-Morales M, Zamorano M (2021) Defining strategies to adopt Level(s) for bringing buildings into the circular economy. A case study of Spain. J Clean Prod 287. https://doi.org/10.1016/j.jclepro.2020.125048
22. Gil A (2002) Como Elaborar Projetos de Pesquisa. 4ª Edição. Editora Atlas S.A. São Paulo

Part XI
CE Supporting Policies and Legal Barriers for Implementation in Buildings

Chapter 23
Circular Economy Supporting Policies and Regulations: The Portuguese Case

Marco Frazão Pedroso and Vanessa Tavares

Abstract Construction is one of the critical sectors in the transition to a Circular Economy due to its contribution to resource depletion, waste, and emissions. Despite its acknowledged limitations (e.g., low productivity), the construction sector has been the focus of policies and regulations to improve its sustainability and circular economy capabilities. This study focuses on circular economy policies and regulations related to the construction sector in the European Union and Portugal, identifying political and regulatory barriers and opportunities. The analysis identified a growing number of publications since 2019 and divided the policies and regulations for the Circular Economy into four areas: Resource and Waste Management, Sustainable Development Goals, Green Public Procurement, and Circular Economy. Four main barriers were identified: policies and regulations harmonization, digital innovation (within the twin transition), support to the transition (e.g. financial and educational), and clear and focused governance models. It then discusses the documents and barriers, analyzes the Portuguese strategy (Portuguese Action Plan for Circularity in the Construction Sector), and proposes a strategy to be followed by other countries. The findings provide a holistic understanding of why policies and regulations fail to support Circular Economy day-to-day practices and provide insights on how to trigger the transition.

Keywords Construction sector · Circular Economy · Policies and regulations

M. F. Pedroso (✉) · V. Tavares
BUILT CoLAB, 4150-003 Porto, Portugal
e-mail: marco-pedroso@sapo.pt; marco.pedroso@builtcolab.pt

V. Tavares
e-mail: vanessa.tavares@builtcolab.pt

M. F. Pedroso
CERIS, IST, Universidade de Lisboa, Av. Rovisco Pais 1, 1049-001 Lisboa, Portugal

V. Tavares
FAUP, CEAU, Universidade do Porto, Via Panorâmica S/N, 4150-564 Porto, Portugal

L. Bragança et al. (eds.), *Creating a Roadmap Towards Circularity in the Built Environment*, Springer Tracts in Civil Engineering,
https://doi.org/10.1007/978-3-031-45980-1_23

23.1 Introduction

Worldwide, the Construction Sector (CS) has been following a linear production and consumption model—based on a "*take/make/dispose*" approach—with the main focus on the economic dimension [1]. When considering the European built environment, the CS is responsible for approximately 50% of natural raw material consumption, 30% of the waste generated, and 30% of greenhouse gas (GHG) emissions [2]. This inefficient approach continues to contribute to the depletion of non-renewable virgin resources, the generation of GHG emissions and waste, and the loss of biodiversity, colliding with the Sustainable Development Goals (SDG) [3].

The transition to a circular CS is imperative to overcome these burdens and contribute to waste and pollution elimination, avoid raw material depletion, and nature regeneration goals [4]. However, the resistance to change, the CS complexity (e.g., long life span, different materials, and numerous stakeholders), and the lack of productivity associated with the resistance to adopting digitization (proven to boost productivity and efficiency [5]), are significant barriers to the transition to a Circular Economy (CE)—based on a "*return/(re)make/(re)use*" approach [6–8]. This context led the European Commission to draw policies and regulations framing CE objectives and goals to support this paradigm shift [9–12], after being translated by different stakeholders (mainly industry leaders and policymakers) in each member-state. The growing literature focuses on the barriers and opportunities for CE adoption in the CS [8, 13, 14], and policies and regulations framing CE objectives through the slowing, narrowing, and closing of material and energy loops. However, these studies [7, 15–18] present a broad scope, simultaneously presenting distinct realities (countries) and neglecting each country's specificities: stakeholder needs, industry leaders, and policymakers.

Focusing on Portugal, this study reviews the EU and Portuguese policies and regulations and identifies the perceived barriers and opportunities. Considering the CS, three main objectives were recognized: *(i)* mapping and relating the main policies and regulations for adopting a CE in the EU and Portugal; *(ii)* identifying the barriers and opportunities; and *(iii)* discussing possible strategies to overcome the barriers to CE adoption. As a result, this study aims to provide an integrated understanding of the main causes of the unsuccessful transition to CE in the CS, namely between the existing policies and regulations and the identified barriers and opportunities. The paper is organized into four sections: Sects. 23.1, 23.2, 23.3 and 23.4.

23.2 Research Methodology

23.2.1 Research Strategy

This research strategy includes two complementary approaches: *(i)* documental research to identify the circular economy (CE) main policies and regulations published by the EU and Portugal; *(ii)* working sessions with Portuguese construction industry stakeholders to recognize policy and regulation barriers and opportunities in CE adoption.

First, the document research categorized and characterized the policies and regulations published by the EU and their influence on the Portuguese context. To draw the regulatory framework, the websites of European and Portuguese organs responsible for policies and regulations were mapped to evaluatethe adoption of a CE.

Afterwards, the working sessions with the main stakeholders allowed to identify, prioritise, and categorise the critical barriers and opportunities felt by Portuguese stakeholders in the Construction Sector (CS) in the CE adoption within policies and regulations. This research strategy was based on design thinking and discussion sessions in which all stakeholders were represented, as discussed in Tavares and Pedroso [19].

23.2.2 Data Collection

Data collection was undertaken between January 2021 and November 2022, considering only the CE documents influencing the CS. The document research started by querying the following sources: European Commission Environment [20], EUR-Lex [21], Journal of the Republic (*Diário da República*) [22], Portuguese Environment Agency (APA) [23] and the National Laboratory for Civil Engineering (LNEC) [24] using the following keywords (also in Portuguese), isolated or combined with synonyms and acronyms: '*Circular Economy*', '*barriers*', '*opportunities*', '*Construction and Demolition Waste*', '*Green Public Procurement*', '*resources*', '*Sustainable Development Goals*', '*sustainable*', and '*decarbonization*'. Although the EU showed concerns about raw materials depletion for many years, the document "Towards a circular economy" [25] was only published in 2014 with the clear objective of adopting CE across the EU and economic sectors. Therefore, 2014 was considered the most relevant starting point.

The sessions, further discussed by Tavares and Pedroso [19], were possible due to the project that funded the execution of the "*Portuguese Action Plan for Circularity in the Construction Sector*" [26]. These eleven sessions took place between January 2021 and November 2022, allowing to: *(i)* identify barriers, *(ii)* prioritize those barriers and *(iii)* identify opportunities. Altogether, the sessions [27] involved more than 800 Portuguese individuals connected to the CS, with the following distribution: Academia and research centres (10%), Contractors and builders (16%), Demolition and CDW management teams (3%), Manufacturers (6%), Government and regulators (14%), Investors, developers, and insurance providers (4%), Designers (17%), Users and owners (30%).

23.2.3 *Data Analysis*

During documental research, relevant data were extracted from each document into an Excel matrix, including the document name, type, publication date, sectorial scope, and a summary of its objectives and goals. This information was analysed, and documents were removed when redundant, old (before 2014), or had limited influence on the CS.

The identified barriers and opportunities were then divided and organized according to priorities into different pillars, following the methodology presented by Tavares and Pedroso [19] and others [8, 28]. This paper focuses on Policies and Regulations.

23.3 Results and Discussion

23.3.1 *Policies and Regulations Linked to CE in the EU and Portugal*

The EU Framework

Table 23.1 presents the results obtained for the EU document search in terms of the main policies and regulations regarding the adoption of circular economy (CE) with a focus on the construction sector (CS), since 2014.

Although CE started to be more relevant in 2014, most CE-related documents were published in 2020, due to the publication of "*The European Green Deal*" in 2019, considered to be the document which started this transition.

After "*The European Green Deal*", several aspects were addressed through different instruments such as green financing (e.g., EU Taxonomy) and major policies (e.g., Renovation wave and NEB) promoting the adoption of sustainability and CE goals in the CS. The most recent documents promote the CS twin transition (green and digital) aiming at increasing productivity, efficiency, decarbonization and circularity of the built environment, as in the "*Circular Economy—Principles for buildings design*" [12]. The growing number of documents published in the last four years shows the importance of CE adoption. When clustered, these documents follow four main topics: Resource and Waste Management (RWM), Sustainable Development Goals (SDG), Circular Economy (CE), and Green Public Procurement (GPP).

Table 23.1 Policies and regulations on CE in the EU (since 2014)

Name I type I date I scope	Summary
Towards a circular economy (COM/2014/0398 final) I Communication I 2014 I CC ⭘	Reduce carbon emissions, increasing energy efficiency, sustainable reindustrialisation of economy, and securing access to raw materials, whilst reducing environmental impacts and GHG emissions. Tackle Construction and Demolition Waste (CDW) challenges and use the "Resource efficiency opportunities in the building sector"
Resource efficiency opportunities in the building sector I Communication I 2014 I CS •	Promote efficient use of resources consumed by new and renovated commercial, residential, and public buildings and reduce their overall environmental impacts throughout the full life cycle
EU action plan for the CE (COM/2015/0614 final) I Communication I 2015 I CC ⭘	Ensure recovery of valuable resources and adequate waste management in the construction and demolition sector and assess the environmental performance of buildings. Develop a monitoring framework for the CE
Green Public Procurement—Buying green! I Book I 2016 I CC ⭘	Help public authorities buy goods and services to lower environmental impact which can also have benefits in terms of CE. Identifies office buildings and linear infrastructures, providing guides for them
A renewed EU Industrial Strategy (COM/2017/0479 final) I Communication I 2017 I CC ⭘	Guidelines to strengthen the industry's ability to adapt and innovate by facilitating investment in new technologies and embracing changes brought on by digitisation. Companies must upgrade the technology base, future-proofing business models, internalise sustainability principles and embrace innovation
Directive (EU) 2018/849 to 852 (PE/9 to 12/2018/REV/1) *All from 2018 I CC ⭘	Guidelines to improve waste management in the EU, to protect, preserve, and improve the quality of the environment, protect human health, ensure prudent, efficient, and rational utilisation of natural resources, promote the principles of the CE, and reducing the dependence on imported resources
Measuring progress towards CE—(COM(2018)29 final) I Communication I 2018 I CC ⭘	Establishes the framework to measure progress and assess the effectiveness of actions towards CE in the EU. Concerns towards the construction sector and the raw materials consumed and waste generated. The recovery rate of construction and demolition waste: mandatory with a recovery target (70% by 2020) under the Waste Framework Directive (2008/98/EC)

(continued)

Table 23.1 (continued)

Name I type I date I scope	Summary
The European Green Deal (COM(2019) 640 final) I Communication I 2019 I CC ⭘	Need for a 'renovation wave' of public and private buildings; enforces the energy performance of buildings; reviews the CPR; ensures that the design of buildings is in line with CE, and lead to increased digitalisation of the building stock. GHG emission targets to at least 50% and towards 55% (by 2030, considering 1990) and climate neutrality by 2050
Sustainable Europe Investment Plan I Plan I 2020 I CC ⭘	The investment pillar of the European Green Deal to achieve its goals. The construction industry referred to contribute to climate neutrality in 2050
New Industrial Strategy (COM/2020/102 final) I Communication I 2020 I CC ⭘	Address the sustainability of construction products and improve the energy efficiency and environmental performance of built assets. The world's first climate-neutral continent by 2050, and creating 700,000 new jobs across the EU by 2030
Taxonomy Regulation (PE/20/2020/INIT) I Regulation I 2020 I CC ⭘	Promotes the financing of the increase in the durability, reparability, upgradability, and reusability of products, but also the reduction of the use of resources through the design and choice of materials, facilitating repurposing, disassembly and deconstruction in the buildings and construction sector
Level(s) I Framework I 2020 I CS •	A common language for the sustainability performance of buildings. Entry point for applying CE principles in the built environment (16 indicators), promoting, among others, the reduction of GHG emissions, raw materials consumption, and CDW generated
A new Circular Economy Action Plan (COM/2020/98 final) I Communication I 2020 I CC ⭘	Promotes CE principles throughout the lifecycle of buildings by: sustainability of construction products (revision of CPR); improving durability and adaptability; developing digital logbooks; integrating LCA in public procurement and the EU sustainable finance; revise material recovery targets for CDW; reduce waste and raw materials consumption; achieve the 2030 SDG and carbon neutrality in 2050
CE—Principles for Building Design I Report I 2020 I CS •	To inform and support stakeholders along the construction value chain about the principles for the circular design of buildings

(continued)

Table 23.1 (continued)

Name I type I date I scope	Summary
A Renovation Wave for Europe (COM/2020/662 final) I Communication I 2020 I CS •	Decarbonisation and integration of renewables; Lifecycle thinking and circularity; High health and environmental standards; Twin challenges of the green and digital transitions. In the construction sector, double the annual energy renovation rate (by 2030) and foster deep energy renovations
New European Bauhaus (COM(2021) 573 final) I Communication I 2021 I CS •	Places, practices, and experiences that are: Enriching; Sustainable, in harmony with nature, the environment, and our planet; Inclusive; Provide citizens access to goods and constructions that are circular and less carbon-intensive, supporting regeneration and biodiversity
Proposed Directive on the energy performance of buildings (COM/2021/802 final) I Communication I 2021 I CS •	Vision for achieving a zero-emission building stock by 2050, focusing on energy efficiency and minimising whole life-cycle GHG emissions of buildings through resource efficiency and circularity. Require calculating the life-cycle Global Warming Potential (GWP) of new buildings. Prevention and high-quality treatment of CDW
Revision of GPP I tba I tba I CS •	Scope expansion to other buildings, in particular schools and social housing. Criteria proposal in line with recent policy developments relating to the Renovation Wave, the Level(s) framework, and the EU Taxonomy
Revision of the Construction Product Regulation (CPR) I tba I tba I CS •	Two objectives: (1) achieve a well-functioning single market for construction products; (2) contribute to the objectives of the green and digital transition
Ecodesign for Sustainable Products Regulation I tba I tba I CS •	Wide range of requirements, including: product durability, reusability, upgradability, and reparability; presence of substances that inhibit circularity energy and resource efficiency; recycled and remanufactured content; carbon and environmental footprints; Digital Product Passport

Caption CC **o**—cross-cutting considering the CS; CS •—dedicated to the CS

The Portuguese Framework

Table 23.2 presents Policies and regulations on CE in Portugal (since 2016). Although Portugal showed concerns towards sustainability since 2006 with Decree-Law 178/2006 (revoked by Decree-Law 102-D/2020) on waste management, only from 2016

Table 23.2 Policies and regulations on CE in Portugal (since 2016)

Name I type I date I scope	Summary
New strategy to Green Public Procurement (Resolution 38/2016) I Resolution I under revision I CC ○	Encourages a Green Public Procurement (GPP) policy for office buildings and linear infrastructures. Public Procurement budget contemplates environmental criteria until 2020, 40% rate in the business sector); and a 60% rate in direct and indirect state administration
Material specifications by the National Laboratory for Civil Engineering (LNEC) I Specification I 2016 I CS •	Establishment of the minimum requirements that a given material must comply with to be used (recycled). Increase the use of recycling and recycled materials in the Portuguese construction industry
Portuguese Circular Economy Action Plan (Resolution 190-A/2017) I Resolution I under revision I CC ○	Increase the introduction of secondary raw materials into the economy. Reducing: waste production; the demand for raw (primary) materials; the emission of GHG; water consumption. Additionally, presents a framework to create an action plan for the construction industry
Legal Regime for Urbanization and Edification (DL 555/99 revision 118/2019) I Decree-Law I 2019 I CS •	It lacks further references to CE principles in the construction sector but addresses the need to treat CDW
Roadmap for Carbon Neutrality 2050 (Resolution 107/2019) I Resolution I 2019 I CC ○	Establishes the trajectory to achieve carbon neutrality by 2050. CS should adopt CDW recycling, use natural materials, and lower the energy consumption. Compensation of emissions through land use and forests. Emissions reduction: 45–55% by 2030, and 65–75% by 2040 (ref 2005)
National Plan to Energy and Climate 2030 (PNEC 2030) (Resolution 53/2020) I Resolution I 2020 I CC ○	Decarbonisation, energy efficiency, security of supply, internal energy market and research, innovation, and competitiveness. Lower GHG emissions through construction and buildings, incorporating renewable energy, energy efficiency, and energy security. Application of the RNC2050
General Waste Management Scheme (RGGR) (DL 102-D/2020) I Decree-Law I 2020 I CC ○	Portuguese general waste management scheme (RGGR). CE in the CS: >70% (by weight) of materials prepared for reuse, recycling, and other forms of recovery, including site filling; >10% of recycled materials or incorporating recycled materials; reduce by 5% (by 2025) the amount of non-urban waste per GDP unit, and (by 2030) by 10% (ref 2018)

(continued)

Table 23.2 (continued)

Name I type I date I scope	Summary
Public Procurement Code (CCP) (DL 18/2008 rectified by DR 25/2021) I Decree-Law I 2021 I CS •	Aligned with the previous GPP strategy. Promotion of CE, short distribution circuits and environmental sustainability. The sustainability and CE concerns can be (sub)factors of the award and tie-breaking criteria
Climate Law (Law 98/2021) I Law I 2021 I CC ⭘	Public policy that addresses climate change goals applied crosswise (including CS). GHG reduction targets (ref 2005) exclude the use of soil and forests: (a) >55% (2030); (b) >65–75% (2040); (c) >90% (2050). Net CO_2 eq sink in the land use and forestry sector >13 Mton (2045–2050)
Action Plan for Circularity in Construction in Portugal (PACCO) I Plan I 2022 I CS •	Guides the CS in the transition from a linear to a CE model with 30 measures and 100+ actions. Increase reuse and recycling to lower the depletion of virgin natural resources; lower GHG emissions by the built environment to contribute to the European 2030 and 2050 goals

Caption CC ⭘—cross-cutting considering the CS; CS •—dedicated to the CS

(as in the EU was from 2014) appeared policies towards embracing CE principles, showing Portugal as an early adopter. Most documents were published or revised since 2019, showing the EU's influence on national policies and regulations. Although the "*Action Plan for CE*" was published in 2017 [29], already identifying specific CE measures towards the CS, the "*Portuguese Action Plan for Circularity in the Construction Sector*" [26] was only published at the end of 2022. As in the EU, Portuguese documents on CE can be divided into the same four scopes: RWM, SDG, CE, and GPP.

The "*Material specifications*", developed by the Portuguese National Laboratory for Civil Engineering, supports designers and other stakeholders in choosing materials (reused, recycled, etc.) and proposes tests (ideally non-destructive) to ensure their performance. However, these documents need to be revised (last publication in 2016). Additionally, the construction-specific legislation needs to be renewed to include reused and recycled materials, and aspects such as disassembly, adaptability, and flexibility.

23.3.2 Political and Regulatory Barriers and Opportunities in Portugal

Although numerous studies have identified CE barriers and opportunities in the CS, the ones identified in this study (Table 23.3) were obtained directly from different Portuguese stakeholders during the previously described sessions.

Since the economic driver in Portugal is crucial [30], a significant number of barriers are related to the lack of financial incentives and the need to create financial policies that promote and recognize sustainability (e.g. LCA and LCC methodologies) and CE efforts in both projects and companies. GPP can promote case studies, showing stakeholders the benefits of the transition to a CE, without compulsory measures.

Other barriers and opportunities are centred on documentation harmonization and interconnection, as well as policies, legislation, and regulations simplification with streamlined access. The definition of certain degrees of obligation and the need to make processes and procedures faster and more efficient (due to harmonization) were identified, with the twin transition being of significant relevance in this aspect.

23.3.3 Discussion and Implications

Policies and regulations mapping (in EU and Portugal) shows that since 2019, there is a rising concern regarding decarbonization and the adoption of a CE in the CS (e.g., by the document "*Circular Economy—Principles for buildings design*" [12]). Several EU documents have been published and influenced Portuguese policies and regulations, with Portugal being able to keep up with the most recent objectives. Nonetheless, Portugal can follow other countries' examples (e.g., The Netherlands) to move forward in terms of CE in the CS, since the existing political and regulatory documents seem to lack clarity and assertiveness to the Portuguese stakeholders.

As described in Sect. 3.1, the EU and Portuguese policies and regulations intended to push this paradigm shift (linear to circular economy) can be divided into four large groups based on their scopes and descriptions: Resource and Waste Management (RWM), Sustainable Development Goals (SDG), Green Public Procurement (GPP), and Circular Economy (CE). This demonstrates the importance of articulation within each document macro-group. GPP can promote and finance case studies. RWM documents can support resources/waste identification and classification, which will support CE principles and procedures and how SDG objectives can be reached.

Barriers and opportunities identified by the Portuguese stakeholders were analysed, and aggregated into four groups: *(i)* **development of policies and regulations** properly integrated, thus supporting the transition through finance, case studies using GPP, and other mechanisms; *(ii)* **new and innovative digital tools and platforms** that allow, for example, to collect large amounts of data and convert it into information, thus supporting the twin transition (green and digital) and the conversion

Table 23.3 Political and regulatory barriers/opportunities identified by Portuguese stakeholders

Barriers
• Complexity and contradiction of some legislation and regulations, blocking the transition • Lack of policies that compel the Bill of Quantities to include the reuse/recycling of materials • Lack of policies and regulations that enforce the accounting of environmental and social costs • Lack of discussion, information and of a schedule for applying Green Public Procurement: case studies • Lack of political support to accelerate the transition using digital tools: Digital Twin; etc. • Lack of incentives and benefits for projects that integrate the principles of sustainability and circularity • Difficulty in establishing a fluid/continuous relationship between the different value chain operators • Lack of political and economic stability: different political parties usually create distinct policies • Lack of benefits in municipal fees when considering sustainable and circular projects • Lag between legislation and state of the art (in content and in time) • Need to integrate Life Cycle Assessment (LCA) and Life Cycle Cost (LCC) in public procurement • Lack of a common classification basis and concrete goals • Lack of robust official statistics and uncertainties regarding the end of life • Need for new material specifications (e.g., LNEC material for reuse and recycled materials) • Lack of risk financing when using recycled (or reuse) materials or even the incorporation of CDW; • Create and promote a green building permit, to reward the environmental performance of companies
Opportunities
• Incentives for new circular business models (e.g., material recycling businesses) • De-bureaucratize processes and procedures making them faster and more efficient • Simplified certification of processes and procedures • Municipalities define fee exemptions/reductions for projects with circularity/sustainability concerns • Specifications adjusted to the new requirements (e.g., including LCA information) • Promote Pre-audit demolitions and the use of the Waste Management Plan as a strategic element • Creation of a digital platform/portal that brings together updated legal information in this area • Creation of new "obligations" or "incentives" to be considered from the beginning of the project • Improve the separation and sorting of material/waste and its classification • Promote the execution of pre-demolition audits and the development of supporting digital technologies • Promote funding mechanisms for training stakeholders in the AEC sector • Dissemination and sharing of knowledge, including the recommendations of the European Union • Taxonomy based on performance quantified by LCA and LCC • Support and encourage entities to apply the LCA (clarifying the economic benefits: LCC) • Legislation should allow/support, based on technical opinions, the options intended to be implemented • Energy efficiency legislation extended to a life-cycle approach

of residues into resources; *(iii)* **education and capacitation** of the different stakeholders, supporting clarity and empowering them to surpass new challenges and promote their resilience; and *(iv)* **governance model** that supports the paradigm shift across the society, based on a continuous flux of information and discussion between the government and the stakeholders, for policies and regulations adapted to the needs identified on the terrain.

The recently published "*Portuguese Action Plan for Circularity in the Construction Sector*" [26] can be an effective starting point to ignite and accelerate the transition, as this document is expected to help the government and other stakeholders to have a more holistic view of the barriers and opportunities specific to the CS and of the reality.

23.4 Conclusions

This study investigated and identified policies and regulations in the EU and Portugal for the Construction sector (CS) and presented the barriers and opportunities identified by Portuguese stakeholders for those policies and regulations. It employed a systematic document review procedure to search, retrieve, evaluate, and extract relevant data from various sources and consultation sessions with stakeholders.

The analysis revealed a growing number of policies and regulations in the EU and Portugal, mainly from 2019. These documents can also be divided into four main areas of interest for the transition to a Circular Economy (CE): Resource and Waste Management (RWM), Sustainable Development Goals (SDG), Green Public Procurement (GPP), and dedicated to CE. The study also identifies four main barriers and opportunities related to policies and regulations found in Portugal: *(i)* **development of policies and regulations**; *(ii)* **new and innovative digital tools and platforms**; *(iii)* **education and capacitation**; and *(iv)* **governance model** adequate to needs. The study then discussed the possible interactions between the documents and barriers and opportunities, showing that the chain reaction mechanisms leading to a CE tend to fail since policies and regulations lag the needs identified in the field by stakeholders. However, the current Portuguese strategy, with a recently published sectorial Action Plan and where all stakeholders took part, can be an important tool to identify the most significant measures to take. This type of strategic document is relevant for any EU member that wishes to embrace a CE in the CS. Despite fulfilling the objectives, the study has some recognized limitations and uncertainties, such as: different countries can show different policies and barriers due to their own realities; a more even distribution of the participating stakeholders, can influence the identified barriers; and the national policies can significantly influence the perceived barriers by the stakeholders.

In this research it was identified further research, such as gathering workgroups with similar contributions from different stakeholder groups, as well as applying a similar research approach to other countries.

Acknowledgements The authors gratefully acknowledge the COST Action CircularB [CA21103] for this opportunity. The authors would also like to acknowledge the project "Circular Agreement with the Construction Industry" funded by the Portuguese Environmental Fund and its workgroup contributions: AECOPS, AICCOPN, APA, CIP, CPCI, IMPIC and PTPC. Marco F. Pedroso would like to acknowledge the Lisbon Regional Operational Programme [LISBOA-05-3559-FSE-000014], and Vanessa Tavares the North Regional Operational Programme [NORTE-06-3559-FSE-000176].

References

1. Benachio GLF, do Carmo Duarte Freitas M, Tavares SF (2020) Circular economy in the construction industry: a systematic literature review. J Clean Prod 260:121046
2. Gallego-Schmid A, Chen H, Sharmina M, Mendoza J (2020) Links between circular economy and climate change mitigation in the built environment. J Clean Prod 260(2020):121115
3. Nazarko J, Chodakowska E, Nazarko Ł (2022) Evaluating the transition of the european union member states towards a circular economy. Energies 15(11):3924
4. Ellen MacArthur Foundation, Report (2015) Growth within: a circular economy vision for a competitive Europe, Amsterdam
5. European Construction Sector Observatory, Analytical report (2021) Digitalisation in the construction sector, Brussels
6. Bertozzi C (2022) How is the construction sector perceiving and integrating the circular economy paradigm? Insights from the Brussels experience, City. Cult Soc 29(2022):100446
7. Govindan K, Hasanagic M (2018) A systematic review on drivers, barriers, and practices towards circular economy: a supply chain perspective. Int J Prod Res 56:278–311
8. Wuni I (2022) Mapping the barriers to circular economy adoption in the construction industry: a systematic review. Build Environ 223:109453
9. Fidélis T, Cardoso AS, Riazi F, Miranda AC, Abrantes J, Teles F, Roebeling P (2021) Policy narratives of circular economy in the EU—assessing the embeddedness of water and land in national action plans. J Clean Prod 288:125685
10. Calisto Friant M, Vermeulen WJ, Salomone R (2021) Analysing European Union circular economy policies: words versus actions. Sustain Prod Consum 27(337–353):337–353
11. Sáez-de-Guinoa A, Zambrana-Vasquez D, Fernandéz V, Bartolomé C (2022) Circular economy in the european construction sector: a review of strategies for implementation in building renovation. Energies 15(4747):4747
12. Comission E, Economy C (2020) Principles for buildings design. JRC, Brussels
13. Ababio B, Lu W (2023) Barriers and enablers of circular economy in construction: a multi-system perspective towards the development of a practical framework. Constr Manag Econ 41(1):3–21. https://doi.org/10.1080/01446193.2022.2135750
14. Giorgi S, Lavagna M, Wang K, Osmani M, Liu G, Campioli A (2022) Drivers and barriers towards circular economy in the building sector: stakeholder interviews and analysis of five European countries policies and practices. J Clean Prod 336:130395
15. Masi D, Kumar V, Garza-Reyes J, Godsell J (2018) Towards a more circular economy: exploring the awareness, practices, and barriers. Prod Plan Control 29:539–550
16. Liu J, Wu P, Jiang Y, Wnag X (2021) Explore potential barriers of applying circular economy in construction and demolition waste recycling. J Clean Prod 326:129400
17. Hjaltadóttir R, Hild P (2021) Circular economy in the building industry European policy and local practices. Eur Plan Stud 29(12):2226–2251. https://doi.org/10.1080/09654313.2021.1904838
18. Rahla K, Mateus R, Bragança L (2021) Implementing circular economy strategies in buildings—from theory to practice. Appl Syst Innov 4(26):26

19. Tavares V, Pedroso MF (2023) Barriers and opportunities in the transition to a circular construction sector in Portugal. In: Bragança L (eds) Workshop "Creating a Roadmap Towards a Circular Built Environment", to be published, Skopje, Macedonia
20. European Commission Environment Homepage. https://environment.ec.europa.eu/topics/circular-economy_en. Accessed 29 Dec 2022
21. European Commission EUR-Lex Homepage. https://eur-lex.europa.eu/homepage.html. Accessed 28 Dec 2022
22. Portuguese Parliament—Journal of the Republic Homepage (in PT). https://dre.pt/dre/en, Accessed 22 Dec 2022
23. Portuguese Environment Agency (APA) Homepage (only in PT). https://apambiente.pt/desafios#residuos. Accessed 12 Dec 2022
24. National Laboratory for Civil Engineering (LNEC) Homepage. https://www.lnec.pt/en/. Accessed 10 Dec 2022
25. European Commission, COM (2014) 398 final—Towards a circular economy: a zero waste programme for Europe, Brussels
26. Fonseca P, Pedroso M, Tavares V, Laranjeira L (2022) Portuguese action plan for circularity in construction (PACCO) [in PT]. BUILT CoLAB/Fundo Ambiental, Porto, Portugal
27. Portuguese Action Plan for Circularity in Construction (in PT) Homepage. https://circularidade.builtcolab.pt/#inicio. Accessed 29 Dec 2022
28. Kirchherr J, Piscicelli L, Bour R, Kostense-Smit E, Muller J, Huibrechtse-Truijens A, Hekkert M (2018) Barriers to the circular economy: evidence from the European Union (EU). Ecol Econ 150:264–272
29. Portuguese Ministry of Environment—Leading the transition—Action plan for circular economy in Portugal Homepage. https://circulareconomy.europa.eu/platform/sites/default/files/strategy_portuguese_action_plan_paec_en_version_3.pdf. Accessed 29 Dec 2022
30. Klein N, Deutz P, Ramos T (2022) A survey of circular economy initiatives in portuguese central public sector organisations: national outlook for implementation. J Environ Manage 314:114982

Chapter 24
Measuring Circularity in Brazilian Social Housing

Mayara Regina Munaro and Vanderley Moacyr John

Abstract The incorporation of circular economy (CE) principles in social housing (SH) is a not widespread agenda in government public policies. If, on the one hand, the growing demand for housing puts pressure on public services and the construction value chain, on the other hand, the reduction in the extraction of virgin materials and waste generation is essential to mitigate the effects of global climate change. This article presents a case study of how the Brazilian government is addressing circularity in the country's National Housing Plan. The study presents the proposal for circularity indicators to be implemented, both in new and existing buildings, aiming to reduce material demand and waste generation in SH. The adoption of circular indices in the input flow of materials has greater applicability in SH and promotes incentives to the secondary construction materials market. Circularity in the outflow of materials requires greater maturity in the application of CE principles and a systematic change in the housing design stage. The study aims to guide decision-makers around the consistent measurement of circularity in housing, promoting improved performance and the shift towards a more sustainable built environment.

Keywords Circular economy · Indicator · Resources · Construction and demolition waste

24.1 Introduction

The world is on the brink of a climate catastrophe, and to limit warming to 1.5 °C above pre-industrial levels, global greenhouse gas (GHG) emissions must decline by 43% by 2030, falling to net zero by 2050 [1]. This is a challenge because more than half the world's population lives in cities, and accounts for more than 70% of global GHG emissions [1]. Cities are drivers of economic growth, contributing more than 80% of global GDP, however, the rapid and poorly planned urbanization

M. R. Munaro (✉) · V. M. John
Department of Civil Construction Engineering, São Paulo University, São Paulo, Brazil
e-mail: mayara.munaro@lme.pcc.usp.br

L. Bragança et al. (eds.), *Creating a Roadmap Towards Circularity in the Built Environment*, Springer Tracts in Civil Engineering,
https://doi.org/10.1007/978-3-031-45980-1_24

leads to many challenges, including a shortage of affordable housing, high rates of residents living in slums or informal settlements, insufficient infrastructure, limited open spaces, unsafe levels of air pollution, and increased climate and disaster risk [1].

Access to adequate housing has been a challenge for a significant part of Brazilian residents. In 2019, the housing deficit in Brazil was estimated at 5.9 million households [2]. Of the 70.6 million Brazilian households in 2019 [3], 44% were framed with a deficit or unsuitable housing [4]. In addition to the housing deficit, it is necessary to plan the housing provision for a population estimated at 232 million inhabitants in 2040, which means 17 million more inhabitants than in 2022 [5].

Since 2009, with the creation of the National Housing Plan (PlanHab), Brazil has had the Minha Casa, Minha Vida Program (MCMV), which supported the production of around 4.5 million housing units for the population with income of up to 10 minimum wages [6]. The MCMV was a recurrent theme in the Brazilian political scenario and represented a break with previous practices by bringing the issue of housing to the center of the government agenda [7]. The program enables access to housing for the low-income population, historically excluded from financing for home ownership [7]. However, it presents challenges that need to be overcome due to the standardized and large-scale production of housing units, poorly inserted and isolated from the city, contributing to the maintenance of social and spatial inequalities [7, 8].

The development of the MCMV brought other challenges to environmental and urban planning. Apart from equipping houses with solar heaters, justified mostly as a strategy to reduce electricity bills for poor families, and the introduction of sewage collection and treatment in the housing project, MCMV did not follow any environmental policies, since the PlanHab 2008–2023. Housing projects introduced a temporary increase in the consumption of building materials—cement, steel, and bricks made up the bulk of material consumption—with a broad environmental impact. At the city level, it also increased sharply the construction and demolition waste (CDW) generation, causing a significant impact on the environment and even in city-related waste management costs and directly impacting public health management and urban drainage [7]. Additionally, because the housing projects were located mostly on the outskirts of cities it introduced a new vector of urban development, requiring municipalities to reassess and expand the network of community facilities and public services [6], increasing the construction-related impacts.

Nowadays the Brazilian government preparing the PlanHab 2040, which aims to the double challenge, improving housing conditions and urbanization meanwhile mitigating environmental impacts. Besides climate-related policies, circularity can become an important aspect for PlanHab since housing makes up the bulk of construction materials consumption, which represents more than 50% of natural resources and generates about 60% of Brazilian urban solid waste [8, 9]. CE is considered instrumental even for the mitigation of GHG emissions and indispensable for maintaining resource security [9].

The preparation of PlanHab 2040 seeks to create a sustainability agenda for social housing (SH) in Brazil, with objectives and strategies to reduce environmental impacts, increase resilience, guarantee performance, optimize costs, and combat social inequality. For the Brazilian construction industry, these proposals aim to minimize the use of resources and waste generation through the more efficient use of resources, seeking to promote the circularity of materials in the construction value chain. To operationalize these strategies, actions, and indicators are being developed to monitor the progress of the proposed goals, throughout the next PlanHab cycle.

From the perspective of how public policies can encourage the adoption of circular principles in the construction of the built environment, this study presents the proposal of circular indicators in the new 2040 Brazilian Housing Plan. Circular performance indicators are analyzed, aiming to reduce the consumption of construction materials and the generation of CDW. The development of consistent circular indicators will serve as a tool for the formation of benchmarks and performance standards to support decision-making and improve environmental performance during the life cycle of housing units and the quality of life of residents.

24.2 Background

24.2.1 Sustainability and Circularity in Social Housing Production

The adoption of the circular economy (CE) is one prerequisite for sustainability. Sustainable housing must present adequate performance, that is, it must be safe, healthy, and comfortable, meanwhile presenting initial and running costs compatible with long-term family income. The total housing cost is frequently neglected in public policies. About 10.4% of the European Union (EU) population cannot pay rent and the costs of housing services (water, gas, electricity) absorb about 40% of income [10]. In Brazil, the excessive rent burden is the main component of the Brazilian housing deficit, corresponding to 52% of the deficit or 3.0 million households [2].

CE is a model that offers the opportunity to reduce resource use, through end-of-life replacement strategies such as reduction, reuse, and recycling of materials in production/distribution and consumption processes. A circular building optimizes the use of resources while minimizing waste throughout its life cycle [11]. It is a business opportunity for innovation and circular value creation for the construction sector [11, 12].

Measuring the actual environmental performance of construction is essential to reducing environmental impacts because it encourages not only building design optimization, and careful selection of material but also procurement of the most eco-efficient supplier for each material improved construction site management or seeking a more efficient building operation scheme [13]. This requires relevant performance indicators, as they represent key components for describing the environmental, social, and economic impacts of buildings [14], coupled with mitigation goals supported by a robust benchmark baseline. It is also required to establishing not only minimum acceptable performance over time but a system to track the aggregated variation of each performance indicator for all housing projects [14].

Environmental performance indicators must have priority, measurability; reliability; comparability, and understandability [13]. Excess of indicators implies a large burden of data collection, and the priority is to reduce the number of indicators that meets the expectations of decision-makers. Measurability to facilitate data gathering at a low cost. Reliability ensures stakeholders trust the indicators, which foster its introduction in actual decision-making. Comparability enables benchmarks and the evaluation performance over time. Comprehensibility is a prerequisite for engagement and for a clear understanding of each indicator by those involved in the construction value chain [13].

24.2.2 Circularity in Social Housing

The literature consulted on social housing refers mainly to the themes of energy efficiency, climate change, and economic development. The European Energy Package 2030 suggested testing initiatives in public housing to achieve EU targets on energy efficiency, energy costs, and environmental sustainability [10]. Few reviewed studies explicitly examine the circular transition of the housing sector. For the Dutch social housing sector, the implementation of CE is in the experimental phase. In interviews with representatives of social housing organizations in the country, none of the represented associations has completed a circular project so far, but almost 80% of them are carrying out circular pilot projects [15].

In Italy, design solutions to improve the energy and environmental quality of public housing have been carried out with the involvement of inhabitants to define new design strategies and to program long-term sustainability [10]. In Milan, a project for the renovation and revitalization of a neighborhood proposed a functional and social mix of housing, with flexible spaces, green design, renewable energy, CE criteria, and continuous maintenance to drive social revitalization and improve equity, safety, comfort, energy efficiency, and sustainability [10].

Recent experiences on the implementation of circular principles in social housing in Denmark have focused on the development of flexible and adaptable housing, considering the intended lifespan of each of the building layers, optimizing the longevity of the construction, and maximizing the recovery of 90% of material at end of life [16]. The Circular Economy Plan for Housing Regeneration in the United Kingdom developed by KLH Sustainability and Clarion Housing [27] considered a set of activities such as demolition for recovery, products with high-recycled content, supply chain integration, and construction waste management, aiming to eliminate and reduce waste before considering conventional management opportunities such as recycling [17].

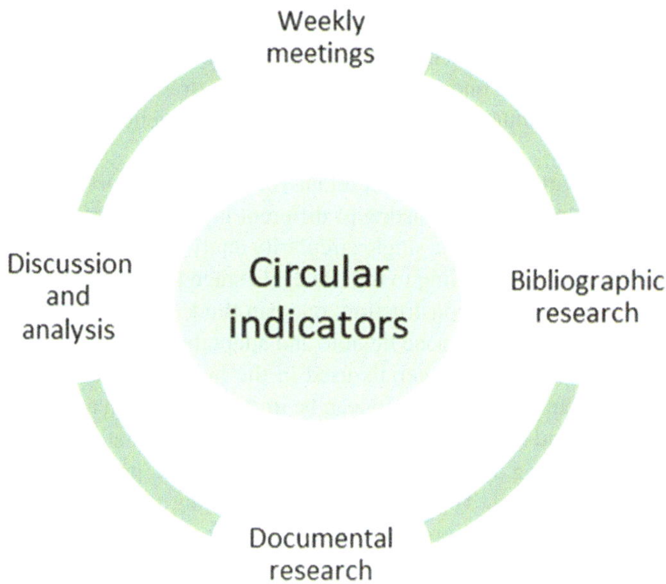

Fig. 24.1 Research strategy adopted for the proposal of circular indicators

24.3 Research Strategy

The research has a qualitative approach; specifically, we used the case study method, which enables researchers to obtain in-depth and comprehensive information about the phenomenon in its natural context [18]. This research describes the case for the proposal of circular indicators to be included in the sustainability axes of the Brazilian National Plan of Social Housing. The process of proposing the circular indicators included weekly meetings with the academic team from interdisciplinary disciplines and government representatives. The authors reviewed relevant literature and documents and discussed their ideas in an iterative and interactive process (Fig. 24.1). The proposed circular indicators were consolidated after discussion and analysis by the team.

24.4 Possible Indicators of Circularity

24.4.1 Aspects Related to Housing

The construction sector has a large potential for CE, given the scale of material use, the value contained in buildings, labor intensity, and the long-term effect of measures adopted in buildings [19]. In long-term resource use and waste generation, minimization can be achieved by longer service lives, which keeps materials and resources in use as long as possible. Therefore, the measures to be adopted are based on the incorporation of secondary materials in the construction of buildings and the reduction of wasted resources in the manufacturing/construction process of building products.

Design criteria are particularly important for the built environment because a building is a complex object consisting of different layers with different life cycles, making it difficult to assimilate a single circularity analysis [19]. The building circularity index can change according to the life cycle stage [20]. Whether a building is in the design stage or under construction, one can think of following circular principles, making its design and space flexible and adaptable. This facilitates the reuse or recycling of products and materials used in the building when their useful life ends. For existing buildings, circularity can be improved through requalification and choice of products and materials that are reused, recycled, or that allow better energy efficiency in the building. If a building is at the end of its life, products, and materials can be reused or recycled. Therefore, even if a building is not circular during its lifetime, it can influence and provide opportunities for other buildings to improve their circularity [20].

The circularity percentage is calculated by taking the weighted average of the circular inflow and outflow of material (Fig. 24.2). Key considerations in measuring this circular flow are monitoring the type of material, the percentage of materials reused and recycled, and the recovery of materials at end of life. There is no standardized methodology and existing circular indicators are being discussed. Therefore, benchmarks, standards, and key performance indicators must be defined [20]. The circular inflow is the mass of non-virgin or sustainably sourced renewable materials used in the building. The circular outflow is calculated by multiplying the potential recovery factor by the percentage of materials recovery after the end of the product/building life [20].

The mass of the materials used in a building is an important consideration in measuring circularity, but not the only one. Additional variables must be considered, such as how materials were transported and how the building was constructed [20].

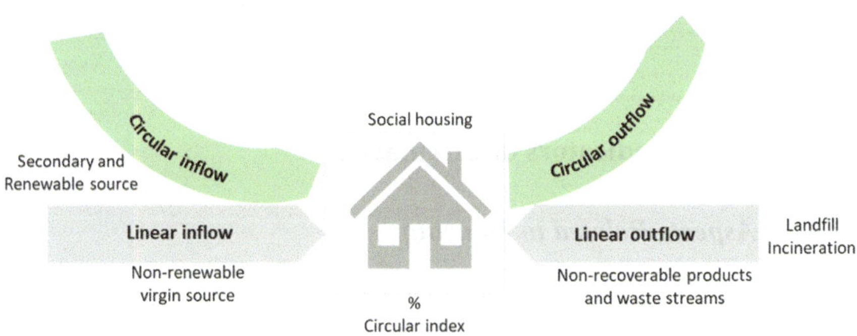

Fig. 24.2 Scheme of linear and circular material flows within a building (adapted by WBCSD [20])

24.4.2 Material Demand Indicators

Material demand (MD) is the sum of the materials necessary to produce a given dwelling, including waste in its entire production chain, divided by the housing unit (*t/HU*) (24.1). MD_t is the total demand for materials; M_i is the mass of primary materials in the product *i*; *HU* is the number of housing units in a project, or it can be the built area of the house/project.

$$MD_t = \frac{\sum iM_i}{HU} \tag{24.1}$$

Within the same construction technology, the environmental impact and the cost are approximately a function of the mass of materials used, so dematerialization improves the performance of this indicator. Lower construction mass will imply less demolition waste. The mass of a material is important information on the production process, and easy to measure, and understand. From the builder's perspective, the main goal of the intensity of materials is to reduce the mass of the building, and, therefore, the indicator can be expressed in kg/m^2 [21]. Compared with a proper benchmark it allows to track of construction dematerialization.

Amount of secondary material in the housing unit (t/HU). Circularity requires substituting virgin raw materials with secondary materials. The use of renewable or secondary (reuse or recycling) and end-of-life materials are an option to reduce the demand for natural resources. Therefore, the total material demand can be expressed as the sum of primary and secondary raw materials, and the total material demand of a housing unit can be expressed as Eq. (24.2).

$$MD_t = \frac{\sum iM_p + \sum iM_s}{HU} \tag{24.2}$$

M_p is the mass of primary materials in the product *i*; M_s is the mass of secondary materials. This indicator is used to monitor progress towards a CE in secondary raw materials. The indicator measures both a country's ability to produce secondary raw materials and its effort to collect waste for recovery [22].

Demand for renewable material used in the housing unit (*t/HU*). Total material demand can also be disaggregated into materials from renewable and non-renewable sources (24.3). Renewable materials comprise cultivated or native biomass from sustainably managed areas. Non-renewable materials include metallic and non-metallic minerals, petroleum (used as a raw material to produce plastic), and native biomass from areas of unsustainable management, that is, areas managed with an intensity greater than that necessary for the recovery of the forest [13].

$$MD_t = \frac{\left(M_{p,ren} + M_{p,nren}\right) + M_s}{HU} \tag{24.3}$$

$M_{p,ren}$ is the mass for renewable primary materials in the product i; $M_{p,nren}$ is the mass for non-renewable primary materials.

Circularity index (%), considers the amount of primary renewable and non-renewable, and secondary material (24.4). Values can range from 0 to 1. Higher values represent a higher roundness rate.

$$IC = \frac{\sum M_{s,} + \sum M_{p,ren}}{\sum M_{p,ren} + \sum M_{p,nren} + \sum M_s} \times 100. \tag{24.4}$$

24.4.3 Waste Generation Indicator

Amount of waste generated at the end of construction of the housing unit (m^3/m^2 or t/m^2) (24.5). WG is the waste generation rate (mass or volume/HU); W_i is the amount of waste generated during construction (mass or volume); HU is the housing unit.

$$WG = \frac{\sum Wi}{HU} \tag{24.5}$$

In Brazil, the generation of waste at the end of a building is regulated by Law No. 12,305/09, which provides for the National Solid Waste Plan [23], and CONAMA Resolution No. 307/02, which establishes guidelines for the management of civil construction waste. In addition to separating waste by class, the creation of a Construction Waste Management Plan is required for large generators, establishing the necessary procedures for sorting, storing, transporting, and disposing of the waste generated in an environmentally appropriate destination.

24.5 Discussion

Consistently measuring circularity and adopting a standardized approach are important to enable stakeholders to understand the level of circularity of their buildings and determine actions to improve their performance, accelerating the shift towards a sustainable built environment [20]. The main advantages of a circular assessment approach are to pay more attention to the renewability of the input resources, focus more on the use phase and the possibility of reapplying products, and introduce the assessment of the potential recoverability of the product at the end of its useful life [19].

The proposed indicators are multiscale and can be used at the level of buildings, suppliers, a city, or globally. They were designed for SH but can be used in other buildings, even if they are not part of public housing policies. In addition, they can be used both for new developments and for the requalification of existing homes and neighborhoods, which is a circular strategy. The application of the indicators suggests a reduction in the mass of construction and waste generation. Dematerialization, reduction of losses and residues, reuse, and recycling increase in the intensity of use and useful life, and use of less energy-intensive materials are strategies for the efficient use of materials and allow reductions in GHG emissions.

The majority of the indicators are related to the inflow of materials and one to the outflow of materials, aiming to obtain a balance of input and output of materials during the construction of a housing (Fig. 24.3). These indicators focus on increasing circularity in the incoming flow of materials, as they are metrics that facilitate the introduction of circular principles and encourage the development of the secondary materials market. In Brazil, CE in the construction industry is still in its infancy, and there is no CE regulation in government urban projects. Still, there is a lack of robust regulation that facilitates the reuse of salvaged building components in new construction projects [15].

Addressing circularity at the 'exit' of the flow of materials, that is, at the very long building service life, increases the uncertainties both in estimating the possibilities of reinserting materials into production chains and in the real benefits of circular practices, such as recycling [9]. In addition, it requires a systemic change in the

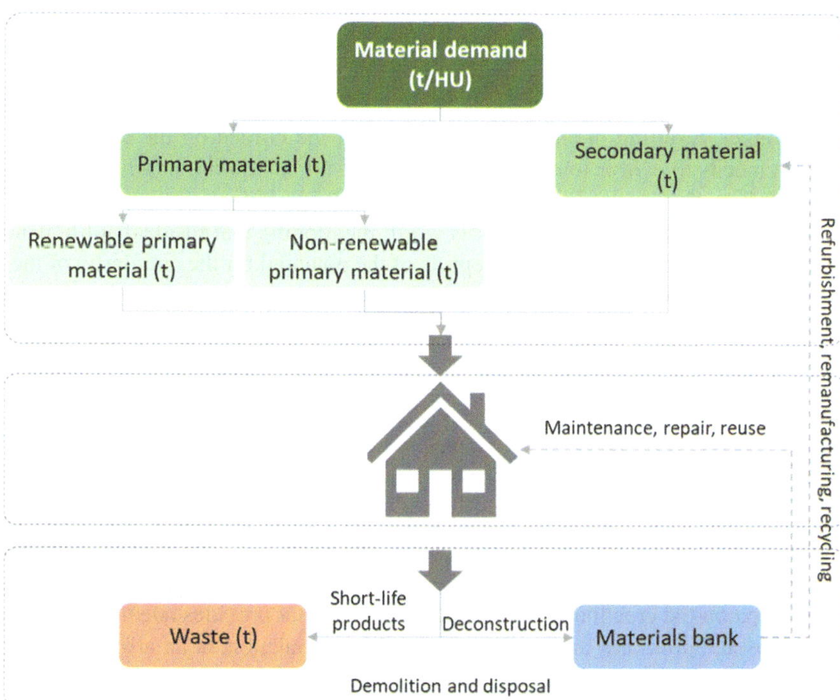

Fig. 24.3 Summary of the input flow of materials and output of waste in a dwelling

housing design stage and design plans for the deconstruction and recovery of materials at the end of the building's service life. The potential environmental benefits will happen in the future if users keep with the building project. Therefore, solutions that increase environmental loads during construction must be well-pondered. An efficient housing project must prioritize the adaptability of spaces to accommodate the different needs of users over time. Strategies such as modular construction, use of materials following the respective technical standards, and design for deconstruction must be an integral part of the projects [24]. Besides, in the construction stage, waste can be avoided through the construction process and integrated and efficient management of the flow of materials and labor at the construction site.

The implementation of circular indicators is important for planning and incorporating efficiency measures in buildings and for consolidating public policies that encourage the construction of more sustainable buildings. Commitment and support from the top management and CE legislation are the main enablers of the development of Dutch social housing [15]. In addition, a lack of incentives or requirements at a national scale can also make it difficult to apply sustainable practices, even where public policy exists SH providers can also be hindered by restrictive or absent funding requirements, inappropriate competing policies, or organizational barriers [25]. It is also necessary to overcome problems arising from the scale both in the supply of products and in the training of human resources, as well as the objective limitations of cost and production chain [26]. Therefore, the responsibilities of each party must be assigned [26].

The quantity of built stock is important, but more so is its quality, as better buildings and infrastructure will last longer, require less maintenance, and serve users' needs longer, resulting in an overall reduction of the burden on the environment. Housing needs to evolve towards a system based on circularity, in which building materials are used, reused, adapted, and rebuilt, considering economic and environmental rationality in decisions. It is worth mentioning that residents of housing units, normally, do not have the perception of the potential for the circularity of their dwellings. Communication projects and user awareness is essential for the effective implementation of the CE, especially in the requalification activities carried out by the residents themselves.

24.6 Conclusions

The study presents the proposal for circularity indicators for social housing in Brazil. The indicators will help to introduce circular principles in the new 2040 Housing Plan, based on the creation of a sustainability axis, that provides adequate housing conditions and promotes the quality of life of the population, at an affordable cost, while reducing impacts on the environment throughout the life cycle of the house.

As the CE gains momentum, governments and construction companies must prepare for their transition based on metrics about their circular performance and associated risks and opportunities. To do this, buildings need a universal and consistent way to measure their circularity. Public governance has shown few efforts on circular policies on social housing and focusing on the energy system. The adoption of these indicators in social housing demonstrates the concern of a social governance policy engaged with sustainable development, in addition to being a way of accelerating the implementation of circular principles in the construction industry.

The proposed circular indicators are directed towards the entry of material flow into housing, favoring the use of secondary and renewable materials in construction chains. In the material flow output, the total waste generation during the construction/renovation of the house was considered. The main issue considered is the efficient use of materials, through dematerialization and efficient management of resources during the housing construction process.

The study has limitations related to the research strategy. The inclusion of different stakeholders of the construction value chain in the meetings could contribute to the circular indicators proposals. In future research, it is suggested to monitor the acceptance and application of the indicators in SH projects, as well as to analyze the creation of new indicators aimed at the output of the material flow with a more interdisciplinary team. The creation of housing scenarios with different circularity indices, through numerical modeling, can be a strategy for the dissemination and awareness of the importance of dematerialization in social housing.

Acknowledgements The authors would like to thank the Superintendência de Gestão Ambiental (SGA) and the Department of Civil Construction Engineering at São Paulo University.

References

1. UN (2022) The Sustainable Development Goals Report. New York, NY, USA
2. FJP (2021) Deficit Habitacional no Brasil 2016–2019. Belo Horizonte, Minas Gerais
3. IBGE (2019) Tabela 6820—Domicílios e moradores, por tipo de domicílio. sidra.ibge.gov.br/tabela/6820#resultado
4. FJP (2020) Inadequação de domicílios no Brasil 2016–2019. FJP, Belo Horizonte
5. IBGE (2023) Projeções e estimativas da população do Brasil e das Unidades da Federação
6. Ministério das Cidades: Caderno 3 Diretrizes e recomendações: Qualificação do projeto urbano. Brasília (2018)
7. Amore C, Shimbo L, Rufino M (2015) Minha Casa... E a Cidade? Avaliação do programa Minha Casa Minha Vida em seis estados brasileiros. Rio de Janeiro
8. Siqueira-Gay J, Gallardo A, Giannotti M (2019) Integrating socio-environmental spatial information to support housing plans. Cities 91:106–115
9. Haas W, Krausmann F, Wiedenhofer D, Heinz M (2015) How circular is the global economy? An assessment of material flows, waste production, and recycling in the European union and the world in 2005. J Ind Ecol 19:765–777
10. Lucchi E, Delera A (2020) Enhancing the historic public social housing through a user-centered design-driven approach. Buildings 10
11. WBCSD (2021) The business case for circular buildings: exploring the economic, environmental and social value
12. UNEP (2019) Global resources outlook 2019: natural resources for the future we want

13. Belizario Silva F (2022) Proposal of life cycle-based environmental performance indicators for decision-making in construction. Universidade de São Paulo
14. Fernandes L, Costa D (2021) Environmental performance analysis for the construction phase of building projects by using the indicators recommended by PBQP-H. Ambiente Construído 21:7–26
15. Çetin S, Gruis V, Straub A (2021) Towards circular social housing: an exploration of practices, barriers, and enablers. Sustainability 13:1–24
16. GHX and Responsible Assets: Circle House—Denmark's first circular housing project, 1st ed. KLS PurePrint, Denmark
17. Marchesi M, Tweed C (2021) Social innovation for a circular economy in social housing. Sustain Cities Soc 71:102925
18. Creswell JW (2014) Research design—qualitative, quantitative, and mixed methods approaches, 4th edn. Sage, United States of America
19. Cottafava D, Ritzen M (2021) Circularity indicator for residential buildings: addressing the gap between embodied impacts and design aspects. Resour Conserv Recycl 164:105120
20. WBCSD (2022) Measuring circular buildings—key considerations
21. Reis Daniel da C (2020) Fluxo de materiais e eficiência no uso de recursos na indústria da construção. Escola Politécnica, Universidade de São Paulo
22. European Comission: Circular material use rate: calculation method. Luxembourg (2018)
23. Brasill: Lei No. 12.305, de 02 de agosto de 2010. Institui a Política Nacional de Resíduos Sólidos. Brazil (2009)
24. Munaro MR, Tavares SF, Bragança L (2022) The ecodesign methodologies to achieve buildings' deconstruction: a review and framework. Sustain Prod Consum 30:566–583
25. McCabe A, Pojani D, van Groenou AB (2018) The application of renewable energy to social housing: a systematic review. Energy Policy 114:549–557
26. MMA (2015) Sustentabilidade urbana: impactos do desenvolvimento econômico e suas consequências sobre o processo de urbanização em países emergentes. Brasília, DF
27. KLH Sustainability. Clarion Housing Group (2023) Available at: https://www.klhsustainability.com/project/clarion-housing-group

Part XII
Circular Business Models

Chapter 25
Circular Economy Business Model Framework Considering Product Sustainability

Bengü Güngör

Abstract With the development of technology and industry, which started in the latter half of the twentieth century, environmental degradation intensified because of the depletion of natural resources, global warming, the disintegration of the ozone layer, acid rain, droughts, and other issues. The Sustainable Development Goals (SDGs) have inspired the creation of a brand-new business, economic, environmental, and social structure model called the Circular Economy (CE) concept, which is an alternative to the current linear economy and contends that all things in nature are in constant transformation with one another. At this point, it is now important to extend the product's useable and functional lifespans as well as the point at which it stops working. This study primarily seeks to give a foundation for developing a circular economy business model considering the product life cycle before outlining how it is integrated. The relationship between sustainability and the circular economy is first and foremost. After that, the circular economy and extended lifespan of product strategies are defined using literature research. All defined actions and decision points are used to design the phases of a model framework. In conclusion, it thinks the suggested framework will help the decision-makers who want to include circular economy principles into their business procedures, especially based on the manufacturing process. No matter how the business model relates to a specific industry, the framework's general behavior will surely help managers decide on the application stages.

Keywords Circular economy · Business model framework · Product sustainability

---

B. Güngör (✉)
Izmir Democracy University, Gürsel Aksel Blv. No: 14, 35140 Karabağlar/Izmir, Turkey
e-mail: bengu.gungor@idu.edu.tr

L. Bragança et al. (eds.), *Creating a Roadmap Towards Circularity in the Built Environment*, Springer Tracts in Civil Engineering,
https://doi.org/10.1007/978-3-031-45980-1_25

25.1 Introduction

The circular economy (CE) is a management approach that promotes the use of waste as resources to create value. It aims to manage resource circularity, efficiency, and optimization [1]. As a strategy for achieving local, national, and global sustainability, the circular economy (CE) has been gaining popularity. Multinational corporations and decision-makers in developed nations have paid it more attention recently. The current and established extract-produce-use-dump linear material and energy flow model of the contemporary economic system is unsustainable. A cyclical alternate flow model is offered to the economic system by the circular economy [2]. Recently, most of the research, particularly in managerial journals, has concentrated on how enterprises' business models might incorporate the circular economy and on the establishment of frameworks designed to comprehend the dynamics of value creation and capture in circular business models [3]. However, established businesses will face hurdles when the economy transitions to a circular one. The value of their current talents, networks, and business models may even be destroyed in some circumstances [4]. Thus, it becomes necessary to redesign current value networks and associated business models as a result of new players or shifting responsibilities. Companies must engage in a circular business model innovation process, which begins with creating the components of business models, to manage these changes. The CE business model incorporates a process of innovation and transformation. Given that many people think the CE is revolutionary, this approach is important for small businesses and numerous larger companies are paying attention. The examination of the CE business model's sustainability is still in its early stages. The boundaries and connections between sustainability and the circular business model are also not investigated. Taking all of these aspects into consideration, the present circular economy business models were examined to complete the literature search for this study.

For all of the aforementioned reasons, this study is driven to fill this research gap in the literature by creating a general CE business model framework to extend product life cycle and ensure sustainable innovation. The objective of the study is to identify the essential elements of a generic circular economy business model framework while also considering applications in several industries. To success these aims, the research questions of the study are constructed as following:

RQ1: What are the common CE business models in the literature?

RQ2: What are the main indicators/components of a CE business models considering manufacturing process?

RQ3: To which stages of the model flow does product life-cycle perspective implemented?

25.2 Circular Economy Concept

The late Pearce and Turner first proposed the idea of a circular economy in 1989, and it tackles the connections between four economic functions of the environment: amenity values, a resource base for the economy, a garbage bin for residual flows, and a life-support system [5]. In a circular economy, waste generation is reduced by thoughtfully designing new products and using an industrial process where resources are continuously recycled in a "closed-loop system". The demand for raw materials is higher than ever, and waste has spread widely throughout the entire earth. According to estimates, about 30% of processed food is wasted once it enters the food supply chain, and about 80% of all materials and consumer goods are discarded [6]. Implementing the circular economy concept promotes social progress, environmental conservation, and sustainable economic growth. A circular economy can improve the creation of new added value while decreasing environmental damage across the board.

The extract-produce-use-dump model (see Fig. 25.1) underlies the current linear production technique, which consumes energy throughout the whole manufacturing process. Instead, a circular economy model (see Fig. 25.2) makes it possible for raw resources to cycle and be changed from one form into another, producing little to no waste in the process. Seeing the detrimental effects of the linear manufacturing approach on the environment has led to the development of the circular economy concept.

There are three fundamental principles covered by this idea. The first is making the best use of natural resources by recycling, restoring damaged products, and reusing production waste. The second involves protecting and enhancing natural capital by managing resource consumption and balancing the supply of renewable energy sources. The third involves increasing productivity by recognizing and eliminating any harmful external consequences, such as those caused by exploiting agricultural land or by harming the air, water, or other elements of the environment. Although the circular economy concept comprises systemic changes based on innovations and leveraging new technical systems, as well as by changing how policies, society, business models, and funding methods are perceived and handled, consumer involvement plays a significant role in its implementation. In this instance, the primary objective is to create a CE model that enables the regeneration of materials, product components, and products in a way that preserves their best value for the longest amount of time. Resources should simultaneously be able to be molded and reintegrated into the economic system or used as fertilizers for the environment.

Fig. 25.1 Linear economy model

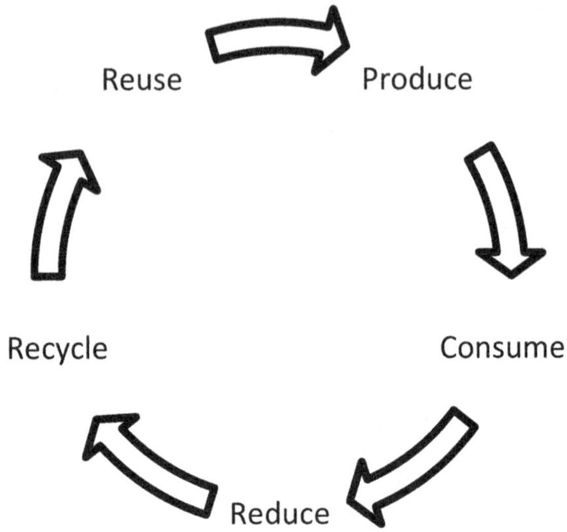

Fig. 25.2 Circular economy model

25.2.1 The Relationship Between Circular Economy and Sustainability

The idea of sustainability predates that of the Circular Economy. It was first introduced in the 1980s with the well-known Brundtland report as an economic development that takes into account the demands of current generations without compromising the potential of future generations to meet their own needs" [8]. This idea has mostly concentrated on anticipating the requirements of the environment and promoting social fairness by taking those demands into account. By promoting the adoption of environmental measures to reduce environmental impacts while achieving economic benefits, these findings have been applied to a single firm level. Additionally, some of the Sustainable Development Goals (SDGs), including SDG-6 (Clean Water and Sanitation), SDG-7 (Affordable Clean Energy), SDG-12 (Reasonable Consumption and Production), and SDG-15 (Life on Land), can be linked to circular economy practices.

A circular economy should aim to reduce resource exploitation and promote waste prevention because it is obvious that global resource depletion (and associated carbon emissions) is accelerating while ironically mountains of waste (and accompanying pollution) are still building up. However, a circular economy should aim to restore and regenerate the environment by supporting sustainability from the whole system perspective of optimizing social, environmental, technical, and economic values of materials and products in society. This is because depletion and pollution pose serious threats to the stability of an environment that is conducive to the thriving of the human species. Concerns about resource depletion and environmental deterioration as a result of ongoing economic growth are the foundation of sustainability research. The pluralist definition of circular economy, whose only shared goal is better resource

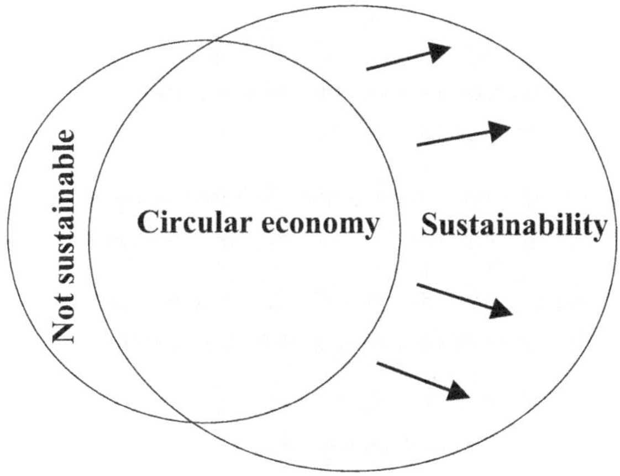

Fig. 25.3 A scheme of circular economy implementation to achieve sustainability (Adopted from [9])

use, results from the integration of many concepts that have their roots in the same founding literature. The prevailing view is that CE is used as a means of achieving sustainability, but CE and sustainability are distinct ideas that are closely related. Businesses of all sizes and industries said that CE can have a positive impact on sustainability (see Fig. 25.3). Notwithstanding the diversity of opinions, organizations typically utilize CE to advance sustainability rather than with circularity as their ultimate objective.

This is following those who view CE as a "toolbox" or "means to an end" rather than as a goal in and of itself, as shown by the arrows pointing to CE in the direction of sustainability in the figure above. This figure indicates that sustainability is directly related to CE integration. As another saying, the CE integration process should consider engaging with sustainability objectives and the CE model should be designed according to three dimensions of sustainability which are environmental, economic, and social. To address the most pressing sustainability concerns, the circular economy has access to a wide range of technologies, viewpoints, and solutions. By utilizing this diversity, circular economy practices may be implemented in a way that is specifically tailored to the local environment.

25.3 Literature Review About Circular Economy Strategies and Business Models

The literature has recently begun to include a number of frameworks that aid businesses in creating CE innovations. By recycling materials and/or energy after use to prevent leakage out of the system, the CE seeks to reduce resource consumption. As one of the primary fields in CE adoption, built environment is being addressed by a growing number of initiatives, laws, and scholarly studies. Systemic innovation across the value chain is necessary for the construction industry to make the transition to a circular economy. Resources (materials or components) recovered at the end of their useful lives must be reincorporated into the value chain. To do this, businesses must gather and recover building materials and components in an economical manner, as well as in adequate quantity and quality. Also, businesses at the start of the value chain must modify building designs to make it possible for construction materials to be recovered at the end of their useful lives and reused in new construction projects. Circular strategies are most effectively used for design optimization in early stages of design. In addition, the authors emphasized the vital importance of lifecycle optimization and policy in enhancing circularity in the built environment. Design for Disassembly (DfD), Design for Modularity, Specifying Reusable and Recyclable Materials, and Design for Remanufacturing are a few examples of common circular design techniques used in the built environment.

Companies that sell sophisticated products with a lot of embedded value may find retaining product ownership (RPO) to be an appealing approach. In order to implement this strategy, businesses may need to make significant investments in their after-sales and maintenance skills, which could be more costly for them and, ultimately, for their clients than a selling and replacing plan. Another strategy that companies use is product life extension (PLE), which focuses on developing items to endure longer and may create opportunities for marketplaces for old goods. For original equipment makers, this could seem like a poor idea as a longer product lifecycle results in fewer purchases over time. PLE can assist businesses in preventing their customers from switching to a competing brand. The final strategy, design for recycling (DFR), enables businesses to rethink their goods and production methods to maximize the number of materials that may be recovered for reuse in new products. This tactic frequently entails collaboration with businesses that have specialized technical knowledge or who may be best able to make use of the materials retrieved.

Instead of concentrating on a single circular loop, the butterfly diagram emphasizes the understanding of how biological and technological closed loops function as a continuous flow of materials across the value circle [10]. Using conceptual and methodological frameworks that sought to circumvent the limits associated with the construction of circular business models for numerous cycles, contributions in this direction have been addressed toward the extension of the literature on business models. In addition, the studies seek to include the 3R principles—reduce, reuse, and recycle—into the business model, as well as to identify new KPIs and two extra blocks (the take-back system and adoption factors) into the traditional depiction of a business model canvas framework to evaluate and clarify the various degrees of circularity in business models [11–13]. Recalling the many patterns found through morphological analysis of the alternative design possibilities is also helpful in terms of repair and maintenance, reuse and redistribution, remanufacturing, recycling, repurposing, and organic feedstock. The literature also largely acknowledges and refers to two more frameworks. The first is the participatory "backcasting and eco-design circular economy" framework, which combines operations and strategic planning. The second is the framework of regenerating, sharing, optimizing, looping, virtualizing, and exchanging, which was put forth to spur innovation in the circular economy among businesses [10].

25.4 Methodology: Circular Economy Business Model Framework Development

The only manner a circular economy business model can remain viable is if the product's value can be economically recovered. Recycling could be accomplished by reusing the product to increase the value of the resources and labor used in its production, or by disassembling it into its parts or raw materials so that they can be recycled for another purpose. According to another proverb, high-value items that are simple to use and process are best for circularity because they don't demand a big shift in the way businesses are run. In a while, it may be simple for companies to legally reclaim particularly large or bulky products and those containing potentially dangerous components, but difficult and expensive to carry and recondition. For this reason, it's crucial to identify the best strategy for the type of product and production process before beginning to design a CE business model.

At this phase, the designed framework (see Fig. 25.4) within this study helps in determining a company's appropriate place in adopting strategic alternatives for developing a circular business model. In this study, the framework behaves as a conceptual assessment tool which is developed according to literature reviews and academic feedbacks. However, it will be verified by more experts' feedbacks, even in sectoral field, and then, the multiple case study approach will be applied to validate the framework for accessing it to final form.

25.4.1 Exploratory Qualitative Research in the Study

This study employs a qualitative methodology that involves theme analysis of literature in the area of CE transition. Because this study is exploratory in nature, a qualitative and inductive methodology based on a review of the published reports, research papers, and documents from eminent institutions, organizations, researchers, and experts is required. For in-depth research and data collection, there are many qualitative research techniques available, including field research, open and in-depth interviews, participatory and nonparticipatory observation, literature analysis, thematic analysis, content analysis, case studies, historical studies, action studies, grounded theory, and other approaches. However, the primary method of analysis is induction. Almost two-thirds of the study in environmental management used qualitative methods, such as the frequent application of qualitative techniques in exploratory studies that also concentrated on the business side, to gather data via social surveys

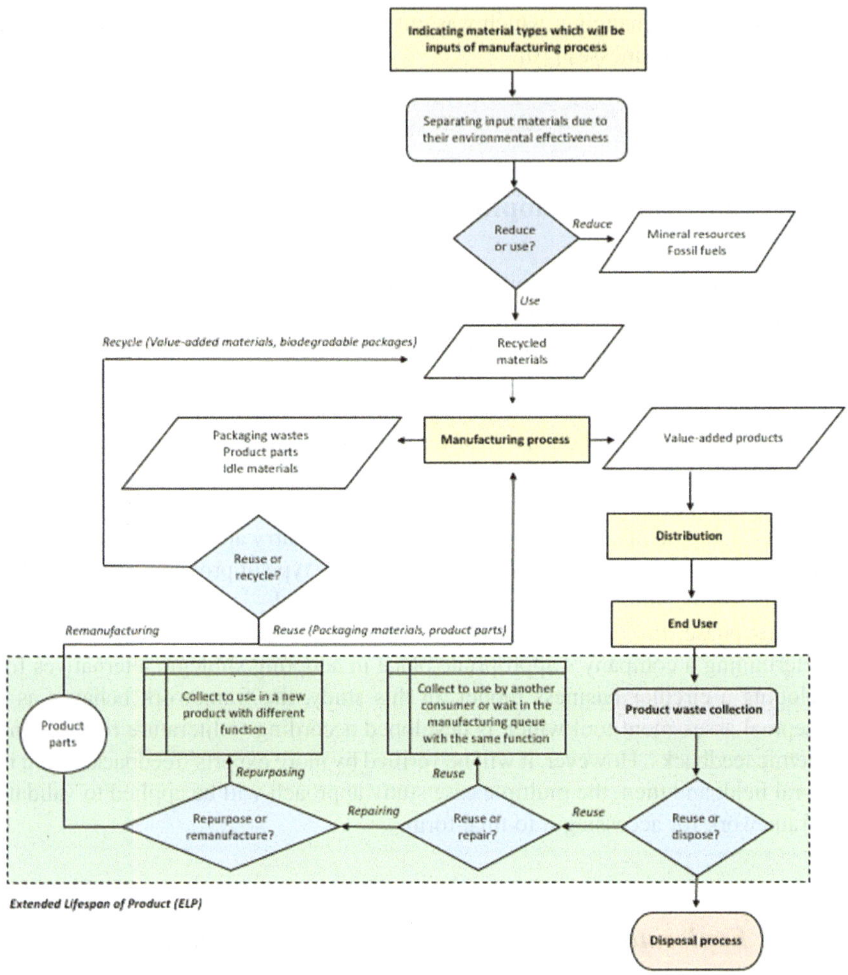

Fig. 25.4 Designed Circular Economy business model framework in the scope of extending product lifespan

and interviews. As a result, it is appropriate for investigating the tools and principles aiding CE transition for enterprises around the world.

Considering the defined research questions in Sect. 25.1, the first research step starts with a thematic analysis of the common CE business models suit for manufacturing processes. In this case, a total of 41 articles (conference proceedings, reports and working papers excluded) were selected by using the keywords: "Circular Economy" AND "Business Models" AND "Manufacturing" via Google Scholar database. Then, each article was classified according to its' CE strategy which affects the business model structure. After this classification, generic indicators were defined that are integrated with business model levels. At this point, it was realized that many studies present a model covering manufacturing operations do not evaluate product life cycle principle. Therefore, extending the product life span strategy was researched under the further literature review as an additional keyword next to the previous ones. In this step, 15 more articles were examined to understand which business model level should be redesigned to prolong product life within the case of specific manufacturing process. At the end, whole model levels were stated as a required process either is effective in the life of the product or not. They were kept as gathered data to design a framework which guides a business decision-making process for implementing product-based CE objectives into their actual business models.

25.4.2 Explanation of the Designed CE Business Model Framework

Raw materials are used as inputs in each manufacturing process. While some raw materials are environmentally friendly, others may cause pollution to the point where natural resources are harmed during manufacturing. As a result, it is critical to categorize production inputs in terms of the environment and sustainability. If a raw material contains harmful components, it is equivalent to the 3R principle's "reduce" action, which is used in production by selecting an environmentally friendly type that can be substituted instead. On the contrary, recycled materials that are generated as a result of a manufacturing process can be reused and transformed into value-added products after manufacturing. Product and part wastes generated after the delivery of value-added products to the end user are also evaluated within the context of the circular economy model. The recycling of product or part wastes generated by products that have completed their product life cycle is handled under the "extended lifespan of product (ELP)" strategy. At this point, it is determined which process of the cycle the wastes separated by type will be included in. The first decision point in this strategy is whether to integrate products or parts that can be "reused" without losing their existing functions in the distribution process to the end user, or to "repaired" ones to be included in the next manufacturing process while maintaining the properties constant. However, a product waste that has been "repaired" can be "repurposed" to have a different feature. For example, if a change in the product is required due to incoming customer demands, this action can enable the conversion of existing wastes to a new raw material type rather than ordering raw materials based on the new product type. All of these processes ensure cyclicity in the manufacturing process as well as product sustainability.

25.5 Conclusion

The goal of Circular Economy Business Models (CEBMs) is to "produce, deliver, and collect value while using circular methods that can lengthen the useful life of goods and parts (e.g., repair and remanufacturing) and close material loops (e.g., recycling)". There are review articles and bibliometric studies dealing solely with CE in the literature, as well as the relationship of CE with different ideas including the built environment, industrial symbiosis, industrial ecology, green and bio-economy, demolition waste sector, and sustainability. The adoption of eco-friendly products and technologies has made the construction industry one of the three with the highest potential for implementing CE policies. The application of the CE principle in the construction sector encourages the use of environmentally friendly products, maximizes material recovery, and reduces waste generation and landfill disposal. Even though the study's goal is to create an industry-independent generic framework, it can be employed particularly in the built environment, such as by prolonging a building product's life cycle. The focus of the aforementioned framework is that it covers the steps and decision mechanisms for extending the product life cycle, distinguishing it from the operational process design-oriented flow charts commonly encountered in the literature. It is crucial to keep in mind that while extended lifespan of product strategy is good at identifying the optimal option based on a certain metric at a particular time, it can occasionally cause us to pursue short-term, individual benefits at the price of long-term, communal ones. At this point, life cycle assessment (LCA) can be used to pinpoint effect "hotspots" within the lifespan of a given solution, such as emphasizing the particularly resource- or pollution-intensive phase of a product's lifecycle, and then help evaluate how well various solutions for that stage of the life cycle mitigate those impacts. Future study is therefore focused on this framework's integration with a product development R&D process that minimizes the use of resources (primarily materials, energy, water, and chemicals), hence dramatically lowering environmental impact.

References

1. Ahmed Z, Mahmud S, Acet DH (2022) Circular economy model for developing countries: evidence from Bangladesh. Heliyon 8(5):e09530
2. Korhonen J, Honkasalo A, Seppälä J (2018) Circular economy: the concept and its limitations. Ecol Econ 143:37–46
3. Ferasso M, Beliaeva T, Kraus S, Clauss T, Ribeiro-Soriano D (2020) Circular economy business models: The state of research and avenues ahead. Bus Strateg Environ 29(8):3006–3024
4. Antikainen M, Valkokari K (2016) A Framework for sustainable circular business model innovation. Technol Innov Manag Rev 6(7):5–12
5. Banaitė D (2016) Towards circular economy: analysis of indicators in the context of sustainable development. Soc Transform Contemp Soc 2016(4)
6. Scheel C, Aguiñaga E, Bello B (2020) Decoupling economic development from the consumption of finite resources using circular economy. A model for developing countries. Sustain 12(4):1–21

7. Bacher, J (1987) The brundtland report: our common future. Peace Mag
8. Walker AM et al (2022) What is the relation between circular economy and sustainability? answers from frontrunner companies engaged with circular economy practices. Circ Econ Sustain 2(2):731–758
9. Ellen McArthur Foundation: Towards The Circular Economy: Economic and business rationale for an accelerated transition. TransitIng Circ Econ, 1, (2013)
10. Ranta V, Aarikka-Stenroos L, Ritala P, Mäkinen SJ (2018) Exploring institutional drivers and barriers of the circular economy: A cross-regional comparison of China, the US, and Europe. Resour Conserv Recycl 135:70–82
11. Geissdoerfer M, Morioka SN, de Carvalho MM, Evans S (2018) Business models and supply chains for the circular economy. J Clean Prod 190:712–721
12. Centobelli P, Cerchione R, Chiaroni D, Del Vecchio P, Urbinati A (2020) Designing business models in circular economy: A systematic literature review and research agenda. Bus Strateg Environ 29(4):1734–1749
13. Heyes G, Sharmina M, Mendoza JMF, Gallego-Schmid A, Azapagic A (2018) Developing and implementing circular economy business models in service-oriented technology companies. J Clean Prod 177:621–632

Part XIII
Criteria and KPIs for Circular Buildings and Materials

Chapter 26
Key Indicators for Evaluating the Energy Efficiency Improvement of the Renovated Building Facades

Liljana Dimevska Sofronievska, Milos Knezevic, Meri Cvetkovska, Ana Trombeva Gavriloska, and Teodora Mihajlovska

Abstract Adopting the circular economy (CE) principles in building sector can reduce the quantity of materials used for the renovation of existing buildings, improve their energy performance and sustainability and minimize harmful emissions embodied in building materials. The main key indicators for energy performance evaluation of buildings, related to CE principles are: transmission losses, heating and electricity energy consumption, greenhouse gas emissions (GHGs), thermal comfort and financial costs for building maintenance. The building stock from the sixties and seventies is still in use, but from the aspect of energy efficiency, it shows a low level. From that reason, all these buildings have to be renovated. The effects of the renovation can be followed by the values of the key indicators. A simulation of a renovated scenario of an existing building was carried out and the results are presented in this paper. The analyzed building was built only in nature concrete without any facade thermal insulation. One of the renovation conditions was the appearance of the building should not be changed. An aerogel thermal plaster, which is nanomaterial with high thermal properties, was applied on the building facade. The results shows that the energy performance of the building is significantly improved in terms of reducing the heating energy consumption by 65%, electrical energy consumption by 40%, CO_2 emissions by 55%, PM10 particles by 46%, and the financial costs by 49%. According to the key indicators, it is found out that the renovation with appropriate material can significantly improve the building functionality.

Keywords CE · Key indicators · Energy efficiency · Emissions · Renovation

---

L. D. Sofronievska (✉) · M. Cvetkovska · A. T. Gavriloska · T. Mihajlovska
Ss. Cyril and Methodium University, Faculty of Architecture, Skopje, North Macedonia
e-mail: dimevska.liljana@arh.ukim.edu.mk

M. Knezevic
University of Montenegro, Faculty of Civil Engineering, Podgorica, Montenegro

L. Bragança et al. (eds.), *Creating a Roadmap Towards Circularity in the Built Environment*, Springer Tracts in Civil Engineering,
https://doi.org/10.1007/978-3-031-45980-1_26

26.1 Introduction

Buildings are responsible for 40% of annual energy consumption and 36% of annual greenhouse gas (GHG) emission in EU. From this reason, improving the energy efficiency and sustainability of the building stock is critical for meeting EU climate targets [1]. Circular economy (CE), especially in the building sector, strive to reduce the pollution, extend the building's lifespan, reduce the material waste and favorite the use of long-lasting building materials and products.

Proper renovation by using sustainable materials with low embodied energy will lead to the fulfillment of both goals, namely energy efficiency and circularity. The key indicators for evaluation the energy efficiency improvement of the buildings such as energy consumption, greenhouse gas emissions and costs are very important, especially in the process of renovation of existing buildings. In order to assess the efficiency of the measures taken to improve the energy performance of old buildings that were built without thermal insulation and whose function and architectural appearance are compromised, a dynamic energy simulation of the renovated scenario was made. For the simulated scenario, a sustainable and low embodied energy thermal insulation material was applied on the façade walls and the values of the key indicators were registered. The key indicators show a significant improvement of the energy performance of the building.

26.2 Key Indicators

26.2.1 Energy Consumption

Buildings energy consumption prediction is essential to achieve energy efficiency and sustainability [2]. Buildings energy consumption is mostly highly dependent on buildings characteristics such as shape, orientation, envelope and building materials. The comparison of the energy consumption for the original building and the improved scenario, is one of the most important key indicators for the buildings energy efficiency improvement. Different types of the energy consumptions, such as energy consumption for heating, electricity energy consumption for: cooling, lightning, equipment etc., are analyzed and presented in this paper.

26.2.2 Greenhouse Gases

Building construction and operations account for 36% of global final energy use and 39% of energy-related carbon dioxide CO_2 emissions [3]. These emissions from building operation arise from the energy used for heating and/or cooling, hot water supply, ventilation and air conditioning, lighting, and from the embodied energy for the production of building materials [4].

Cutting the GHGs in the building sector is a key indicator for not only the energy efficiency improvement, but also it is much more important from aspect of the climate changes and CE measures in the building sector. In order to assess the improvement that the application of the new façade material has on the building energy performance, the CO_2 emissions and PM10 particles in case of the original building and in case of the improved scenario of the selected building are defined and the comparison of both scenarios are presented in this paper.

26.2.3 Financial Costs for Maintenance

Maintenance and operation costs are part of the buildings life cycle costs [5]. The maintenance of the analyzed building is highly dependent of the heating and cooling conditions and the corresponding bills are responsible for the high financial costs of the building. The reduction of the bills for heating and cooling is a key indicator for the improvement of the energy efficiency of the building.

26.3 Façade Material

Aerogel-based building products are currently considered to be promising insulation materials mostly due to the fact they have high thermal performances with limited thickness. Furthermore, they have a very low embodied energy, lower than traditional insulation products [6–12]. In order to keep the original appearance of the building and at the same time to improve the thermal comfort, energy efficiency and costs, aerogel thermal plaster is used as a facade insulation material in the improved renovated scenario. The aerogel thermal plaster has a thermal conductivity of 0.028 W/mK and even applied in a small thickness has a great insulating effect as a result of its nano porous structure [7]. Due to the composition and method of application, aerogel plasters perfectly mimic the texture of natural concrete, while the original material remains preserved under the plaster. The cost of the aerogel is still high, which prevents its intense use in construction.

26.4 Energy Simulation

A dynamic energy simulation of the original building and the improved scenario in which the facade is renovated by applying the aerogel thermal plaster has been carried out by using Design Builder and EnergyPlus software [13]. The goal was to

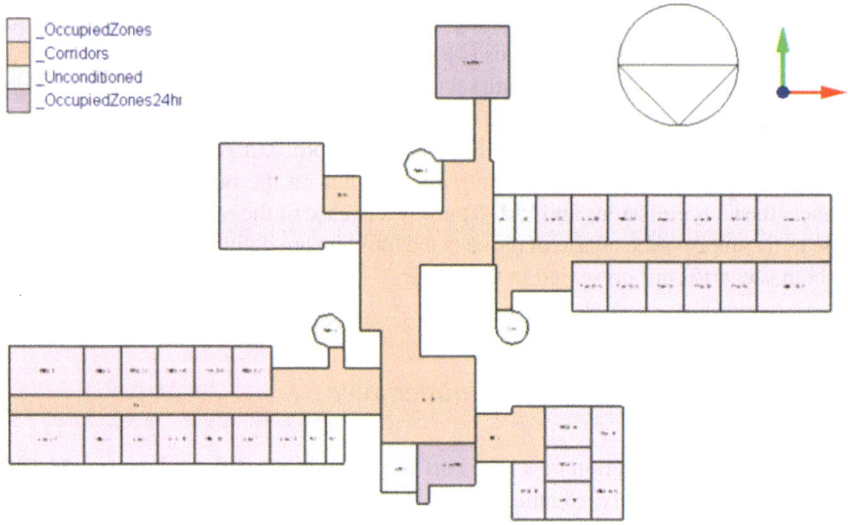

Fig. 26.1 Thermal zones division—ground floor

evaluate the energy efficiency improvement of the building by comparison of the key indicators for both cases, the existing building and the improved scenario.

The selected case study building is an office building, considered as a cultural heritage from the post-earthquake period in Skopje. The structure and the facade are designed are built in concrete, with no insulation and the appearance of the facade is untreated natural concrete. The selected building floor plan is shown in Fig. 26.1, where the principle of dividing the building into thermal zones can be find out. The building is divided into 140 thermal zones.

Each of the zones has its own design temperature, orientation, number of people, lighting, electrical equipment and appliances, type of heating, cooling, ventilation, glazing area, etc. The designed room temperature in the offices is 20 °C, and in the halls and corridors is 15 °C. The outdoor climate data are defined by appropriate measurements. The building general information such as gross area, volume, openings etc. are presented in Table 26.1.

26.5 Results

26.5.1 Heating Energy Consumption

The simulation results of the existing condition of the building show that the building is a large energy consumer during winter time. This is due the lack of thermal insulation of the building's envelope. This implies large financial costs for maintaining the

Table 26.1 Thermal zones summary

Zone summary	Area [m^2]	Volume [m^3]	Above ground gross wall area [m^2]	Underground gross wall area [m^2]	Window glass area [m^2]	Total openings area [m^2]
Total conditioned area (m^2)	2647.7	8085.9	2468.9	9.69	658.8	712.3
Total unconditioned area (m^2)	694.5	2103.8	694.7	325.57	43.3	49.3

thermal comfort. By improving the heating energy consumption which is defined as a key indicator for the energy efficiency evaluation, it can be concluded the energy efficiency is significantly improved in the renovated scenario (see Fig. 26.2), which leads to reduced financial costs for maintenance and improved thermal comfort. Figure 26.2 shows the graphs of the average monthly values for heating energy consumption in kWh, for both, the existing condition and improved scenario, also shown in Table 26.2.

In the actual scenario, or existing condition, the average monthly heating energy consumption is 27 684.9 kWh (see Table 26.1), which means 332 218.8 kWh annually or 125.3 kWh/m2. Scenario 1 showed a reduction of the heating energy by 65%, which means that the average monthly heating energy consumption is 8 765.8 kWh (see Table 26.1). This means that the annual heating energy consumption in the improved scenario is 105 186.6 kWh or 40 kWh/m2.

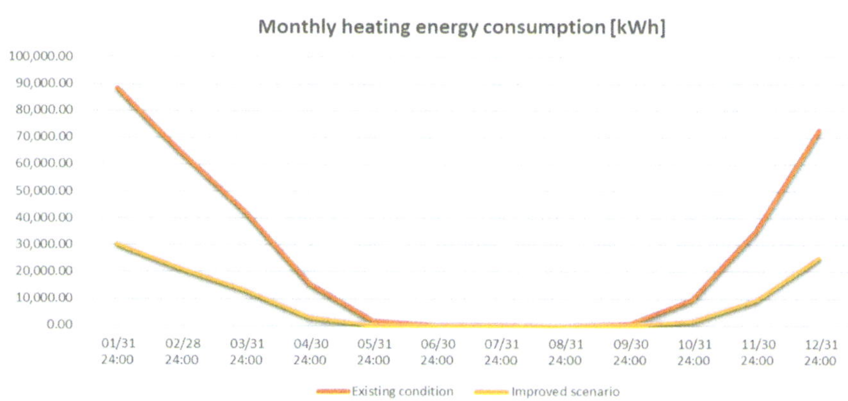

Fig. 26.2 Comparisons of monthly heating energy consumption between actual scenario and improved scenario

Table 26.2 Key indicators summary

Key indicators	Existing condition	Improved scenario
Heating energy [kWh]	27 685	8 765.7
Electricity heating energy [kWh]	6 133.5	2.96
Electricity cooling energy [kWh]	865.5	646.2
Total electricity energy [kWh]	16 157	9 736
CO_2 emissions [kg]	14 022.5	7 017.9
PM10 particles [kg]	1.3	0.7

26.5.2 Electricity Energy Consumption

The total electricity energy consumption is divided into electricity for additional heating (electric heaters); electricity for cooling (air conditioners); electricity for lighting and electricity from electrical appliances and equipment. The results show that apart from high consumption of thermal energy for heating, the building also consumes electricity for heating, which indicates the poor thermal insulation of the building, that despite the high consumption of heating energy, the heating system in the coldest months does not satisfy the thermal comfort and additional electrical heating is used. In addition, the simulations show high-energy consumption for cooling during the summer, which again indicates the poor thermal characteristics of the building envelope.

The average monthly total electricity energy consumption (heating, cooling, appliances and lighting) for the existing condition is 16 154 kWh, i.e. 193 848 kWh annually or 73.1 kWh/m^2 (see Fig. 26.3 and Table 26.2). The simulations of the improved scenario show an improvement in the consumption of total electricity (See Fig. 26.3 and Table 26.2) and also in both, electricity for heating and electricity for cooling the building (See Table 26.2 and Fig. 26.4). The average monthly total electricity energy consumption (heating, cooling, lighting equipment) is reduced by 40%, or 9 736 kWh, i.e. 116 832 kWh annually or 44 kWh/m^2.

The electricity energy consumption for heating, cooling and maintenance of the building, is key indicator for evaluating the energy efficiency improvement, thermal comfort and financial costs of the building.

26.5.3 CO_2 Emissions and PM10 Particles

Figure 26.5 shows the comparisons of the monthly CO_2 emissions of the building between actual scenario (existing condition) and the improved scenario. In the existing condition the monthly CO_2 emissions are 14 022.5 kg. The improved scenario shows much lower CO_2 emission i.e. the average monthly emissivity is 7 017.9 kg, which means that there is a reduction of the CO_2 emissions by 50%. (See

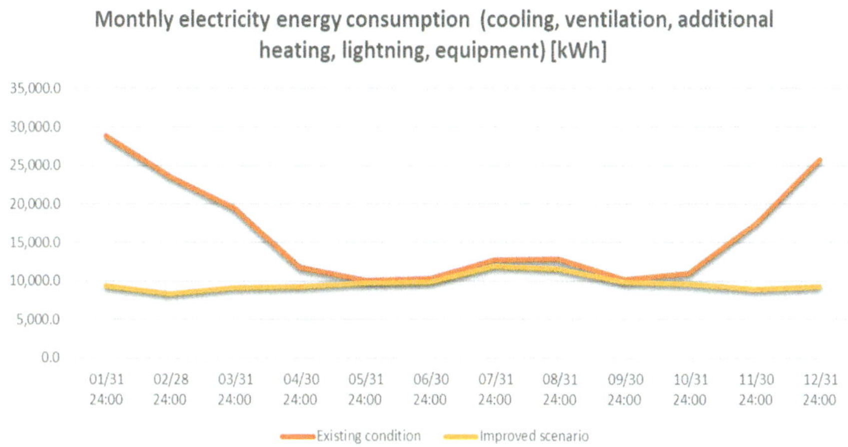

Fig. 26.3 Comparisons of total monthly electricity energy consumption between actual scenario and improved scenario

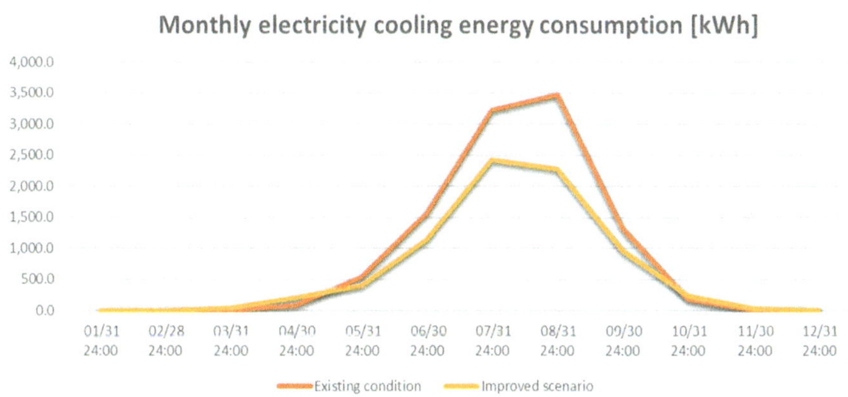

Fig. 26.4 Comparisons of monthly electricity energy consumption for cooling between actual scenario and improved scenario

Table 26.2 and Fig. 26.5). Reducing the CO_2 emissions is not just a key indicator for energy improvement evaluation, but also an indicator for CE implementation by proper buildings renovation. The same situation is with the PM10 particles reduction. From Fig. 26.6 and from Table 26.2 it can be concluded that great reduction of PM10 particles in the improved scenario is achieved. In the existing condition, the building emits an average of 1.3 kg monthly, or 15.6 kg annually. The improved scenario shows lower emission of PM10, i.e. an average of 0.7 kg monthly or 8.4 kg annually. It can be concluded that by adding insulation on the building envelope, the PM10 emissivity is reduced by 46.1% compared to the actual scenario. The PM10 emission is a key indicator for reducing the air pollution.

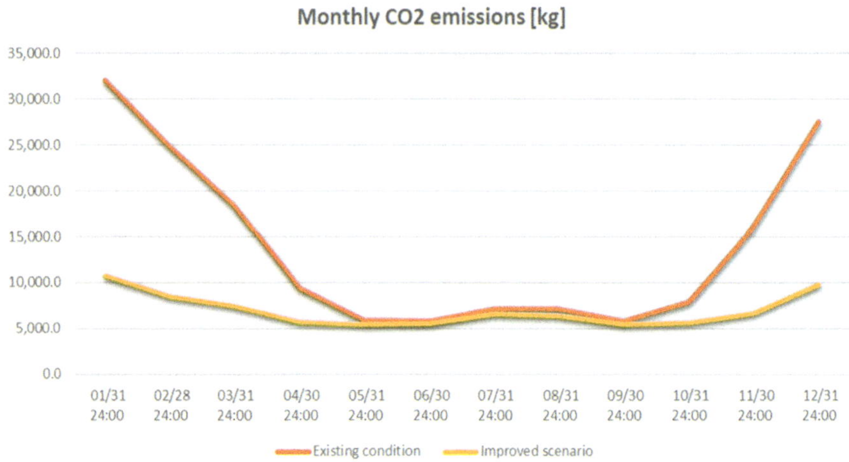

Fig. 26.5 Comparisons of monthly CO_2 emissions between actual scenario and improved scenario

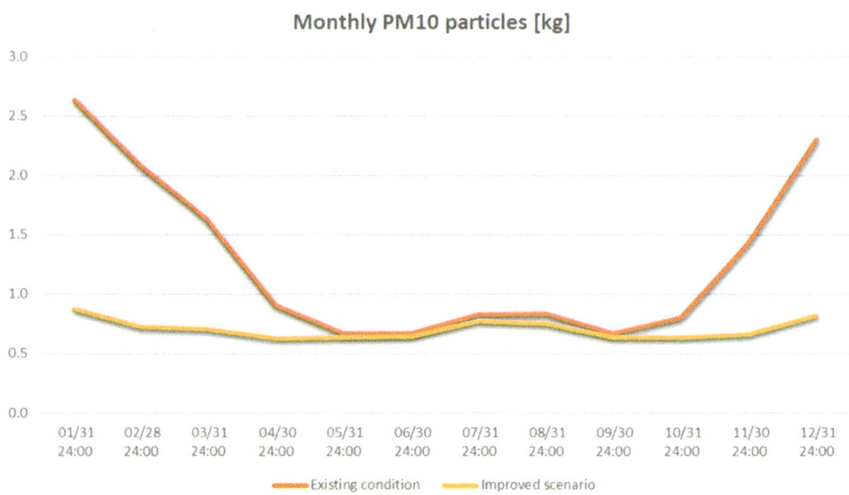

Fig. 26.6 Compariosons of monthly PM10 particles between existing condition and improved scenario

26.5.4 Financial Costs

Finally, a financial analysis are carried out for existing situation improved scenario, (See Fig. 26.7). It can be seen that the annual building's maintenance costs (heating and cooling) are reduced by 49% in the improved scenario compared to the existing condition. The highest costs are observed during the winter months, while the lowest during May, June and September, when the outside temperature is closest to the inside

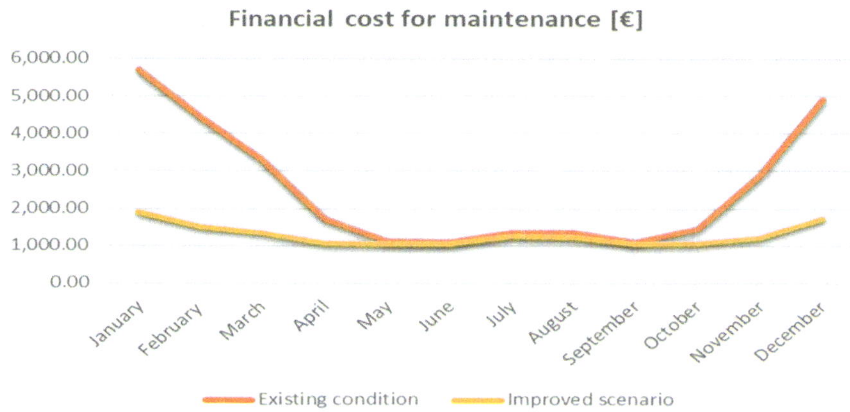

Fig. 26.7 Comparisons of monthly PM10 particles between actual condition and improved scenario

temperature. This analysis proves the role of the thermal insulation of the envelope in the reducing of the building's maintenance costs.

26.5.5 Key Indicators Summary

The key indicators that play the most important role in the evaluation of the energy efficiency improvement of the building are summarized in Table 26.2. By com-paring the indicators of the actual scenario (existing condition) and the improved scenario, it can be concluded that the proper building renovation can significantly reduce energy consumption, emissions, financial costs, and improve the general energy performance of the building.

26.6 Conclusion

Reducing the building's energy consumption and greenhouse gases, lowering the financial costs, improving the thermal comfort and lifespan of the building, are both energy efficiency and CE key indicators which must be analyzed before the renovation of existing buildings. All of the above mentioned key indicators are analyzed in this paper in order to evaluate the improvement of the energy performance of an existing building after its renovation. For that purpose, a simulation of both, the existing condition of the building and the improved renovated scenario with new façade material application were made. The results showed that the buildings energy efficiency is significantly improved in terms of reducing the heating energy consumption by 65%, electrical energy consumption by 40%, CO_2 emissions by 55%, PM10 particles by 46%, and the financial costs by 49%. It can be concluded that the key indicators play a big role in the energy efficiency and CE improvement evaluation.

References

1. EEA (2022) Building renovation: where circular economy and climate meet. Eur Environ Agency
2. Amal A. Al-Shargabi, Abdulbasit Almhafdy, Dina M Ibrahim, Manal Alghieth, Francisco Chiclana (2022) Buildings' energy consumption prediction models based on buildings' characteristics: Research trends, taxonomy, and performance measures, J Build Eng, 54
3. Abergel T, Dean B, Dulac J, Hamilton I, Wheeler T (2018) Global status Report—Towards a zero-emission, efficient and resilient buildings and construction sector
4. Röck M Ruschi M, Saade M, Balouktsi M, Rasmussen Nygaard F, Birgisdottir H, Frischknecht R, Habert G, Lützkendorf T, Passer A (2020) Embodied GHG emissions of buildings—The hidden challenge for effective climate change mitigation, Appl Energy, 258
5. Krstic H, Marenjak S (2020) Analysis of buildings operation and maintenance costs, UDK 624.012.004.5:69.003.12, Gradzevinar 4
6. Curto DD, Cinieri V (2020) Aerogel-based plasters and energy efficiency of historic buildings. Literature review and guidelines for manufacturing specimens destined for thermal tests. Sustain 2020, 12, 9457; https://doi.org/10.3390/su12229457
7. Stahl Th, Brunner S, Zimmermann M, Ghazi WK (2016) Thermo-hygric properties of a newly developed aerogel based insulation rendering for both exterior and interior applications. Energy Build, 44, pp 114–117
8. Handojo DU, Xiaodong L, Eng TJN (2022) Sustainable production in circular economy: aerogel upscaling production. Environ Sci Pollut Res 29:20078–20084, Springer
9. Carty L (2017) Analysis of the effects of aerogel insulation on the thermal performance of existing building envelopes, Edinb Napier Univ, 09013398
10. Ganobjak M (2019) Aerogel materials for heritage buildings: Materials, properties and case studies. J Cult Herit, Elsevier
11. Castro-Díaz M, Osmani M, Cavalaro S, Parker B, Lovato T, Needham P, Thompson J, Philippe K, Ruiz F (2022) Impact of circular silica aerogel on plasterboard recycling. loughborough university. In: Conference Contribution. https://hdl.handle.net/2134/21435765.v1
12. Ibrahim MP, Biwole HP, Achard P, Wurtz E, Ansart G (2015) Building envelope with a new aerogel-based insulating rendering: Experimental and numerical study, cost analysis, and thickness optimization. Appl Energy 159:490–501
13. Strand KR, Crawley BD, Pedersen OC, Liesen JR, Lawrie KL, Winkelmann FC, Buhl WF, Huang YJ, Fisher ED (2016) EnergyPlus: A New-Generation Energy Analysis and Load Calculation Engine for Building Design, Conference: ACSA Technology Conference. Massachusetts, Cambridge

Part XIV
Circular Building Assessment Models, Tools and Frameworks

Chapter 27
SDGs and the SBTool^{PT} Urban Assessment Method: A Conceptual Analysis of Urban Sustainability Indicators to Assess the Circular Economy in the Built Environment

Genesis Camila Cervantes Puma and Luis Bragança

Abstract The Sustainable Development Goals (SDGs) and the SBTool^{PT} Urban assessment method can be used together to assess the circular economy in the built environment and promote urban sustainability. In this conceptual analysis, the focus is on developing urban sustainability indicators that can help to assess the circular economy in the built environment SBTool^{PT} Urban includes a set of tools that are designed to help to align operations and activities with the SDGs. These indicators should be aligned with the SDGs and the SBToolPT Urban assessment method, which provides a comprehensive framework for assessing sustainability in urban areas. Using these indicators, the SBToolPT Urban can become an assessment method for projects that are wished to be implemented based on the SDGs. Through these indicators, urban areas can be assessed on their sustainability and circular economy performance and opportunities for improvement can be identified as well. This can help to promote sustainability and circular economy in the built environment and contribute to the achievement of the SDGs related interconnected and provide a comprehensive framework for achieving sustainable development and addressing the world's most pressing environmental, social, and economic challenges.

Keywords Circular economy · Sustainable development goals · Urban sustainability · SBTool^{PT} urban

---

G. C. Cervantes Puma · L. Bragança (✉)
University of Minho, ISISE, ARISE, Department of Civil Engineering, Guimarães, Portugal
e-mail: braganca@civil.uminho.pt

G. C. Cervantes Puma
e-mail: cami_cervantes@hotmail.com

L. Bragança et al. (eds.), *Creating a Roadmap Towards Circularity in the Built Environment*, Springer Tracts in Civil Engineering,
https://doi.org/10.1007/978-3-031-45980-1_27

27.1 Introduction

It is well known that Circular Economy indicators are not fully defined because the scope and magnitude of their approach are extremely broad this leaves more freedom to generate criteria with the basic concepts of Circular Economy. Circular Economy has three main principles. The first is based on preserving and enhancing natural capital by creating a balance in the flow of renewable resources. The second is to optimize the circular use of resources by calculating their highest level of usefulness, both technically and biologically. Finally, the third one consists in stimulating the efficiency system by trying to exclude all external negative points on the first two points [1].

This study took into account these three parameters to study the sustainable development goals (SDGs) and to take those that are compatible with SBTool^{PT} Urban [2] to establish this new scope of the evaluation. SBTool^{PT} Urban is a Sustainable Building Assessment Tool that promotes sustainable design, construction, and operation of buildings and urban areas. SBTool^{PT} Urban evaluation parameters must be nalysed based on Sustainable Developments Goals (SDGs) [3] since they are general objectives at a global level. With this, it is possible to have a much greater scope of an evaluation system, and in SBTool^{PT} Urban, that is why this research tries to build upon an already created evaluation system, generate new categories to broaden the scope and make its use much easier at a global level. It has to be taken into account that innovation in these issues is implemented progressively, this will always leave a gap to be able to implement new parameters in any evaluation method.

The urban circular economy is a concept that focuses on applying the principles of the circular economy to urban environments, intending to promote sustainable and resilient cities. In the context of the circular economy, urban areas are seen as hubs of economic activity and innovation, where resources can be used more efficiently and waste can be minimized through circular systems of production, consumption, and waste management.

The urban circular economy can be seen as a key strategy for promoting sustainable urban development and achieving several of the SDGs, by integrating environmental, social, and economic considerations into urban planning and management. Overall, the urban circular economy can contribute to more sustainable and livable cities, by promoting resource efficiency, reducing waste and pollution, and creating new economic opportunities and jobs.

In addition, it is known that after eight years of implementation of the 2030 agenda [4] with three fundamental principles Inclusion, Participation, and Accountability, several countries are making efforts to incorporate new sustainable development practices. Therefore, UN DESA [5] and UNDP [6] have developed a theoretical and analytical framework to enable all governments and stakeholders to generate a greater scope with essential facilities that play a critical role in supporting countries in achieving sustainable development goals, promoting human development, and addressing critical global challenges.

27.2 Literature Review

27.2.1 Urban Circular Economy

There are research that present an overview of CE strategies in different scopes and from different perspectives, and importance. [7]. Urban circular economy is an approach to resource management in urban areas that aims to create a more sustainable and resilient economy. It involves designing out waste and pollution, keeping products and materials in use for as long as possible, and regenerating natural systems. The concept is based on the principles of the circular economy, which is an economic model that promotes the use of resources in a more sustainable and efficient way.

Urban circular economy can be applied to a wide range of economic activities in cities, including construction and demolition, transportation, energy, food production and consumption, and waste management. Some of the key elements of urban circular economy include Resource Efficiency, Reuse and Recycling, Regenerative Systems and Collaboration and Innovation.

Urban circular economy can play a crucial role in achieving the SDGs by promoting sustainable resource use, reducing waste, creating green jobs and economic opportunities, and reducing the negative impact of economic activities on the environment and society in urban areas. It can help create more livable and sustainable cities and contribute to the transition towards a more sustainable and equitable global economy.

27.2.2 Sustainable Developments Goals (SDGs)

The Sustainable Development Goals (SDGs) are a set of 17 global goals established by the United Nations General Assembly in 2015 [8] as a universal call to action to end poverty, protect the planet, and ensure peace and prosperity for all by 2030. The SDGs are designed to be a blueprint for a better and more sustainable future, with the aim of creating a world that is more equitable, just, and sustainable for all people and the planet. The scope of the Sustainable Development Goals (SDGs) is broad and comprehensive, covering a wide range of social, economic, and environmental issues that are critical for achieving sustainable development. These goals are interconnected and interdependent, and they address various aspects of sustainable development, such as poverty reduction, health and education, gender equality, environmental protection, and sustainable economic growth. The SDGs are designed to be universal and apply to all countries, regardless of their level of development, and they aim to balance economic, social, and environmental sustainability to create a better world for present and future generations.

27.2.3 SBTool^{PT} Urban

SBToolPT Urban is a sustainability assessment and certification system developed in Portugal for urban areas. It focuses specifically on the sustainability of neighborhoods, districts, and cities, and assesses the performance of these areas against a set of sustainability criteria, such as environmental quality, energy efficiency, mobility and accessibility, water and waste management, social cohesion and inclusion, and economic development and innovation. The goal of SBTool^{PT} Urban is to support decision-making processes related to sustainability and promote sustainable urban development in Portugal. It also supports the implementation of sustainability policies and strategies at the local level, fostering a more sustainable and livable urban environment.

SBTool^{PT} Urban is organized into Dimensions, Categories, and Indicators. These categories have an optimal life cycle analysis based on EN 15,942:2021 [9]. In addition, there are additional criteria for Sustainable Buildings and Environmental Management, which are intended to reward urban areas that possess additional attributes to those already established. This method is intended to clarify the relevance of the scale of intervention to establish an evaluation of urban areas. The evaluation process is graded in 3 stages: the first one where the performance is quantified by indicators, the second one where the performance of categories is graded, and the last one where the development by dimensions is graded. After this, the sustainability certificate can be issued [10].

The categories defined in Dimensions intend to add a set of indicators related to the verification of sustainability in the Environmental, Social, and Economic areas. Where in the Environmental Dimension the contributions to the sustainability of the environment, use of energy, water, metals, and waste, among others, are assessed. The Social Dimension manages the improvements to the aspects related to the users' quality of life like safety, accessibility, and mobility, among others. And the Economic Dimension seeks to identify the economic expenditure to generate implementations reducing costs, promoting the local economy and local employment.

27.3 Materials and Methodology

To evaluate the SDGs and SBTool^{PT} Urban's as a tool for Urban Sustainability to assess a Circular Economy, the first step is the literature review. The research was carried out on the following scientific databases: Web of Science, Scopus, Science Direct, United Nations Goals, and Google Scholar. It addressed the keywords 'Urban Sustainability', 'Sustainability Developments Goals', 'Urban growing', 'Circular Economy', 'Urban Impact on Economy' and 'urban sustainability assessment method'. This information is taken into the following main subjects: Sustainability Development Goals, Circular Economy, Urban Sustainability, Impact of Urban growth on the Cities Economy. Then the second step is to make an in-depth study of the SBTool^{PT} Urban indicators. This analysis focuses on how many indicators could be parameterized by Sustainability Development Goals (SDGs). Based on the literature review, the main potential is to make a combination between the SBTool^{PT} Urban indicators and Sustainability Development Goals (SDGs) to make a better method of asses Circular Economy. Finally, a simplified framework of indicators is proposed according to the Indicators of SBTool^{PT} Urban and Sustainability Development Goals. The indicators that we are going to use are discussed in the following section.

27.4 Results and Discussion

The literature review shows that the objective is to perform an analysis of the parameters of the Urban SBTool^{PT} based on the SDGs, to have a greater scope on the 2030 goals of the United Nations, generating an evaluation method for new SDG projects.

While the SDGs and SBTool^{PT} Urban differ in their scope and focus, they share a common goal of promoting sustainable development. In fact, many of the goals and targets of the SDGs are relevant to the sustainability criteria of SBTool^{PT} Urban, such as sustainable cities and communities, clean water and sanitation, affordable and clean energy, and more. Therefore, the implementation of SBTool^{PT} Urban can contribute to achieving the SDGs at the local level, especially in urban areas.

Although there is a remarkably clear link between SDGs and SBTool^{PT} Urban, there is no study conducted on the linkage of these urban sustainability parameters. On the other hand, there are project initiatives that bring the objectives of the SDGs and urban sustainability methods into use throughout the world. There are several types of urban sustainability assessment, such as LEED (Leadership in Energy and Environmental Design), BREEAM (Building Research'Establishment's Environmental Assessment Method), and DGNB (German Sustainable Building Council), of which the first two focus more on environmental aspects, while DGNB is a method adapted for commercial buildings, and provides performance metrics for commercial buildings [11].

There are several SDGs that work with SBTool^{PT} Urban because this tool promotes sustainable practices in buildings and urban areas, which can significantly contribute to the achievement of some of the Sustainable Development Goals. The reviewed literature has shown that there are seven SDGs, which maintain an entire relationship with SBTool^{PT} Urban.

By assessing the environmental, social, and economic performance of buildings and urban areas, SBTool^{PT} Urban can help to achieve several SDGs, including SDG 6 (Clean water and sanitation), SDG 7 (Affordable and Clean Energy), SDG 9 (Industry, Innovation, and Infrastructure), SDG 11 (Sustainable Cities and Communities), and SDG 12 (Responsible Consumption and Production), SDGs 13 (Climate Action) and SDGs 15 (Life on Land).

For all the above mentioned in the previous analysis, several SDGs were selected that work in the same scope as SBTool^{PT} Urban. The selected SDGs are listed in Table 27.1.

The SBTool^{PT} Urban's dimensions were taken to generate a greater link between each one. The dimensions are divided between Environmental, Social and Economic. After the reviewed literature, SDG 11, SDG 15, SDG 6, and SDG 7 are considered environmental close related because they are all related to protecting and preserving the natural environment, which is essential for sustaining life in the planet. SDG 12 and SDG 13 are considered social goals because they focus on promoting sustainable development and addressing issues related to social and economic well-being, while also taking into account environmental concerns. SDG 9 is considered an economic goal because it focuses on promoting sustainable economic growth and development, while also addressing issues related to infrastructure and innovation.

To justify this division, an analysis of several projects like Solar Heat for Industrial Process towards Food and Agro Industries commitment in Renewables (SHIP2FAIR) aims to foster the integration of solar heat in industrial processes of the agro–food industry [13]; "Calvo Residuo Cero" aims of sustainably valuing all the waste generated in its activity and reducing single-use materials [14], that have already been implemented or are in the process of implementation was carried out, which allowed for the identification of the area of work of each of the SDGs collected.

27.4.1 Environmental Dimension

In the environmental dimension, there are SDGs 11, 15, 7 and 6, due to their impact on urban form, land use, ecology, water, energy and materials and waste. Following, it is intended to show several projects according to each SDGs.

SDG 11 is focused on creating sustainable cities and communities. Is credited with several projects such as: Smart grids and energy management systems such as bus rapid transit (BRT), light rail transit (LRT), and bike-sharing programs, that help reduce energy consumption, while smart traffic management systems reduce traffic congestion and air pollution; Promoting the use of energy-efficient buildings can reduce energy consumption and greenhouse gas emissions and can also save money on energy costs for building owners and tenants.

SDG 15 focuses on protecting, restoring and promoting sustainable use of terrestrial ecosystems, sustainably managing forests, combating desertification, halting and reversing land degradation, and biodiversity conservation: Establishing protected areas and wildlife conservation programs can help protect biodiversity and preserve ecosystems; Restoring degraded land through reforestation, wetland restoration, and other restoration techniques can help promote biodiversity, reduce soil erosion, and restore ecosystem functions.

SDG 6 is focused on ensuring access to clean water and sanitation for all and improving water quality and efficiency. Environmental projects that can help achieve this goal include: Improving sanitation infrastructure and promoting safe

Table 27.1 Chosen sustainable developments goals

Goals	Purpose	Analysis
SDG 6: Clean water and sanitation	Ensure availability and sustainable management of water and sanitation for all	Where energy efficiency and wastewater treatment can be introduced to boost their level of sustainability
SDG 7: Affordable and clean energy	Ensure access to affordable, reliable, sustainable and modern energy for all	The aim is to improve energy use, decrease the use of fossil fuels and implement better energy use to increase the use of renewable energies
SDG 9: Industry, innovation, and infrastructure	Build resilient infrastructure, promote inclusive and sustainable industrialization and foster innovation	Which addresses the implementation of innovative ways of approaching new construction and implementing new industry methods
SDG 11: Sustainable cities and communities	Make cities and human settlements inclusive, safe, resilient and sustainable	The parameters are introduced which help to form new ways of achieving sustainable cities, which are important in the urban growth of these cities
SDG 12: Responsible consumption and production	Ensure sustainable consumption and production patterns	Which introduces us to the responsible use of it for the optimization of Circular Economy strategies, generating less waste and emissions
SDGs 13: Climate Action	Take urgent action to combat climate change and its impacts	Recognizes that climate change is one of the biggest challenges facing humanity, and that addressing it will require action at all levels, from individuals to governments and international organizations
SDGs 15: Life on Land	Protect, restore and promote sustainable use of terrestrial ecosystems, sustainably manage forests, combat desertification, and halt and reverse land degradation and halt biodiversity loss	Focused on protecting, restoring, and promoting sustainable use of terrestrial ecosystems, forests, and biodiversity. The goal recognizes the crucial importance of preserving life on land for both current and future generations

Source United Nations and Author [12]

and hygienic sanitation practices can help improve health outcomes and reduce water pollution; Capturing and using rainwater can help reduce water waste, increase water availability, and promote sustainable water use.

SDG 7 is focused on ensuring access to affordable, reliable, sustainable, and modern energy for all: Building smart grids and investing in energy storage technologies can help integrate renewable energy sources into the grid, reduce energy waste, and increase energy efficiency; Encouraging the development and use of renewable energy sources such as solar, wind, and hydropower can help reduce greenhouse gas emissions and promote sustainable energy production.

27.4.2 Social Dimension

In the Social Dimension, there are SDGs 12 and 13, due to their impact on Exterior Comfort, Safety, Amenities, Mobility, Local and Cultural Identity. Following, it is intended to show several projects according to each SDGs.

SDG 12 is focused on promoting sustainable consumption and production patterns and reducing waste: Implementing circular economy initiatives such as product life extension, product sharing, and material recovery can help promote sustainable consumption and production while also creating new economic opportunities; Encouraging sustainable waste management practices such as recycling, composting, and reducing waste generation can help reduce the environmental impact of consumption and production.

SDG 13 is focused on taking urgent action to combat climate change and its impacts: Implementing measures to reduce greenhouse gas emissions, such as promoting renewable energy, improving energy efficiency, and reducing emissions from transportation and industry; Engaging communities in climate action, such as through citizen science projects, participatory planning, and community-led initiatives.

27.4.3 Economic Dimension

In the Economic Dimension, there is SDGs 9, due to their impact on Employment and Economic Development. Following, it is intended to show several projects according to each SDGs.

SDG 9 is focused on building resilient infrastructure, promoting inclusive and sustainable industrialization, and fostering innovation: Providing support for SMEs, such as access to finance, business development services, and technology transfer, can help promote economic growth and job creation; Investing in infrastructure, such as transportation networks, communication systems, and energy systems, can promote economic growth and improve access to basic services.

Table 27.2 Division of categories according to SDGs

Sustainable developments goals	Dimension	Category
SDG 11: Sustainable cities and communities	Environmental	C1—Urban form C2—Land use and infrastructure
SDGs 15: Life on land		C3—Ecology and biodiversity
SDG 6: Clean water and sanitation		C5—Water
SDG 7: Affordable and clean energy		C4—Energy C6—Materials and waste
SDG 12: Responsible consumption and production	Social	C7—Exterior comfort C8—Safety C9—Amenities C10—Mobility
SDGs 13: Climate action		C11—Local and cultural identity
SDG 9: Industry, innovation, and infrastructure	Economic	C12—Employment and economic development

Source Author

The reviewed literature has shown that it is possible to integrate SBTool^{PT} Urban as an evaluation method for the SDG based projects to be implemented at the international level. SBTool^{PT} Urban is an evaluation method that can generate many benefits when used as an evaluation method not only at the urban scale but also when implementing projects based on improving the urban circular economy of a region. The division of the chosen goals is summarized in Table 27.2.

27.5 Conclusions

The present research sought to improve the ranking of circularity indicators and ensure a complete evaluation. Combining the Sustainable Development Goals (SDGs) with SBTool^{PT} Urban can bring many benefits to individuals, organizations, and communities. By aligning with the SDGs and using SBTool^{PT} Urban, individuals and organizations can demonstrate their commitment to sustainable development and responsible business practices. This can enhance their reputation and credibility and help them attract and retain customers, investors and employees who value sustainability.

Also, the combination of the SDGs and SBTool^{PT} Urban can create opportunities for collaboration and partnership with other individuals, organizations and communities that share a commitment to sustainable development. This can lead to greater knowledge sharing, resource pooling and innovation, and create a collective impact greater than the sum of its parts. SDGs projects that are in the process of implementation can be a good study to validate their usefulness with SBTool^{PT} Urban, this achieve to generate an evaluation of projects of various SDGs to be implemented properly with an analysis of their impact in their various fields. Although the implementation of SBTool^{PT} Urban as a method for evaluating SDG implementation projects needs a more detailed study, it is clear the benefits that would arise by being able to maintain standards and quantify the impact of each of them in society.

The assessment process can provide a clear picture of an urban area's sustainability performance and highlight opportunities for improvement, such as increasing the use of renewable energy, improving waste management, or enhancing social cohesion and inclusion. Finally, there are many benefits to using the SBTool^{PT} Urban to evaluate projects in the implementation phase, from enhancing reputation and credibility to contributing to a more sustainable and livable world. However, further research into the SBTool^{PT} Urban's scope is needed for its use at the local level to be more widely adopted for application in conjunction with urban circular economy development projects. The limitations that exist in these evaluation methods leave us a way to continue developing and creating new applications for these evaluation methods. Considering the importance of the SDGs, it is necessary to continue applying these evaluation methods to generate a wider scope and greater number of applications to generate more support within the emergence of the new concepts of circular economy.

Acknowledgements This research received no external funding. However, the authors acknowledge the support of CYTED Network ECoEICo—Circular Economy as a Strategy for a More Sustainable Construction Industry and the COST Action CircularB—Implementation of Circular Economy in the Built Environment.

References

1. de Oliveira CT, Oliveira GGA (2023) What Circular economy indicators really measure? An overview of circular economy principles and sustainable development goals. Resour, Conserv Recycl, 190. https://doi.org/10.1016/j.resconrec.2022.106850
2. Braganca L, Mateus R (2018) Guia de avaliação SBToolPT Urban, V 2018; iiSBE Portugal. Sch Eng: Guimaraes, Portugal. ISBN 978–989–20–8969–0
3. THE 17 GOALS United Nations, https://sdgs.un.org/goals, Last Accessed 2023/02/18
4. The 2030 AGENDA for Sustainable Development, https://globalcompact.pt/index.php/pt/agenda-2030, Last Accessed 2023/02/18
5. Williams L, Tahtinen L, Costa N, Karst L, Proden E (2020) Participación de Actores en la Agenda 2030 para El Desarrollo Sostenible. Naciones Unidas
6. The SDGS in Action UNDP Org, https://www.undp.org/sustainable-development-goals, Last Accessed 2023/02/18
7. Kalioujny B, Ermushko J, Zhavoronok A (2016). Establishment of a strategy of circular economy increasing the well-being of society: comparison of two national policies. https://doi.org/10.1051/shsconf/20162801050
8. United Nations Sustainable Development Summit (2015), https://sustainabledevelopment.un.org/post2015/summit, Last Accessed 2023/02/14
9. CEN EN 15942:2021. Sustainability of construction works—environmental product declarations—communication format—Business to Business, Brussels (2021)
10. Mateus R, Braganca L (2011) Sustainability assessment and rating of buildings: Developing the methodology SBToolPT–H. Building and Environment, 1962–1971, 46(10)
11. Ferreira A, Pinheiro MD, Brito J de, Mateus R (2023) A critical analysis of LEED, BREEAM and DGNB as sustainability assessment methods for retail buildings. J Build Eng, 66. https://doi.org/10.1016/j.jobe.2023.105825

12. Global indicator framework for the sustainable development goals and targets of the 2030 agenda for sustainable development goals and targets (from the 2030 Agenda for Sustainable Development) Indicators, https://unstats.un.org/sdgs/indicators/Global, Last Accessed 2023/02/18
13. SHIP2FAIR, http://ship2fair-h2020.eu/#, Last Accessed 2023/02/19
14. Calvo residuo cero, grupo calvo global compact, https://www.grupocalvo.com/2022/02/24/calvo-residuo-cero-premio-goods/, Last Accessed 2023/02/1

Chapter 28
Implementing Nature-Based Solutions for a Circular Urban Built Environment

Rocío Pineda-Martos, Nataša Atanasova, Cristina S. C. Calheiros, Ranka Junge, Samaneh S. Nickayin, Teresa A. Paço, Laura Dominici, Elena Comino, Maria-Beatrice Andreucci, Dimitra Theochari, Bernhard Pucher, Aránzazu Galán González, Pedro N. Carvalho, and Guenter Langergraber

Abstract This short review outlines the implementation of nature-based solutions in the urban built environment which can contribute to a circular economy as well as the multiple benefits related to the ecosystem services they can provide. The novel Circular City framework on the mainstreaming of nature-based solutions for

R. Pineda-Martos (✉)
University of Seville, School of Agricultural Engineering, Urban Greening and Biosystems Engineering Research Group (ETSIA-US), Ctra. de Utrera, Km.1, 41013 Seville, Spain
e-mail: rpineda@us.es

N. Atanasova
University of Ljubljana, Faculty of Civil and Geodetic Engineering, Jamova Cesta 2, 1000 Ljubljana, Slovenia

C. S. C. Calheiros
Interdisciplinary Centre of Marine and Environmental Research (CIIMAR/CIMAR), University of Porto, Novo Edifício Do Terminal de Cruzeiros Do Porto de Leixões. Avenida General Norton de Matos S/N, 4450-208 Matosinhos, Portugal

R. Junge
Zurich University of Applied Sciences, Institute of Natural Resource Sciences, Grüntal 14, 8820 Waedenswil, Switzerland

S. S. Nickayin
Agricultural University of Iceland, Faculty of Planning and Design, 311 Borgarbyggð, Hvanneyri, Iceland

T. A. Paço
Universidade de Lisboa, Instituto Superior de Agronomia, LEAF—Linking Landscape, Environment, Agriculture and Food Research Center, Associated Laboratory TERRA, Tapada da Ajuda, 1349-017 Lisboa, Portugal

L. Dominici · E. Comino
Politecnico Di Torino, Department of Environment, Land and Infrastructure Engineering, Corso Duca Degli Abruzzi 24, 10129 Turin, Italy

M.-B. Andreucci
Sapienza University of Rome, Faculty of Architecture, Via Flaminia 72, 00196 Rome, Italy

L. Bragança et al. (eds.), *Creating a Roadmap Towards Circularity in the Built Environment*, Springer Tracts in Civil Engineering,
https://doi.org/10.1007/978-3-031-45980-1_28

the enhancement of urban resource management, which was developed within the COST Action CA17133, is presented. Urban circularity challenges addressed by nature-based solutions are assessed in the built environment following three different levels of implementation—i.e., green building materials, systems for the greening of buildings envelope, and green building sites as vegetated open spaces and water-sensitively designed. Considering the possibilities of implementing nature-based solutions in the built environment, we also highlight the circularity processes that can take place through the integration of nature-based solutions at some or all of the proposed scales towards the achievement of at least one of the seven urban circularity challenges. A collection of representative actual case studies exemplifying the development and implementation of nature-based solutions towards circular cities is also included.

Keywords Nature-based solutions · Circular economy · Built environment

28.1 Introduction

As defined by Langergraber et al. [1, 2], nature-based solutions (NBS) are approaches that not only bring nature into cities but also contribute towards solving or mitigating environmental and societal problems. In many cases, this includes ideas for urban design that are inspired or derived from nature [3]. The specific focus of this short review is to present the advances on the implementation of NBS within urban ecosystems—i.e., the built environment, towards Circular Economy (CE), which were proposed within the COST Action CA17133 Circular City: Implementing nature-based solutions for creating a resourceful circular city (2018–2023) [4].

Beside performing their specific and nominal intended design and use—such as water management (e.g., urban drainage), air quality, greening of the city, … — introducing or enhancing NBS within existing urban infrastructure provides multiple benefits, such as climate change mitigation and adaptation, reduction of the urban heat island effect (e.g., via evaporative cooling), wastewater treatment, while enhancing

D. Theochari
MERA Landschaftsarchitekten mbB, Griegstraße 75, Haus 24B, 22763 Hamburg, Germany

B. Pucher · G. Langergraber
University of Natural Resources and Life Sciences Vienna (BOKU), Department of Water, Atmosphere and Environment, Institute of Sanitary Engineering, Muthgasse 18, 1190 Vienna, Austria

A. G. González
Université Libre de Bruxelles, Department of Building, Architecture and Town Planning, Avenue Franklin Roosevelt 50, 1050 Brussels, Belgium

P. N. Carvalho
Aarhus University, Department of Environmental Science, Frederikborgvej 399, 4000 Roskilde, Denmark

human wellbeing and liveability of cities [5], as well as biodiversity, and resource recovery [6]. All these aspects find expression in the design of urban spaces and buildings—what is concerned to as the built environment [3].

Over the last fifteen years, the concept of NBS appears to encompass contemporary landscapes and architecture design solutions, where natural and living materials—as well as policies strategies, measures and actions on participatory planning and governance promoting their applications—are leveraged to come across societal challenges posed by the urban built environment [3, 6, 7]. The urban green infrastructure and the related NBS are connected to the reduction of these issues, interconnected with physical phenomena occurring in cities [3, 6].

Another significant aspect when managing contemporary urban systems is the concept of circularity. Adopting the Circular Economy (CE) model constitutes an evolving umbrella representing multiple definitions and internal complexities, such as the one introduced by Langergraber et al. [1] as an economic system that aims at minimizing waste and making the most of resources (water, nutrients, materials, food, energy) by keeping these in circulation and reprocessing within the city [7]. In a circular system, resource inputs and outputs (as waste or emissions), as well as energy loss are minimized by closing and optimizing material cycles and energy cascades, including their economic and environmental efficiency [2, 7, 8].

This paper describes the urban challenges related to circularity that can be addressed through NBS, and the pathways and case studies on how NBS can be included in the built environment [2, 7, 8].

28.1.1 Urban Circularity Challenges

To comply with the principles of CE, substantial amendments in the infrastructures' management and design along with promoting new or hybrid systems are needed by applying a multisectoral and multidisciplinary approach [7, 8]. NBS encompass an extraordinary potential to address several urban challenges and generate multiple co-benefits while delivering several ecosystem services.

A set of seven urban circularity challenges (UCCs) that can be addressed with NBS was postulated by Atanasova et al. [7].

1. Restoring and maintaining the water cycle;
2. Water and waste treatment, recovery, and reuse;
3. Nutrient recovery and reuse;
4. Material recovery and reuse;
5. Food and biomass production;
6. Energy efficiency and recovery; and
7. Building system recovery.

28.1.2 The Circular City Framework

The Circular City Framework, as defined by Langergraber et al. [2], aims at mainstreaming the use of NBS in urban environment. It is a framework for addressing UCCs with implementation of NBS, and includes:

- The updated definitions of all NBS that clear up hitherto confusing and overlapping terminology in this area;
- The catalogue of technologies for providing/recovering resources with NBS that comprises a set of 39 NBS units (NBS_u), 12 NBS interventions (NBS_i), and 10 Supporting units (S_u);
- The analysis of input and output (I/O) resource streams required for NBS units and interventions (NBS_u/i); and,
- A guidance tool—as a decision support system for the NBS implementation in cities.

28.2 Levels of Implementation

Pearlmutter et al. [3] proposed a series of different scales for NBS implementation in the built environment:

- **Green building materials**: They result from organic materials extracted from low environmental impacts biological cycle—e.g., water, carbon, and energy; in constructing the built environment [3]. It is considered a beneficial reuse of other resource streams to prevent harmful residues and guarantee a user-friendly living environment concerning indoor air quality and climate [3, 8]. Ideally, the material processing cycle should be designed to circle back the nutrients safely to the ecosystem at the end of its usage cycle [3, 8].
- **Green building systems**: These include green building-integrated systems such as extensive and intensive green roofs (GRs), vertical greening systems (VGSs), house trees, building integrated treatment wetlands, and building integrated agriculture (BIA) [2, 3, 8–11]; to optimize, among others, the energy efficiency of the buildings:
  - *Green roofs:* systems that are implemented on a constructed structure comprising several layers, playing different functions, with vegetation on top.
  - *Vertical greening systems:* consist of vegetation planted in soil or in artificial or organic substrates as part of suspended panels on the wall surface where plants are grown.
  - *House trees:* are planted next to or within the building infrastructure – for instance, Bosco Verticale in Milan and Hundertwasser Haus in Vienna. Such traditional European features of buildings notably influence the buildings' energy efficiency, thanks to the cast shadow in summer and light pass during wintertime.

- *Building integrated treatment wetlands*: are systems designed to collect and hold rain runoff or to treat wastewater from the building with high pollutant removal efficiency. If designed properly, GRs and VGSs can also function as treatment wetlands and treat rainwater or greywater [9, 10, 12, 13].
- *Building integrated agriculture*: is a practice to synergize the built environment and agriculture [11, 14]. The result are mixed-use buildings with a farming system, using the local source of water and energy to produce food. It is possible to integrate a rooftop greenhouse with the building below, and thus optimize the metabolism of the building (cooling/heating and gas exchange (CO_2/O_2)) [15].

- **Green building sites**: these may be open land spaces or parcels next to the buildings that offer spaces for establishing nature in cities and provide multiple ecosystem services. Such spaces could be essential in the management of urban blue-green infrastructure and water resources [2, 3, 8, 9].

Following their approach [3], we consider these three scales of implementation—i.e., from the building materials, systems for the greening of buildings, and green urban sites (Fig. 28.1) in subsequent analyses [9, 16].

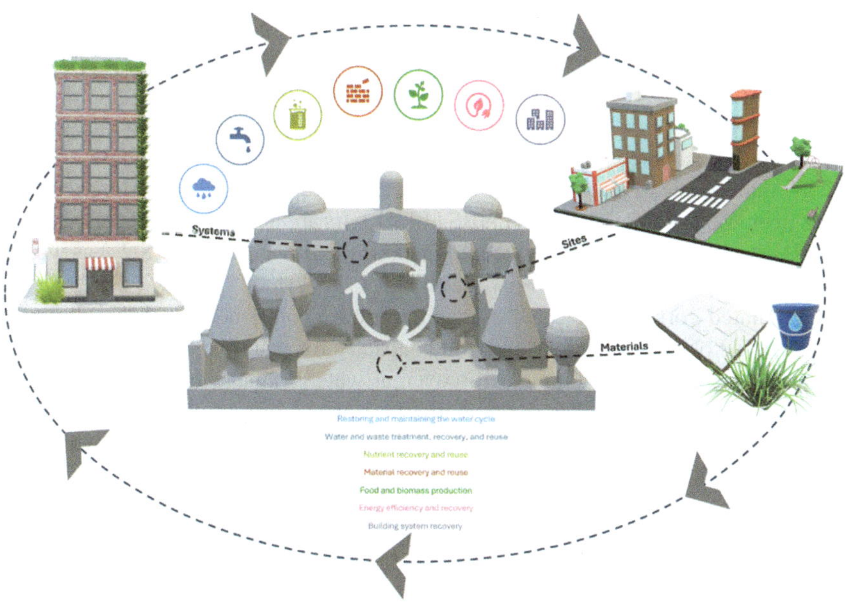

Fig. 28.1 Nature-based solutions' scales of implementation in the built environment: green building materials, systems, and sites; and associated urban circularity challenges

Fig. 28.2 Green roofs and cork as green building material: **a** cork oak tree; **b** cork green roof; **c** granulated cork; and **d** insulation cork board. Original images from Calheiros, C. S. C

28.3 Case Studies

28.3.1 Cork as Green Building Material for Green Roof Systems

The interest in materials considered to optimize GRs' structural layers performance at the building level, as delivering different services and promoting circularity of the built environment is rising [17].

Cork is the bark of the cork oak tree (*Quercus suber* L.) which is periodically harvested from the trees (Fig. 28.2a); and it has been long used in the built environment—mainly due to its capacity of thermal and acoustic insulation. More recently, it has been tried as a component of NBS [17, 18]. Its application has been considered in GRs' systems (Fig. 28.2b) as a component of the substrate (Fig. 28.2c) as well as drainage and protection layer (Fig. 28.2d).

Granulated cork was selected as main substrate component by Monteiro et al. [19] because of its low weight and good capacity for water adsorption, being also adequate for plant establishment. Moreover, as water drainage and protection layer, the insulation cork board has the potential to replace conventional synthetic materials—e.g., extruded polystyrene [20]; and thus, contribute towards negative carbon footprint.

The sustainability of materials deployed in GRs still needs to improve. Life cycle and carbon footprint assessments can support and highlight the best direction [20, 21].

28.3.2 Vertical Greening Systems for Indoor Air Quality Using Recycled Substrates

VGSs are increasingly adopted as a nature-based strategy to improve air quality in interiors. Many investigations demonstrated potential of air phytoremediation in ornamental plant species on main indoor air pollutants [22, 23]. Other studies focused on the importance of growing media composition specially to increase the efficiency of active botanical biofiltration of volatile organic compounds (VOCs) [24]. Despite the importance of cultivation substrate for functional VGSs, scant research is carried out to assess the influence of alternative growing media produced with recycled components on the indoor phytoremediation efficiency.

Organic-rich substrate along with coconut fiber, perlite and vermicompost, obtained from by valorization of organic waste, showed promising results as growing medium for VGSs and contributed to improved indoor air quality [25]. De Lucia et al. [26] obtained good prospects when testing the influence of recycled rice husk-based substrate in modular VGSs on plant health status and thermal comfort.

Focusing on innovation and production of alternative growing media, recycled organic and inorganic by-products and waste derived from local supply chains, can be a promising way to increase circularity and improve the sustainability of VGSs.

28.3.3 Vertical Greening Systems for Greywater Treatment

The use of treated greywater can contribute to restoring and recreating the hydrological cycle in a circular and sustainable way; as well as the reuse and recovery of nutrients in urban environments [7, 9]. Sustainable water management in cities can be addressed through the implementation of potential NBS, such as the case of VGSs; providing, in addition to circularity strategies, additional benefits for the inhabitants and their environment [2, 7, 9].

Proper selection of plants and filter media used in a VGS play an essential role in maximizing the performance of greywater treatment. In addition, in these complex integrated treatment systems, both components influence water requirements, aesthetics, hygienic-sanitary conditions, and system maintenance; fundamental factors for the success of its large-scale implementation [12, 13].

The experimental system installed at BOKU University (Fig. 28.3) demonstrated its multifunctionality on a large scale by identifying its water demand, maximizing local thermal reduction by evapotranspiration [13]. Plants were evaluated for their adequacy to different irrigation conditions and types of water [13].

28.3.4 Building Integrated Agriculture, the Case of Rroofood Project

In climates with dry, hot summers, GRs microclimate can be very harsh for plants; and use of locally adapted wild plants present a solution of increasing agricultural production in urban environments.

Fig. 28.3 Vertical greening system at the University of Natural Resources and Life Sciences Vienna (BOKU), Vienna, Austria. The system has a total size of 6m x 4m and consists of four individual systems with their own irrigation system. Monitoring included air and substrate temperature, substrate water content, precipitation, solar radiation, irrigation water volume and system output [13]

The Rroofood project [27, 28] evaluated the suitability of wild edible plants to find enhanced sustainable production solutions for urban farming in GRs. A collection of plant species was sown in a GR at Instituto Superior de Agronomia (Universidade de Lisboa, Portugal) campus (Fig. 28.4), with the selected following three criteria: (i) used in traditional gastronomy; (ii) from areas with environmental conditions of equal or greater demand than the Lisbon area; and (iii) species with resilience traits. Such traits minimize the need for external resource inputs such as water and energy, fertilizers and pesticides.

Fifteen species from the genera *Amaranthus*, *Beta*, *Cakile*, *Chenopodium*, *Chrysanthemum*, *Nigella*, *Papaver*, *Petroselinum*, *Rumex*, *Scolymus*, *Tragopogon*, and *Viola* were selected. Most of these germinated and developed successfully. Two species reseeded naturally and another group of four was still present in the next spring. From those, a selection of five species with potentially interesting features, combining plant physiology traits and gastronomic aptitude, for further work, was made [27].

Fig. 28.4 **a** Green roof built at the Instituto Superior de Agronomia campus—as part of *the green roof lab*, Universidade de Lisboa, Portugal; **b** detail of *Nigella damascena* plant. Original images from Paço, T. A

28.4 Conclusions

The search for less energy intensive and resource-demanding, and thus more CO_2-neutral solutions for societal challenges has been intensified recently—having in consideration the alignment of the European policies and strategies and the global policies related to the United Nations Agenda 2030 and The Paris Agreement. The present short paper aims to highlight potential contributions of nature-based solutions to the development of COST Action CA21103 CircularB framework of a circularity rating tool build from the state-of-the-art and best practices of circular economy in the construction of the built environment, coupled with the European Union Circular Economy Action Plan. Several case-studies are presented, at the level of green building-integrated systems, illustrating the potential of nature-based solutions to cope with circularity processes, either via structures or materials, towards a circular urban built environment.

Acknowledgements The authors are grateful for the support from the COST Action CA17133 Circular City, https://circular-city.eu/). C.C.S.C. is thankful to Fundação para a Ciência e Tecnologia (FCT) regarding UIDB/04423/2020 and UIDP/04423/2020 projects.

References

1. Langergraber G, Pucher B, Simperler L, Kisser J, Katsou E, Buehler D, Garcia Mateo MC, Atanasova N (2020) Implementing nature-based solutions for creating a resourceful circular city. Blue-Green Systems 2(1):173–185
2. Langergraber G, Castellar JAC, Pucher B, Baganz GFM, Milosevic D, Andreucci M-B, Kearney K, Pineda-Martos R, Atanasova N (2021) A framework for addressing circularity challenges in cities with nature-based solutions. Water 13(17):2355
3. Pearlmutter D, Theochari D, Nehls T, Pinho P, Piro P, Korolova A, Papaefthimiou S, Garcia Mateo MC, Calheiros C, Zluwa I, Pitha U, Schosseler P, Florentin Y, Ouannou S, Gal E,

Aicher A, Arnold K, Igondová E, Pucher B (2020) Enhancing the circular economy with nature-based solutions in the built urban environment: green building materials, systems and sites. Blue-Green Systems 2(1):46–72
4. COST Action CA17133 Circular City - Implementing nature-based solutions for creating a resourceful circular city, https://circular-city.eu/, Last Accessed 2023/02/20
5. de Haan FJ, Ferguson BC, Adamowicz RC, Johnstone P, Brown RR, Wong THF (2014) The needs of society: A new understanding of transitions, sustainability and liveability. Technol Forecast Soc Chang 85:121–132
6. Pineda-Martos R, Calheiros CSC (2021) Nature-based solutions in cities—Contribution of the Portuguese National Association of Green Roofs to urban circularity. Circ Econ Sustain 1:1019–1035
7. Atanasova N, Castellar JAC, Pineda-Martos R, Nika CE, Katsou E, Istenič D, Pucher B, Andreucci MB, Langergraber G (2021) Nature-based solutions and circularity in cities. Circ Econ Sustain 1:319–332
8. Langergraber G, Castellar JAC, Andersen TR, Andreucci M-B, Baganz GFM, Buttiglieri G, Canet-Martí A, Carvalho PN, Finger DC, Griessler Bulc T, Junge R, Megyesi B, Milošević D, Oral HV, Pearlmutter D, Pineda-Martos R, Pucher B, van Hullebusch ED, Atanasova N (2021) Towards a cross-sectoral view of nature-based solutions for enabling circular cities. Water 13(17):2352
9. Pearlmutter D, Pucher B, Calheiros CSC, Hoffmann KA, Aicher A, Pinho P, Stracqualursi A, Korolova A, Pobric A, Galvão A, Tokuç A, Bas B, Theochari D, Milosevic D, Giancola E, Bertino G, Castellar JAC, Flaszynska J, Onur M, Garcia Mateo MC, Andreucci MB, Milousi M, Fonseca M, Di Lonardo S, Gezik V, Pitha U, Nehls T (2021) Closing water cycles in the built environment through nature-based solutions: The contribution of vertical greening systems and green roofs. Water 13(16):2165
10. Cross K, Tondera K, Rizzo A, Andrews L, Pucher B, Istenič D, Karres N, McDonald R (eds) (2021) Nature-based solutions for wastewater treatment: A series of factsheets and case studies. IWA Publishing, London, United Kingdom
11. Caplow T (2009) Building integrated agriculture: Philosophy and practice. In: Urban futures 2030: Urban development and urban lifestyles of the future, pp 48–51. Herausgegeben von der Heinrich-Böll-Stiftung, Berlin, Germany
12. Pradhan S, Al-Ghamdi SG, Mackey HR (2019) Greywater recycling in buildings using living walls and green roofs: A review of the applicability and challenges. Sci Total Environ 652:330–344
13. Pucher B, Zluwa I, Spörl P, Pitha U, Langergraber G (2022) Evaluation of the multifunctionality of a vertical greening system using different irrigation strategies on cooling, plant development and greywater use. Sci Total Environ 849:157842
14. Gould D, Caplow T (2012) Building-integrated agriculture: A new approach to food production. In: Metropolitan sustainability, pp 147–170. Woodhead Publishing
15. Pons O, Nadal A, Sanyé-Mengual E, Llorach-Massana P, Cuerva E, Sanjuan Delmàs D, Muñoz P, Oliver-Solà J, Planas C, Rovira MR (2015) Roofs of the future: rooftop greenhouses to improve buildings metabolism. Procedia Eng 123:441–448
16. Canet-Martí A, Pineda-Martos R, Junge R, Bohn K, Paço TA, Delgado C, Alenčikienė G, Skar SLG, Baganz GFM (2021) Nature-based solutions for agriculture in circular cities: Challenges, gaps, and opportunities. Water 13(18):2565
17. Calheiros CSC, Stefanakis AI (2021) Green roofs towards circular and resilient cities. Circ Econ Sustain 1:395–411
18. Talhinhas P, Ferreira JC, Ferreira V, Soares AL, Espírito-Santo D, do Paço TA (2023) In the search for sustainable Vertical Green Systems: An innovative low-cost indirect green façade structure using Portuguese native ivies and cork. Sustain 15, 5446
19. Monteiro CM, Calheiros CSC, Martins JP, Costa FM, Palha P, de Freitas S, Ramos NMM, Castro PML (2017) Substrate influence on aromatic plant growth in extensive green roofs in a Mediterranean climate. Urban Ecosyst 20:1347–1357

20. Carbon footprint of the insulation cork board (2017) Tártaro A. S., Mata, T. M., Martins A. A., Esteves da Silva, J. C. G. J Clean Prod 143:925–932
21. Tams L, Nehls T, Calheiros CSC (2022) Rethinking green roofs- natural and recycled materials improve their carbon footprint. Build Environ 219:109122
22. Kim KJ, Shagol CC, Torpy FR, Pettit T, Irga PJ (2020) Plant physiological mechanisms of air treatment. In: Soreanu, G., Dumont, E. (eds.) From biofiltration to promising options in gaseous fluxes biotreatment—Recent developments, new trends, advances, and opportunities, pp 219–244. Elsevier
23. Mata TM, Martins AA, Calheiros CSC, Villanueva F, Alonso-Cuevilla NP, Fonseca Gabriel M, Ventura Silva G (2022) Indoor air quality: A review of cleaning technologies. Environ 9:118
24. Pettit T, Irga PJ, Surawski NC, Torpy FR (2019) An assessment of the suitability of active green walls for NO_2 reduction in green buildings using a closed-loop flow reactor. Atmos 10(12):801
25. Kazemi F, Rabbani M, Jozay M (2020) Investigating the plant and air-quality performances of an internal green wall system under hydroponic conditions. J Environ Manage 275:111230
26. De Lucia M, Treves A, Comino E (2021) Rice husk and thermal comfort: Design and evaluation of indoor modular green walls. Dev Built Environ 6:100043
27. do Paço TA, Arsénio P, Pinto da Costa F, Raymundo A, Prista C, Magalhães A, Espírito-Santo D (2022) The wild food rooftop—Traditional edible plants in a green roof of Lisbon area. In: International Web Conference on Food Choice and Eating Motivation, 19–20 May. Online event
28. The Rroofood project, https://tapaco4.wixsite.com/rroofood, Last Accessed 2023/02/20

Part XV
CE Criteria in Sustainability Frameworks for Buildings

Chapter 29
Developing of Comprehensive Circular Rating System for Office Buildings: A Review of Existing Tools and Case Study

Nika Trubina, Jan Pešta, and Antonín Lupíšek

Abstract As the world moves toward a more sustainable future, the concept of circular economy in the building industry and its evaluation is attracting more and more attention. This study reviews existing building assessment systems from a circular economy perspective, such as LEED, BREEAM, DGNB and SBToolCZ. Since these programs were developed using different approaches, the idea of developing a single comprehensive rating system was put forward. Thus, based on the analyzed criteria of the rating systems, we selected the most relevant ones for our study with a focus on building systems and materials from a circular economy perspective and developed a universal set of key indicators for office building design. This set of criteria was tested on a specific building case study, evaluating its circularity and the effectiveness of the proposed design. The building received a score of 18.1 out of a possible 40 points. In the end, improvements were suggested, after which the overall credit score could be increased by 25%. This study underscores the importance of promoting circularity in the construction industry by requiring a comprehensive rating system that addresses the unique challenges and opportunities of the sector.

Keywords Circular economy · Assessment tool · Circular indicators · Certification · Office building

N. Trubina · A. Lupíšek
Faculty of Civil Engineering, Czech Technical University in Prague, Thákurova 2077/7, 16629 Prague, Czech Republic

N. Trubina (✉) · J. Pešta · A. Lupíšek
University Centre for Energy Efficient Buildings, Czech Technical University in Prague, Třinecká 1024, 27343 Buštěhrad, Czech Republic
e-mail: nika.trubina@cvut.cz

L. Bragança et al. (eds.), *Creating a Roadmap Towards Circularity in the Built Environment*, Springer Tracts in Civil Engineering,
https://doi.org/10.1007/978-3-031-45980-1_29

29.1 Introduction

The construction industry is one of the main anthropogenic factors affecting the environment. For instance, more than 40% of all waste in the Czech Republic comes from the construction industry [1]. In the entire European Union, it accounts for over 35% of all waste produced [2]. The amount of construction and demolition waste and the use of primary resources can be reduced by adopting circular economy principles [3].

However, there is still no standard definition of circular economy [4], with more than 100 different definitions currently in use and so even its implementation into building design is not yet settled [5]. The concept of circular economy has been developed as an update of the economy's linear model, which was recognised as insufficient for a world with limited resources and society with a high consumption rate [6]. One of the most comprehensive analysis of the circular economy definitions was conducted by Kirchherr, Reike, and Hekkert [5], who concluded that the circular economy is an "economic system that replaces the 'end-of-life' concept with reducing, alternatively reusing, recycling and recovering materials in production/distribution and consumption processes. It operates at the micro level (products, companies, consumers), meso level (eco-industrial parks) and macro level (city, region, nation, and beyond), with the aim to accomplish sustainable development, thus simultaneously creating environmental quality, economic prosperity and social equity."

The building industry has established prominent green building certification programs and sustainability standards aimed at implementing focused sustainability strategies. These standards consider also various factors such as material efficiency, waste reduction, emissions, indoor comfort, building design, water management, and more. The most widely known of building sustainable assessment tools are the LEED (Leadership in Energy and Environmental Design) [7], the BREEAM (Building Research Establishment Environmental Assessment) [8], or the DGNB (Deutsche Gesellschaft für Nachhaltiges Bauen) [9]. The Czech Republic also has its national certification tool. SBToolCZ is a regional certification scheme that was adapted for local conditions [10]. It was based on the Sustainable Building Tool developed by The International Initiative for a Sustainable Built Environment (iiSBE) [11].

However, these certificates have been developed with different approaches and lack consistency in terms of indicators related to the circular economy. Each certificate's set of criteria differs from one another [12]. Therefore, the aim of this paper is to identify relevant criteria for our research and develop a universal set of key indicators for designing buildings with a focus on building systems and materials from the perspective of the circular economy. This research is focused on office buildings. To achieve this, the paper conducts a theoretical review of the literature and existing rating systems to identify similarities and gaps. The paper emphasises the importance of functional adaptability, which refers to the ability to change or replace the structural system and reuse building elements and materials to extend the life cycle of the building and its components. Thus, the main objective is to provide a comprehensive overview of the indicators that can be used to design circular office buildings that are sustainable, efficient, and adaptable.

29.2 Materials and Methods

The first part of the study was a literature review and analysis focused on the evaluation of building design from the perspective of the circular economy in accordance with existing rating systems and certification tools. Based on it, new criteria that are necessary for the assessment of office buildings were developed. Further, the newly developed criteria were applied to the case study and assessed with the help of a Bill of Quantities (BoQ) of the building elements and other construction documentation. The circularity calculation was done analytically. On this basis, the building was assessed, and possible constructive optimisations and other measures were proposed.

29.2.1 Overview of Certification Tools from a Circular Economy Perspective

In this paper, the 4 most common certification schemes for the Czech Republic were considered for the evaluation of the design of the office building from the point of view of the circular economy: BREEAM, LEED, DGNB, and SBToolCZ. A list of criteria and indicators will be selected for the assessment, which corresponds to the principles of CE. The manuals used were selected for such typology as administrative buildings, a new construction in the Planning and Definition project phase, when design can still be influenced. Further information on the evaluation criteria is described with reference to the following manuals: BREEAM International New Construction Version 2016 [8], DGNB System—New buildings criteria set 2020 [9], LEED v4 Building Design and Construction 2021 [7] and SBToolCZ manual for the evaluation of administrative buildings in the design phase 2022 [10].

To make the assessment, all manuals were analysed, all criteria names were tabulated, and each was given a brief description and assessment of whether it applies to the circular economy. Items related to optimal water management, or the location of the object and its surroundings were excluded. Special attention was paid only to those indicators related to the design of the building itself, its construction, and the use of materials (Fig. 29.1).

Unlike the others, DGNB certification provides bonuses for the application of circular economy principles in sustainable construction. In the evaluation manual [9] is a list of additional criteria, when applying which extra points will be obtained [9]. In other assessment tools, there is no such list of criteria, but there are separate

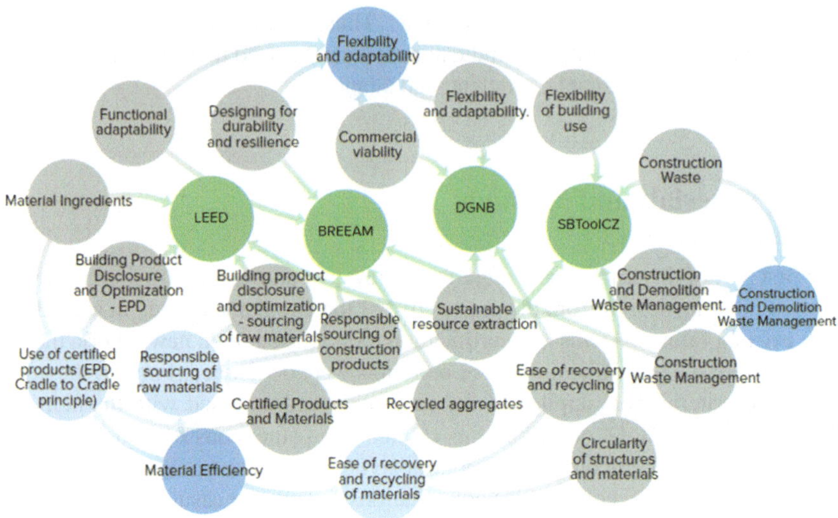

Fig. 29.1 Indicators related to the design of the structural system and materials from chosen rating certificates

indicators that correspond to and complement them. In general, it is possible to highlight key topics that support certificates almost equally (Fig. 29.1).

29.2.2 Proposed Evaluation Criteria

After analysing and comparing the assessment indicators, the criteria that focus on the structural system and materials for assessing an office building concerning circularity requirements were selected. In particular, the focus is on functional adaptability—the possibility of changing or replacing the structural system and on reusing building elements and materials to extend the lifespan of the building and its elements as long as possible. On this basis, further universal indicators have been developed as a sum of all parameters from the certification tools linked to the above-mentioned circular economy criteria.

29.2.2.1 Extending the Life of the Building with a Focus on Functional Adaptability

Objective: To extend the lifespan of the building as far as possible by allowing for continued intended use or possible future changes of use with a focus on functional adaptability—ease of replacement of structural elements, change of layout and renovation, adaptation to changing tenant needs, and market conditions.

Indicator: Credit rating of the adaptability of the building and the resulting sum of credits for each of the adaptability aspects, converted into points (Tables 29.1–29.4).

Assessment method: Evaluation of design concepts with adaptability aspects according to SBTooLCZ certification tools [10].

Simplicity of replacement of structural elements—assembly construction. Note: In the case of evaluation using this criterion, this refers to partitions in the common area, excluding technical and sanitary areas.

Overall evaluation of the criterion: The total credit score is calculated by summing all the credits according to the tables.

Table 29.1 Evaluation of the building's structural support system—column span widths and design heights

Building structural system Supporting system/structural height (m)	Credits C1			
	3.7	3.8	3.9	4.1
Wall—spans up to 6 m	0	2	3	5
Wall—spans over 6 m	3	5	6	7
Combined—spans up to 6 m	3	5	6	7
Combined—spans over 6 m	5	7	8	9
Skeleton—spans up to 6 m	5	7	8	9
Skeleton—spans over 6 m	7	8	9	10

Table 29.2 The ease of replacement of structural elements—partitions

Partitions	Credits C2
Non-dismountable structures requiring demolition work	0
Demountable (plasterboard etc.)	2
Mobile, easily relocatable or can be reused in the same form	4
No partitions	5

Table 29.3 The ease of replacement of structural elements—envelope

Envelope	Credits C3
The non-demountable structure required for demolition works is part of the main load-bearing structure (e.g., monolithic with additional treatment)	0
Demountable (e.g., double-replacement of one shell does not affect the other shell, the space between the two shells can be used for maintenance and repair of both shells)	3
Easily demountable and independent of the main structure, has the possibility of dismantling individual panels and elements (curtain walling)	5

Table 29.4 The ease of access to the heating, ventilation and air conditioning (HVAC) systems

HVAC design	Credits C3
The HVAC systems installed in the building shall be designed to allow the smallest part corresponding to 50% of the total usable area to be changed without affecting the operation of the rest of the building	0
The HVAC systems installed in the building shall be designed to allow the smallest part corresponding to 20% of the total usable area to be changed without affecting the operation of the rest of the building	3
The HVAC systems installed in the building shall be designed to allow a change in the smallest part corresponding to 5% of the total usable area without affecting the operation of the rest of the building	5

Criteria limits: Conversion of credit ratings to points to combine into one final indicator. The maximum number of credits is 25 and points are awarded on the basis of linear interpolation on a scale from 0 to 10.

29.2.2.2 Reduction of Materials Used

Objective: To reduce the absolute quantity of construction materials used.

In the design of the construction system and assembly structures, preference must be given to the proposed option with the lower absolute quantity of construction materials used, which leads to one of the principles of the circular economy—Reduce.

Indicator: Total weight of building materials used per building expressed in tones.

Method of assessment: The weight of materials used will be calculated according to the bill of quantities. When developing multiple design options, the highest score is given to the option that has the least amount of total weight.

Criteria limits The maximum number of credits is endless, and points are awarded based on linear interpolation on a scale from 0 to 10.

29.2.2.3 Enabling Future Reuse of Building Elements and Materials

Objective: To enable the future reuse of building elements, components and parts in their original form at the end of the building or product life cycle, with a focus on avoiding or producing less construction waste and saving materials.

Indicator: Percentage of products or systems for which reuse in its original form without recycling is enabled (%).

Method of assessment: The total weight of building materials and construction elements used for which easy replacement and reuse is possible shall be assessed in relation to the total weight of all materials used (to be reported in %).

Criterion limits: Since the evaluation is done as a percentage, the maximum value is 100% and points are awarded on the basis of linear interpolation on a scale from 0 to 10.

29.2.2.4 Reduction of Construction Waste

Objective: To prevent construction waste and promote resource efficiency at the end of the life cycle of the building or product.

Indicator: Weight of construction waste from used building materials (t) and weight of building materials for which recycling is enabled (t) and its share per building. (%).

Method of assessment: The total weight of used building materials and construction elements that are part of construction and demolition waste at the end of the life cycle of the building or product and in relation to the total weight of all materials used shall be assessed (to be reported in %).

Criterion limits: Conversion of percentage scores into points to combine into one final indicator. Since the evaluation is done as a percentage and in this case, the evaluation will be reversed (the less waste—the better), the maximum value is (best result) 0%, the minimum (worst result)—100%, and the points are awarded based on linear interpolation on a scale from 0 to 10.

29.2.2.5 Usage of Materials with Recycled Content

Objective: To use recycled waste and secondary raw materials, thereby reducing the demand for virgin raw materials and optimising material efficiency in the construction sector. The use of products or systems containing recycled components in the design of construction systems and assemblies is positively evaluated.

Indicator: Percentage of products or systems containing recycled components (%).

Method of assessment: The total weight of construction materials used made from secondary raw materials/recycled waste in relation to the total weight of all materials used shall be assessed (to be indicated in %).

Criterion limits: Conversion of percentage scores into points to combine into one final indicator. Since the evaluation is done as a percentage, the maximum value is 100% and points are awarded on the basis of linear interpolation on a scale from 0 to 10.

29.2.3 Case Study

As a case study, a 4-level office building in the Project Planning and Definition phase was chosen (Fig. 29.2). According to the structural solution, the office building is designed as a skeleton frame system made of S235 steel and based on reinforced concrete footings. The ceiling slabs are composite made of reinforced concrete castings on top of a profile steel base and placed on IPE220 ceiling beams. The vertical load-bearing structures consist of HEB450 steel columns. All the beam-to-column and column-to-column connections are welded. The spatial rigidity of the structure is ensured by transverse, longitudinal and truss stiffeners. The building envelope consists of an aluminum window system with glass facades and high-performance insulated panels. The building has a flat impassable green roof, which is formed of the composite board as well. The space features a double acoustic floor and a ceramic flooring layer designed for the reception, kitchen, and welfare facilities. The floor and facade structures are made partly from secondary materials. Partitions are designed with aerated concrete in technical and sanitary areas. The ceiling in the building is formed by hanging a plasterboard ceiling.

A feature of this building is the shift of each floor by 1.5 m. On the fourth floor, the building has a 4.5-m cantilever suspended on metal cables.

29.3 Results

The original version of the building performed well in the assessment (Table 29.5).

High scores were obtained in the functionality section: the office building being evaluated has a structural height of 4.2 m and a span width ranging from 5 to 7 m in the horizontal part and from 4.3 to 6 m in the sloping part, so it gets an almost maximum score in this criterion (Fig. 29.2). Partitions in the common spaces in the project are not provided, and the space will be made by Shell & Core—without prefabricated interiors. In terms of circular economy, this is a beneficial solution, allowing tenants

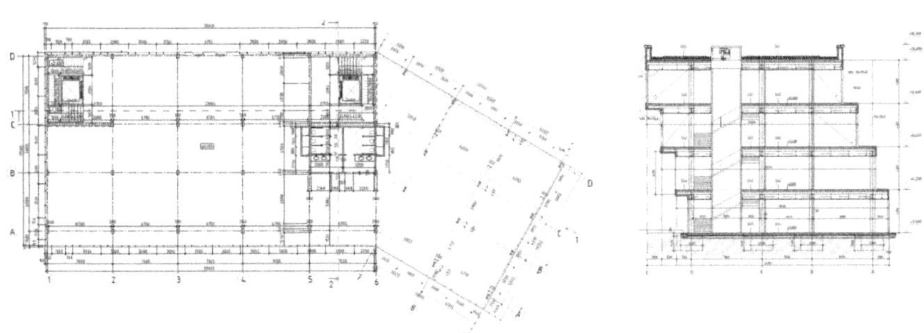

Fig. 29.2 Basic floor plan and section drawing of the reference office building

Table 29.5 The case study's assessment results

Name	Credits	Points
Extending the life of the building with a focus on functional adaptability (total)	16.5	6.6
–Building structural system	9	
–The ease of replacement of partitions	2.5	
–The ease of replacement of envelope	5	
–The ease of access to the HVAC system	0	
Reduction of materials used	2885.2 t	–
Enabling future reuse of building elements and materials	14.3%	1.5
Reduction of construction waste	98%/2%	10
Usage of materials with recycled content	0.02%	0
Total		18.1

to decide on the interior layout of workspaces, creating open spaces or using movable partitions. The use of steel constructions, in this case, is more advantageous compared to the concrete one and lies in the lower dead weight of the structure and more significant variability in the form of structures. Some of the elements used supported sustainable construction and circular economy principles. There were, for example, glass curtain walling, sandwich panels, and double floors which ensure efficient cable routing through the double floor, as well as easy accessibility to installations. These elements could easily be dismantled at the end of their life without damaging other elements.

The lowest scores were for ease of access to the HVAC system and the use of materials containing recycled components. Using EPD, recycled components were found in glass curtain walling and in parts of the floor structure, which amounts to 0.2% of the total weight of the building. This criterion scored 0 points.

29.4 Discussion

Although the original design showed good results in the evaluation, it could be suggested several optimisation measures with an example of potential changes to achieve better assessment results.

Implementing design optimisation measures will allow to recieve the highest credit in the evaluation of building systems, as well as greater flexibility in changing the tenant's layout or the building's purpose. For example, to increase the span between columns above 6 m, to design structural elements for universal reuse using standardised-size, multiples of 0.5 m or 1 m, and design easy-opening doors in areas for access to the installation shaft in case of necessary repairs or replacement of pipes and systems HVAC [13].

The replacement of standard natural aggregates with recycled materials in foundations and their underlayment will reduce the consumption of raw materials [14]. The development of scenarios for designing the further use of the load-bearing structure and its dismantling will allow considering the options of disassembling the skeleton and using "dry joints"—bolted connections instead of welding the main beams [15]. Since a third of the total weight of the structure is the ceiling slab construction, an alternative solution is to use prefabricated panels or lightweight versions of them. Instead of a monolithic trapezoidal slab and monolithic concrete, prefabricated panels will be assembled for possible reuse of this structure. The prefabricated floor structure could be secured to the beams with high-strength structural bolts—mandrels [16]. Although economic factors were excluded from the study, the use of bolted connections is generally more expensive than welded connections. Therefore, new strategies need to be developed in a further study to demonstrate the benefits of circular solutions to project owners.

Together with above-mentioned measures and the replacement of the floor composition with a lightweight vinyl floor with an easy-to-use mounting-dismantling system that can be recycled into new material at the end of its life, such measures could save about 500 tons of construction materials. Additionaly, replacing the existing façade with prefabricated wood-based panels, combined with the above measures, can increase the possible future reuse of the elements by 40% compared to the original version.

Also, the criterion that evaluates the reduction of construction waste has been considered simplistically, as there are several approaches to distributing these types of waste in real life. Each material has several treatment options, e.g., it can be sent directly to a landfill, or separated from other materials and sorted for sending to a landfill or for further recycling or sent for recycling immediately [17]. Therefore, this study considered the most favorable scenario of recycling materials after the end of their life cycle.

29.4.1 Limitation of the Study

The scope of this study was limited in that the structural system and materials of the building were its primary emphasis when evaluating the structure's circularity in terms of comprehensiveness. The circular use of the site is not taken into account in this paper (its terrain, whether brownfields have been used). It also does not evaluate circularity in terms of the project's finance, or whether the building was done using circular business models in some way. Depending on how the building was constructed, the materials or the source of the environmental data were used, the evaluation's results may vary. The BoQ in this paper was based on design documentation from 2D drafting, which was another limiting factor. Using building information models (BIM) could facilitate the process of obtaining BoQ and be used as efficient management and demolition/deconstruction tool. Integration of assessment tool into the BIM environment could provide real-time support for designers and accelerate the building evaluation [18].

29.5 Conclusion

While many rating systems have emerged to assess the sustainability of buildings, there is a need to develop universal criteria for assessing the circularity of buildings. It is crucial to ensure consistency in measuring circularity and enable effective comparison of building performance across different regions and rating systems. Furthermore, a universal set of criteria can facilitate the integration of circularity into building codes and regulations, leading to the broader adoption of circular building practices. This study developed a basic set of criteria for assessing the circularity of office buildings, and the new criteria have been tested in practice in a case study. The building received a score of 18.1 out of a possible 40 points. After the assessment, improving measures were proposed. Using basic circular economy strategies such as prefabrication, unifying the dimensions of structural elements, using recycled materials, and other, in this case, could help to save up to 500 tons of materials. The overall credit score could be increased by 25% by increasing the ability to reuse building elements and materials in the future and recycled materials. The evaluation results can already be applied at the initial stage of the construction preparation in the decision-making process for selecting the most suitable design and material options.

This study contributed to the comparison of existing rating systems for office buildings. We have attempted to achieve a balance between the aforementioned instruments, thereby aligning the criteria with the requirements and priorities of the principles of the circular economy. Research in this area has great potential and still a scope for development by other criteria related to location, water, or circular business models.

References

1. Praha (2021) Odbor statistiky obchodu, dopravy, služeb, cestovního ruchu a životního prostředí: Generation, recovery and disposal of waste for the period 2019
2. Brussels (2015) EC: Closing the loop—An EU action plan for the Circular Economy—COM(2015) 614 final
3. United Nations (2015) The sustainable development goals. United Nations: New York, NY, USA
4. Reike D, Vermeulen WJV, Witjes S (2018) The circular economy: New or Refurbished as CE 3.0?—Exploring controversies in the conceptualization of the circular economy through a focus on history and resource value retention options. Resour Conserv Recycl 135:246–264. doi:https://doi.org/10.1016/j.resconrec.2017.08.027
5. Kirchherr J, Reike D, Hekkert M (2017) Conceptualizing the circular economy: An analysis of 114 definitions. Resour Conserv Recycl 127:221–232. https://doi.org/10.1016/j.resconrec.2017.09.005

6. Ellen MacArthur Foundation (2012) Towards the circular economy: Economic and business rationale for an accelerated transition
7. USGBC (2021) LEED v4.1 building design and construction
8. BREEAM (2016) BREEAM international new construction 2016—Technical manual
9. DGNB (2020) DGNB system new construction, buildings criteria set—Version 2020
10. SBToolCZ (2022) https://www.sbtool.cz/
11. International Initiative for a Sustainable Built Environment (2022)
12. de Oliveira CT, Dantas TET, Soares SR (2021) Nano and micro level circular economy indicators: Assisting decision-makers in circularity assessments. Sustain Prod Consum. 26:455–468. https://doi.org/10.1016/J.SPC.2020.11.024
13. Peltokorpi A, Olivieri H, Granja AD, Seppänen O (2018) Categorizing modularization strategies to achieve various objectives of building investments. Constr Manag Econ 36:32–48. https://doi.org/10.1080/01446193.2017.1353119
14. Pešta J, Pavlů T, Fořtová K, Řepka J (2021) Recycled concrete for foundation structure: LCA case study. IOP Conf Ser Mater Sci Eng 1196:012012. https://doi.org/10.1088/1757-899X/1196/1/012012
15. Webster M, Costello D (2005) Designing structural systems for deconstruction: How to extend a new building's useful life and prevent it from going to waste when the end finally comes (4) (PDF). Greenbuild Conference, Atlanta, GA
16. Girão Coelho AM, Lawson M, Lam D, Yang J (2020) Guidance on demountable composite construction systems for UK practice. Silwood Park, Ascot, Berkshire, SL5 7QN UK
17. Doka G (2009) Life cycle inventories of waste treatment services
18. Carvalho JP, Bragança L, Mateus R (2020) A systematic review of the role of BIM in building sustainability assessment methods. Appl Sci 10:4444. https://doi.org/10.3390/app10134444

Chapter 30
Analysis of Adaptability Requirements Against Their Implementation in Level(s) Framework

Rand Askar, Luís Bragança, and Helena Gervásio

Abstract The adaptability of buildings is considered an essential criterion of sustainability and circularity of the built environment. Change is inevitable in our modern life. Therefore, designing buildings for adaptability and adaptive reuse is urgently necessary to save resources and prevent waste produced by arbitrary demolition activities. The circular economy recognises DfA "Design for Adaptability" as a key strategy to achieve the circularity of buildings, counting on the concept's ability to optimise the effectiveness of other strategies such as design for disassembly (DfD) and promote the waste hierarchy (reduce, reuse, recycle). The recognition is reflected in the EU framework for sustainability assessment Level(s), which embraces four circularity indicators in Macro Objective 2. The paper identifies adaptability requirements building on multiple adaptability and circularity assessment models. In light of these requirements, Level(s) consideration of DfA is examined, leading to multiple possible improvements to more inclusive and objective adaptability and circularity assessment.

Keywords Design for adaptability · Circular economy · Level(s) framework

30.1 Introduction

Design for adaptability (DfA) is a Circular Economy (CE) strategy of intentionally designing buildings with consideration of change along their lifecycle. By considering the end-of-life (EoL) options, DfA facilitates implementing circularity

R. Askar (✉) · L. Bragança
University of Minho, ISISE, ARISE, Department of Civil Engineering, 4804-533 Guimarães, Portugal
e-mail: rand.askar@hotmail.com

H. Gervásio
ISISE, Department of Civil Engineering, University of Coimbra, 3030-788 Coimbra, Portugal

L. Bragança et al. (eds.), *Creating a Roadmap Towards Circularity in the Built Environment*, Springer Tracts in Civil Engineering,
https://doi.org/10.1007/978-3-031-45980-1_30

strategies in buildings, such as design for disassembly (DfD), material reuse and upcycling.

Building adaptability has been defined as "the capacity of a building to accommodate change in response to the emerging needs or varying contextual conditions, therefore prolonging its useful life while preserving the value for its users over time" [1]. Accommodating change requires having a broader design perspective than the one defined by the immediate need and actual context. Similarly, thinking of time allows a building to be perceived as a dynamic system, interacting with rising demands for change [2]. Buildings are usually designed to be fixed objects of fixed use, ignoring their context's temporal reality. Having the capacity to adapt creates continuous possibilities over time, allowing a building to maintain an extended useful life which cannot happen with a one-time solution. A useful life can only be realised when there is persistent value for users, who are a constant in the equation as buildings change to accommodate their needs, provide better service, and ensure their comfort and protection. However, the value for users is often lost in conventional building design as unavoidable mismatches develop over time between the space supply and demand [1].

The massive amount of construction and demolition waste (CDW) resulting from recurrent arbitrary demolition and rehabilitation activities proves a critical deficiency in how buildings are designed, operated and renovated. The one-time solution of defined form and function approached by the predominant design culture of our buildings leads to one end-of-life option—demolition. As a critical driver for change, physical obsolescence is the primary cause of demolition activities. Still, many drivers for change are external. Building obsolescence usually results from changing operational conditions, emerging user needs, and other varying environmental and external factors. Those include social and local factors (e.g., user's preferences, cultural demands, existing materials), environmental motives (e.g., natural hazards, climatic changes such as heat waves), technical requirements and functional performance (e.g., to embrace technological improvements, energy efficiency requirements), economic factors, legislative issues (e.g., regulations and policies, building codes), and stakeholders' interests [1].

30.2 Materials and Methods

The methodology draws on the analysis of multiple adaptability and circularity assessment models performed in [3]. The analysis studies eight adaptability assessment models and examines the implementation and role of adaptability in multiple circularity frameworks. As a result, change enablers and adaptability requirements are identified and then analysed against their implementation in the common European framework for sustainability—Level(s). The DfA indicator in Level(s) is examined, defining its objective, levels and aspects of assessment, and interrelation with other CE indicators in Macro Objective 2. The shortcomings and possible improvements are then identified in light of the requirements previously defined.

30.3 Results

30.3.1 Change Enablers and Adaptability Requirements

The notion of "Design for Adaptability" originated from enhanced resilience of the built environment against vulnerabilities and disruptions on the one hand and the efficient implementation of sustainability in the built environment on the other. The capacity of a building to respond to emerging changes stems from the early design decisions in the first place [2]. Applying adaptability, therefore, requires transforming the usual focus of form and function determined by immediate needs towards employing a context and time-based perspective for long-term uses [2]. By this meaning, the adaptable capacity of a building determines its future potential [4].

Adaptability requirements are identified using the review and analysis performed in [3] of multiple adaptability assessment models and circularity frameworks that consider adaptability an essential criterion in their methodology. The identified change enablers and adaptability requirements are presented in Fig. 30.1.

Design Factors and Strategies. Seven strategies are identified as essential facilitators for design for adaptability (DfA).

System Separation. The functional Independence of major building systems is identified as a critical adaptability enabler in literature studies. Two main concepts are recognised in adaptability assessment models and criteria: Support-infill Separation and The Shearing Layers. Support-infill Separation distinguishes between two types of elements; "Base building" and "Fit-out". The base building is the durable static

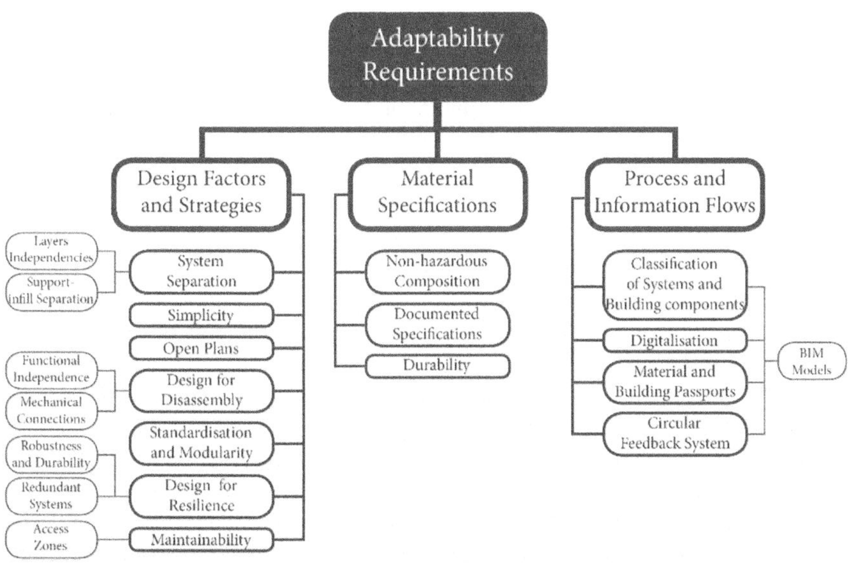

Fig. 30.1 Adaptability requirements

part represented by structural support elements. The fit-out is the flexible infill subject to occupiers' recurrent modifications with minimal interface problems, allowing the building to adapt to the user's needs over time. The idea is brought by the "Open Building" concept of Habraken [5], delivering insights into adaptable design and process flexibility, particularly in residential buildings. Layers Independencies introduced by the shearing layers concept of Brand [6] help to change the general perception of buildings from fixed objects to dynamic systems. Brand's model suggests that a building comprises six layers with different timescales (site, structure, skin, services, space plan, and stuff). The layers should be functionally separated, each containing similar lifespan elements. Understanding a building's composition and identifying the temporal layers of its components is an important strategy to enable its adaptive capacity by allowing flexibility to shorter-life layers or elements and durability to longer-life layers and elements [1].

Simplicity. Designing simple structural systems (e.g., repeating layouts and grids, larger but fewer components) creates easily understood load paths, reducing the uncertainty for the designer working on adaptable solutions. Moreover, the absence of complex systems is vital for the continued operation of the building.

Open plans. Designing layouts free of structural, mechanical and other obstructions allows easy reconfiguration of space plan components to suit changing functional requirements. Open plan layouts grant facilitated adaptation of interior spaces with minimised impact on the existing structure and systems.

Design for Disassembly (DfD). DfD implies that all materials and products used at every level in a building can be neatly disassembled and recovered. By this means, building materials and components have the potential to be reused to their highest extent [1]. DfD goes beyond a building service life by addressing the destination of its materials and components, accounting for the end-of-life (EoL) scenarios at the early design stages [7]. DfD firmly adheres to adaptability as a strategy in which particular components can be changed in response to external factors [1]. By this meaning and considering the EoL in mind, DfD calls for adaptability in its methodology [1] and vice versa since extending the life of a building should think that multiple elements need to be replaced at different stages. DfD relies on achieving *Functional Independence of layers and components* using Mechanical Connections instead of chemical ones. It also provides easy separation of elements and materials without force, minimising contamination of materials and damage to elements during deconstruction and adaptation.

Standardisation and Modularity. Designing modular components and standardised building products facilitates the process of disassembly and reuse, therefore, adaptability. Standardisation can be achieved at three levels: material, component and connection. However, each level has a distinct advantage. For example, standardised materials allow for more efficient recycling, while standardised components create specific conditions for connections between those [4]. Since connections are recognised as essential change facilitators, their standardisation enhances modular design efficiency by ensuring easy removal and replacement of components and exempts those from being standardised. Furthermore, using standardised grids and modularisation also facilitates component interchangeability which is seen as a great enabler of adaptability.

Design for Resilience. Resilient design is achieved using Redundant Systems with overcapacity to support change scenarios, making buildings more adaptable. Examples include clear story height and cavity floors that meet floor-height requirements of different uses. Robustness and Durability also ensure design resilience by granting the ability to withstand the impacts of stressors and external disturbances without major damage or functional failure. In addition, robustness ensures the structure's durability, securing its strength to cater to multiple uses and loading scenarios. Combining the adaptability of use with the required structural robustness under certain conditions contribute to high levels of durability and longevity.

Maintainability. It is the accessibility for assessment to inspect the functionality of varying lifespan elements, especially the shorter-life ones for replacement or repair. Maintainability can be achieved by creating specified Access Zones in the functional layers of systems and elements. Many adaptability mechanisms rely greatly on maintainability and how easily elements are reachable for safe removal and replacement.

Material Specifications. Proper materials selection can significantly impact adaptability values [7]. For example, buildings containing contaminating materials (e.g., asbestos) have less potential for adaptive reuse because of the high risk and rising costs associated with extraction or containment [7]. Conversely, Durable, Non-toxic materials are essential enablers for adaptive reuse projects where they contribute to the prolonged functional life of a building and the reuse of its components in other projects [1]. Moreover, keeping records of materials and Documenting their Specifications facilitates the process of disassembly and appraisal of the most advantageous adaptability scenarios.

Process and Information Flow. Adaptability at its highest level in buildings is a complex procedure that requires an efficient collection of high-quality data on applied materials and products, their characteristics, supply chain information and other important features that facilitate future scenarios [4]. Material and Building Passports are concepts that emerged in the light of CE to grant efficient reporting of all materials and components composing a building describing all characteristics, including composition, value for recovery and recycling, reuse potential, and Eol financial value. Moreover, establishing a common language for data sharing among stakeholders along the value chain facilitates adaptability and disassembly mechanisms. The use of Classification Systems for building systems and elements delivers vigorous means to produce detailed descriptions and categorisation of all building-related data. Furthermore, to ensure facilitated and smart resource reuse and management, a complete knowledge of the building over its entire life span, including all the changes it has undergone is necessary. Studies suggest that Building Information modelling (BIM) could efficiently compile and interpret data and information needed. BIM allows for establishing a Circular Feedback System using a centralised model with all information important to all stakeholders in a Digital Form. BIM

Models enable tracking the components' geometric and mechanical characteristics, composition, recycling value, reuse potential, and expected life cycle.

30.3.2 Level(s) Framework

Description. Level(s) framework is developed to be a common EU framework for a holistic approach towards sustainability assessment in new and existing office and residential buildings. The framework is designed considering the CE action plan, which involves a lifecycle thinking from cradle to cradle, reflected in the utilisation of value and risk rating system [8]. The 16 core sustainability indicators constituting the six Macro-objectives of the Level(s) framework mainly focus on the environmental performance of buildings along their lifecycle. The framework is structured to meet the EU target areas, including energy, resource use, waste production water, and indoor comfort in addition to lifecycle costs. The framework relies on reporting by involved actors on building performance at multiple project stages, from planning and design through implementation, completion and operation to the projected end of life. To achieve this objective, the framework supports three levels of performance assessment: common, comparative and optimised, allowing a progression in accuracy and expertise.

The consideration of the circular economy of building manifests in Macro Objective 2 "Resource efficient and circular material life cycles" which includes 4 core indicators:

- 2.1 Bill of quantities, materials and life spans
- 2.2 Construction and demolition waste and materials
- 2.3 Design for adaptability and renovation
- 2.4 Design for deconstruction, reuse and recycling

Design for Adaptability and Renovation Indicator. This indicator considers addressing future emerging needs by ensuring a building's design capacity since the early project stage to keep fulfilling its function and extend its service life. DfA in Level(s) counts on more efficient use of space throughout the life cycle of a building to increase its longevity and improve its operational performance, contributing to minimising its environmental impacts over the lifecycle [9].

Levels of assessment: The DfA indicator provides a continuous view into the flexibility of design aspects through consequent life cycle stages by promoting three levels of assessment as follows:

- Level 1 of the assessment introduces design concepts in the form of two checklists, one for residential buildings with 8 design aspects and one for office buildings with 12 design aspects helping architects and structural engineers appraise design factors that facilitate future adaptability to emerging needs.
- Level 2 of the assessment relies on detailed design drawings to help configure spatial dimensions and access zones to services during construction. Design

alternatives can be assessed and compared at this level, allowing informative decision-making.

- Level 3 of the assessment evaluates the value of design features in the use phase for potential future adaptations.

Assessment Aspects. The multiple aspects considered in the three levels of assessment are presented in Table 30.1.

Interaction with other Level(s) indicators. Adaptability scenarios have different impacts on input and output flows along a building's lifecycle. The framework, therefore, establishes connections between the DfA indicator and other indicators. The considerations taken by other indicators influencing DfA indicator are presented in Table 30.2.

The input–output relationship among the previous indicators enables comparisons in terms of resource efficiency, allowing users to define advantages and barriers for each and identify potential trade-offs.

Table 30.1 Assessment aspects in the three levels of indicators 2.3

Aspects	Level 1	Level 2	Level 3
Scope	Structure, internal layouts and technical services		
System boundry	Building spatial and structural design features for new and renovation projects		
Reference standards	EN 15643-3, EN 16309, ISO 20887		
Reporting format	Manually filled-in forms		
Lifecycle stage	Planning and initial design	Design and construction	Operation/In use
Measurement unit	N/A	Dimensionless	Dimensionless
Involved actors	Concept architect, structural and service engineers	Architect, structural and service engineers Sustainability consultants	Architect, contractor, structural and service engineers, Sustainability consultants
Calculation method and weighting system	N/A	The calculation method recommended is based on existing methods and criteria. However, using other methods is possible as long as the scores correspond to the same design aspects identified in these assessment levels	
Scoring system	Checklists	Semi-quantitative	Semi-quantitative

Table 30.2 Interrelation of Level(s) indicators with indicator 2.3

Indicator	Consideration of aspects related to DfA
1.2 Life cycle global warming potential	• The potential of the building design to adapt to changing user needs and market conditions in the future to extend the life of the building • The options to extend the lifespan of essential building components
6.1 Life cycle costs	• Cost estimation for adaptation and refurbishment • Annual and periodic costs estimation for maintenance, repair and replacement • Lifecycle cost for adaptation scenarios
2.1 Bill of quantities, materials and lifespans	• Service lives of components and systems • Materials efficiency, durability, repairability, disassembly, reuse, recovery • Minimum material discards and waste
2.2 Construction and demolition waste and materials	• Increase the use of prefabricated elements to promote reuse options • Safe recovery of materials and demolition audits before rehabilitation to avoid damage before adaptation
2.4 Design for deconstruction, reuse and recycling	• Independency of building elements • Use of mechanical, non-destructive connections • Accessibility to elements for easy repair and removal • Standerdised specification of elements • Modular systems to ease updgade and replacement • Design for ease of reconfiguration for different uses

30.3.3 Analysis of Adaptability Requirements in Level(s) Framework

This section addresses the analysis of adaptability requirements against their implementation. The framework's shortcomings and possible improvements are also analysed, particularly for indicator 2.3. Table 30.3 presents the consideration of adaptability requirements in DfA indicators and others addressed by other Level(s) indicators.

Observations and Shortcomings and Improvements. Some of the adaptability requirements identified in this study are split between indicators 2.2, 2.3 and 2.4. Therefore, assessing one of these indicators results in an uncomprehensive outcome. For example, material specification is not addressed in indicator 2.3 but in 2.2. This argument reinforces the necessity for a single circularity index that englobes all

Table 30.3 Implementation of adaptability requirements in Level(s) indicators

Adaptability requirements	DfA indicator	Other indicators
1. Design factors and strategies		
1.1 System separation		
Support-infill separation	× (partially)	
Layers independencies	× (partially)	
1.2 Simplicity	× (partially)	×(2.4)
1.3 Open Plans	×	
1.4 Design for Disassembly		
Functional independence	× (partially)	×(2.4)
Mechanical connections		×(2.4)
1.5 Standardisation and modularity		×(2.4)
1.6 Design for resilience		
Robustness and durability	×	
Redundant systems	×	
1.6 Maintainability		
Access zones	×	×(2.4)
2. Material specifications		
2.1 Non-hazardous composition		×(2.2, 2.4)
2.2 Documented specifications		×(2.2)
2.3 Durability		×(2.2, 2.4)
3. Process and information flows	Not addressed	

aspects in the interrelated indicators of Macro Objective 2, allowing making beneficial trade offs between indicators (e.g., between DfA and DfD) to reach higher circularity values. Also, introducing a circularity index building on Macro objective 2 indicators allows benchmarking circularity values in buildings and identifying added value to sustainability (using indicators 1.2 and 6.1), enabling decision-making regarding the most beneficial circularity aspects to all sustainability dimensions.

The scope of analysis, particularly the layers of consideration in the different indicators in Level(s), is not coherent. For example, in indicator 2.3, the scope considers the structure, façade, space plane and services, whereas in the other indicators of Macro objective 2 the layers are shell, core and external elements. Layering and classifying systems and components using a common methodology recognised by all stakeholders ensure coherent evaluation and more straightforward trade-offs between circularity options. A proper classification must be established since the creation of the bill of quantities and materials, then adopted in the other interrelated indicators. Adopting a coherent layering also allows for assessing each layer's impact on the overall project.

Although the framework mentions that structure and façade have the largest share of the environmental impact of a building project, the sum of weights given to this category in the office buildings weighting system (10.5) is equal to the category of space plan (10.5) and less than the services category (12). The reason is either the design aspects are not attributed to their proper categories or the design aspects for structure and façade are not inclusively addressed. In both cases, the weighting system should be revised considering a coherent attribution of elements to proper categories and allocation of objective weights. Moreover, the weighting and scoring systems in Levels 2 and 3 for indicator 2.3 are only developed for office buildings, with no systems for residential buildings. Therefore, the next version of the framework should address creating a weighting and scoring system for residential buildings. Also, changes from one use to another (from residential to office and vice versa) should be considered.

In its current status, the framework lacks the integration of recent information management tools such as material passports and digitalised information and calculation of indicators. Automating the indicators using a BIM environment could deliver more efficient assessment and centralised management where all stakeholders can interact and provide insights on circularity, testing multiple alternatives and appraising best options.

30.4 Conclusion and Future Work

This paper addresses adaptability requirements drawing on a review study analysing renowned assessment models and frameworks for adaptability and its role in enhancing building circularity. The requirements are analysed against their implementation in Level(s) framework, allowing for improvements to be introduced for more practicality and objectiveness. The conclusion can be made that although Level(s) embraces circularity as a modern-day requirement for sustainability in a well-established and accessible-to-all framework, it lacks some coherence in its methodology. Since Level(s) is still undergoing developments and updates, some shortcomings are addressed and further improvements are proposed to ensure a more inclusive and objective assessment focusing on Macro Objective 2. A potential future work may consider analysing the rest of the circularity requirements against their implementation in Level(s) to identify an overall circularity index. Testing and refining should also be performed using case studies. Future developments should also address the digitalisation of information requirements and automation of indicators and circularity index for interactive assessments.

Acknowledgements This research was financially supported by the Portuguese Foundation for Science and Technology (FCT)/MCTES, under grant number PD/BD/150400/2019, using national funds (PIDDAC). This support was provided through the R&D Unit Institute for Sustainability and Innovation in Structural Engineering (ISISE), with reference UIDB/04029/2020, and the Associate Laboratory Advanced Production and Intelligent Systems (ARISE), under reference LA/P/0112/2020. Additionally, the study received backing from the CYTED Network Circular Economy as a Strategy for a More Sustainable Construction Industry (ECoEICo) with reference 322RT0127, and the COST Action Implementation of Circular Economy in the Built Environment (CircularB) with reference CA21103.

References

1. Askar R, Bragança L, Gervásio H (2021) Adaptability of buildings: a critical review on the concept evolution. Appl Sci 11:4483. https://doi.org/10.3390/app11104483
2. Schmidt R, Eguchi T, Austin S, Gibb A (2010) What is the meaning of adaptability in the building industry?. In: Proceedings of the proceedings of the 16th international conference on open and sustainable buildings. Labein Tecnalia, Bilbao, Spain, pp. 227–236
3. Askar R, Bragança L, Gervásio H (2022) Design for adaptability (DfA)—Frameworks and assessment models for enhanced circularity in buildings. ASI 5:24. https://doi.org/10.3390/asi5010024
4. Geldermans RJ (2016) Design for change and circularity—accommodating circular material and product flows in construction. Energy Procedia 96:301–311. https://doi.org/10.1016/j.egypro.2016.09.153
5. Habraken NJ (1972) Supports: An alternative to mass housing. Praeger Publishers, New York, NY, USA. ISBN 978-0-85139-625-5
6. Brand S (1994) How buildings learn: What happens after they're built. Penguin Books, New York, NY, USA. ISBN 978-0-14-013996-9
7. Ross BE, Chen DA, Conejos S, Khademi A (2016) Enabling adaptable buildings: results of a preliminary expert survey. Procedia Eng 145:420–427. https://doi.org/10.1016/j.proeng.2016.04.009
8. Dodd N, Donatello S, Cordella M (2021) Level(s)—a common EU framework of core sustainability indicators for office and residential buildings, user manual 1: Introduction to the level(s) common framework (Publication Version 1.1)
9. Dodd N, Donatello S, Cordella M (2021) Level(s) Indicator 2.3: design for adaptability and renovation user manual: introductory briefing, instructions and guidance (Publication Version 1.1)

Part XVI
Digitalization and BIM for Enhanced Circularity of Buildings and Building Materials and Products

Chapter 31
Using City Information Modelling to Evaluate Urban Circularity Through Sustainability Indicators

Adriana Salles, Maryam Salati, and Luís Bragança

Abstract While promoting a holistic view of cities, sustainability assessment methods and the circular economy concept have gained attention among urban planners and policymakers. Those methods associated with information technologies can enable intelligent solutions to accelerate sustainability goals. City Information Modelling (CIM) can facilitate the assessment of urban sustainability and circularity. Continuing a previous study, this article examines whether traditional sustainability assessment tools may be adapted to CIM while promoting circular economy practices. Furthermore, the relationship between the most prioritized sustainability indicators and primary urban circularity concerns is examined through theoretical analysis. Therefore, a correlation matrix is proposed and indicators associated with each circularity concern are identified. In total, 24 indicators out of 48 are directly related to urban sustainability. Accordingly, this article discusses how urban circularity concerns can be determined using CIM, based on their correlation with the prioritized indicators. The findings indicate that a CIM-oriented strategy could be used to evaluate urban circularity concerns through sustainability indicators.

Keywords City information modelling · Sustainability assessment methods · Circular economy

31.1 Introduction

Cities can play a significant role in sustainable development by providing opportunities and serving as an important instrument for change and quality of life improvement. In addition, to minimize cities' environmental footprints, the circular economy provides symbiotic ways to design circular urban systems and optimize resource and energy consumption [1]. A circular economy could also help cities achieve urban

A. Salles (✉) · M. Salati · L. Bragança
University of Minho, ISISE, ARISE, Department of Civil Engineering, 4804-533 Guimarães, Portugal
e-mail: adfsalles@gmail.com

L. Bragança et al. (eds.), *Creating a Roadmap Towards Circularity in the Built Environment*, Springer Tracts in Civil Engineering,
https://doi.org/10.1007/978-3-031-45980-1_31

sustainability [2]. A growing focus on urban sustainability has led to the development of indicators, assessment methods, tools, and rating systems [3, 4]. It is possible to use sustainability assessment methods to identify alarming weaknesses in the environment, socio-economic conditions in communities, and inadequacies or problems related to the built environment [5]. To evaluate sustainable aspects, assessment systems use indicators, which serve many purposes, including simplifying, quantifying, and analyzing complex data. They also provide a better way to predict outcomes, monitor progress toward targets, and reinforce them. Nevertheless, one single issue can be assessed using an array of indicators and addressed from multiple perspectives [6].

It is necessary to use more innovative and efficient urban planning and management methods in order to meet the growing demands of cities and the challenges of sustainable development [7]. Therefore, City Information Modeling—CIM—can be viewed as a paradigm for the intelligent modeling of urban spaces by integrating different elements of the urban environment. The use of an intelligent city model enables city planners and urban designers to analyze the demands of cities and make more effective, efficient, and sustainable urban plans [8]. City Information Modelling has received significant attention in the literature over the last decade. CIM is a novel concept that has spawned several approaches in the literature. In general, the concept is based on urban digital models that include rich geospatial information and a complete and up-to-date database [9–12]. CIM has been defined as the "latest advancement from BIM" by Dall'O et al. [13], where data can be viewed in a 3D environment, including various city components. Using city data, all stakeholders can collaborate and make informed decisions. A key feature of CIM, according to Wang and Tian [14], is its close relationship with smart cities, integrating various technologies such as the Internet of Things (IoT), Geographic Information Systems (GIS), and Building Information Modeling (BIM). Despite the fact that CIM is currently under discussion and no widely accepted concept has emerged, many authors propose integrating GIS and BIM to compose a CIM platform [14–18]. As with BIM, CIM could be applied throughout the entire life cycle of a city, from the design and planning stages through the construction and maintenance stages [16]. Furthermore, sustainable and circularity criteria such as energy use, water and waste management, mobility and transport, can be incorporated into CIM's urban database. Using data from BIM and CIM, Dantas et al. [19] evaluated how CIM could be used to obtain data to support the indicators of ISO 37120 to improve public services and quality of life. In their study, Sabri et al. [20] proposed a 3D geospatial platform for assessing the processing, publishing, and visualization of urban environments in 3D. According to the authors, the platform is intended to support urban planning decisions from an intuitive perspective, so that planners can interact with the lives of people directly. In a case study using BIM for urban planning and management, Tao and Qian [21] proposed simulation models to analyze and evaluate solar radiation, ventilation, and energy consumption. Beirão [22] discusses an approach to utilizing a CIM platform, suggesting that it should be open, accessible, and interactive for all users, such as urban planners and citizens. Despite the wide variety of approaches, a

CIM platform typically has elements of interactivity, collaboration, interoperability, information sharing, as well as BIM and GIS integration.

In a previous study, Salles et al. [23] analyzed internationally recognized systems and identified the prioritized issues and aspects of urban sustainable assessment techniques, demonstrating existing sustainability concerns. Considering the Portuguese methodology SBTool^{PT} Urban as a baseline, the study compared it with the similar indicators of the following methods: Sustainable Neighborhood Tool (SNTool) from iiSBE, BREEAM Communities (BREEAM-C), and LEED for Neighborhood Development (LEED-ND). Also, the alignment of the indicators with the Sustainable Development Goals (SDGs), ISO 37120 standards, and Level(s) was considered.

Accordingly, this study aims to discuss how CIM can be used to implement circular economy strategies in the built environment. After that, it is evaluated how the method proposed in the previous study might help achieve urban circularity.

31.2 Materials and Methods

As stated previously, Salles et al. [23] proposed a method to assess urban sustainability using CIM. By evaluating the method, this paper aims to discuss the feasibility of using it for urban circularity. The results obtained by Salles et al. [23] showed that the averaged indicators of most evaluation methods promote similar sustainability concerns. As a result, 48 indicators have been identified as being essential to be assessed in every urban context, with 41 indicators being covered by the Portuguese methodology SBTool^{PT} Urban, while the remaining seven are uncovered or partially covered. Then the article analyzes the feasibility of each indicator being calculated using CIM. A total of 52 parameters are used in the calculation of the 41 indicators presented in the SBTool^{PT} Urban methodology. As part of the assessment process, each indicator was validated based on the calculation criteria and methodology and their ability to be integrated into a digital information model. In this regard, the indicators are categorized as 'YES', 'NO', or 'PARTIALLY' meaning, respectively, that indicators can or cannot be assessed using CIM, while others can only be partially assessed using CIM. Data type, the feasibility of modeling, and availability of relevant information and similar cases in the literature determine whether a parameter is considered 'YES'. If a parameter cannot be modeled, is not found in any related case study, or is not likely to benefit greatly from CIM, then it is classified as 'NO'. 'PARTIALLY' refers to the fact that most of the criteria can be assessed using the model, however, one or two criteria cannot be assessed. In the absence of a calculation methodology for the new indicators, they were considered 'NON-APPLICABLE'. Therefore, the CIM model can be used to calculate at least 28 parameters, and seven more can be partially calculated by CIM.

The method proposed by Salles et al. [23] uses CIM to assess urban sustainability. Considering that CIM is the integration of BIM and GIS, the study proposed the use of Autodesk Revit, as a BIM tool along with CADMAPPER, for the incorporation of GIS information into the model, and Dynamo, for the calculation of the results.

As a Revit plug-in, Dynamo allows users to create routines to execute activities and calculations, it is also a versatile and adaptable tool to calculate indicators. Once the platform is created, the modeling phase is the following step, adding to the model all the necessary data. CADMAPPER allows the urban area to be inserted into the model, then the area has to be characterized by the material parameters in the model, the topography, and the shared parameters. Afterward, calculation routines can be created using Dynamo to quantify the indicators. Each indicator has a different calculation routine, according to its parameters.

The literature review pointed out a relationship between circular economy and urban sustainability [1, 2, 24, 25]. Dong et al. [1] explored the synergies between urban sustainability and circular economy. According to the authors, applying the circular economy to urban areas enables a systematic approach to exploring symbiotic relationships. This includes, for example, urban industrial symbiosis and community waste separation and recycling, which can be used to create circular urban systems, optimizing the metabolism of cities while maximizing resource efficiency and reducing environmental impact. In this regard, the concept of urban sustainability can be described as an increase in the efficiency of urban resources. By integrating circular economy principles, the urban space, industrial facilities, and infrastructure can be designed in a closed-loop environment.

Furthermore, Schaubroeck et al. [26] proposed the use of a 3D city model to store building joints. According to the authors, building joint data can be used to apply circular strategies at an urban scale. In this regard, city models can store data and serve as databases. Moreover, the digitalization of the building stock can contribute to reaching various sustainability targets.

Accordingly, based on a literature review, this study evaluates the synergies between sustainability indicators and the circular economy in the built environment. In addition, it is discussed the feasibility of using CIM for urban circularity.

31.3 Results and Discussion

Analyzing the prioritized sustainability indicators identified in the previous study by Salles et al. [23], and comparing them with the main concerns of urban circularity, this study finds out that 24 out of 48 indicators are directly connected to urban circularity strategies. The results are presented in the correlation matrix shown in Fig. 31.1. The main concerns about urban circularity identified in this paper are screened and selected based on a literature review. A general finding is that applying circular economy concepts in the urban environment can contribute to urban sustainability. At least, half of the sustainability indicators are directly related to urban circularity concerns. And most of the circularity concerns are related to more than 6 indicators. This indicates that the implementation of circular economy concepts in an urban environment could have a positive impact on multiple aspects of sustainability, including the economy, environment, and society. The matrix also reveals that it is possible to focus on circularity concerns to address a significant number of sustainability indicators, which could lead to improved outcomes for the city.

Circularity is the concern that has devoted the highest number of indicators. This concern represents the extent to which circular economy principles have been applied. Increased circularity leads to increased recycling ratios, greater reuse, and reduced resource consumption. Reduce, Circular business model, and Industrial symbiosis are subsequently second, third, and fourth, addressing 15, 13, and 12 indicators, respectively. Reuse and Recycle are related to 11 indicators. In the sequence, Extend lifetime

	Urban circularity concerns												
Prioritized Sustainability Indicators	Prioritize renewable resource	Extend lifetime of products	Reduce	Reuse	Recycle	Circularity	Industrial symbiosis	Circular business model	Team up to create joint value	Design for the future	Application of digital technology	Knowledge creation	Education
Uses Density and Flexibility		■				■							
Reuse of Urban Land		■	■	■	■	■							
Building reuse		■	■	■	■	■				■			
Technical infrastructure		■	■	■	■	■				■			
Energy Efficiency	■		■			■	■				■		
Renewable energies	■						■	■		■	■		
Centralized energy management	■		■				■	■					
District heating and cooling	■		■				■	■					
Efficient water consumption			■	■	■		■	■					
Effluent management			■	■	■		■	■					
Centralized management			■	■	■		■	■					
Low impact materials	■	■	■	■	■	■	■	■		■			
Energy embodied in construction materials	■		■	■	■	■	■			■			
Construction and Demolition Waste		■	■	■	■	■	■						
Solid waste management			■	■	■	■	■	■					
Local food production						■	■	■					■
Public transport						■							
Cycle path network						■							
Valuing heritage		■	■	■	■	■							
Social inclusion						■		■	■			■	■
Economic viability						■		■					
Local economy						■		■					
Sustainable buildings						■				■			■
Environmental management								■			■		■

Fig. 31.1 Relationship between sustainability indicators and urban circularity concerns. *Source* Authors

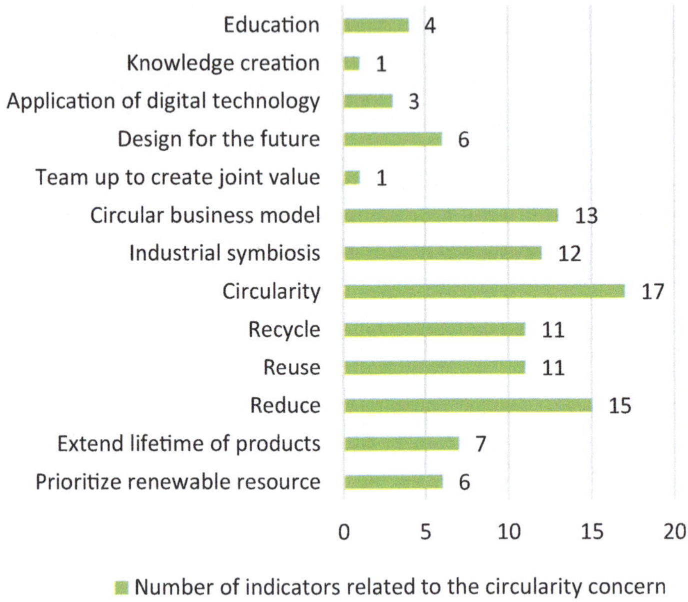

Fig. 31.2 Number of indicators related to the circularity concern. *Source* Authors

of products is related to 7 indicators, followed by Prioritize renewable resources and Design for the future, with 6 indicators each. Education, Application of digital technology, Team up to create high value and Knowledge creation are the latest ones, with 4, 3, and 1 indicators, respectively. The results are shown in Fig. 31.2.

Evaluating the method proposed by Salles et al. [23] the results indicate that CIM can be used to access urban sustainability. The SBTool^{PT} Urban methodology promotes 41 of the 48 prioritized indicators. To calculate these indicators, SBTool^{PT} Urban Guide developed a set of calculation parameters. Among the 41 indicators, there are 52 calculation parameters, and at least 28 parameters can be calculated using the CIM model, and another 7 can be determined in part. The classification criteria (YES, NO, or PARTIALLY) are related to the calculation parameters and the information needed to quantify them. Therefore, 24 indicators have a direct relation to urban circularity concerns from the 48 prioritized indicators. This helps to identify which indicators contribute the most to overall circularity performance, as well as which have the greatest potential for improvement.

Considering the relationship between sustainability indicators and the circularity concerns, presented in Figs. 31.1 and 31.2, this study has identified the number of indicators classified as YES, NO, PARTIALLY, or NON-APPLICABLE, for each circularity concern. Figure 31.3 presents the results. According to the findings of the study, CIM cannot be used to calculate the indicators of only three circularity concerns. The majority of concerns have indicators that can be assessed by CIM. In five circularity concerns, more indicators were able to be determined through CIM

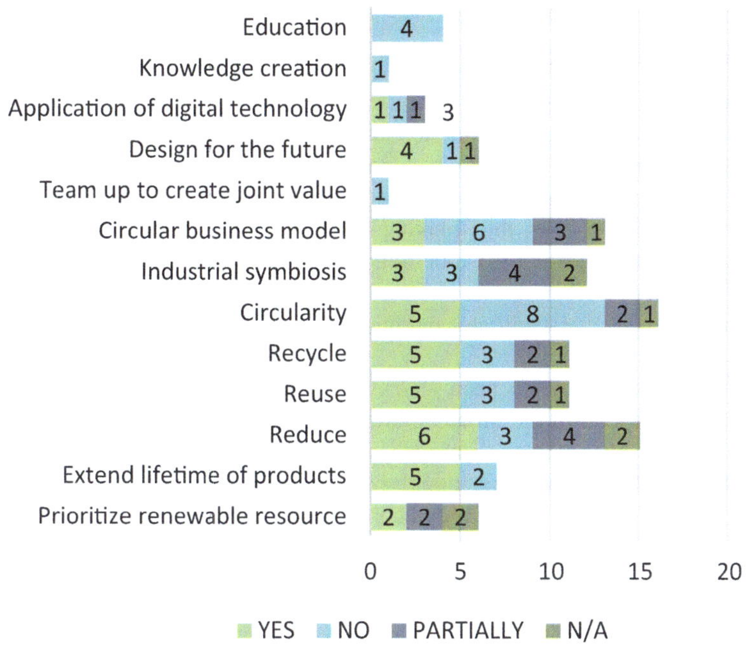

Fig. 31.3 Number of indicators classified according to the calculation using CIM for each urban circularity concern. *Source* Authors

than in those that could not. There are only two concerns in which the number of indicators that cannot be assessed by CIM exceeds the number that can be assessed. These results suggest that CIM is a viable tool for measuring sustainability indicators for most circularity concerns.

Nevertheless, the use of CIM to evaluate urban circularity concerns is highly dependent on quantification. As such, the accuracy of CIM evaluations is limited by the quality of available data. Furthermore, the use of CIM for urban circularity analysis involves the consideration of multiple variables. In this study, circularity concerns were related to sustainability indicators, previously evaluated regarding the feasibility of calculation using CIM. Since SBTool^{PT} Urban sustainability indicators are typically evaluated through quantitative assessments, CIM can provide a quantitative basis for assessing urban circularity concerns. It also can be used to identify and quantify urban circularity concerns, and to evaluate the effectiveness of proposed strategies for urban sustainability.

31.4 Conclusion

The pursuit of urban sustainability has led to advances in research as well as the development of indicators. Information and communication technologies have also contributed to sustainable urban development by providing tools and systems that enhance urban management and operations efficiency. In addition, they help cities transition to digitalization, thus making them more 'intelligent' and sustainable. Accordingly, City Information Modelling can aid in the implementation of sustainability strategies in urban spaces. Moreover, the CIM model can provide a multidisciplinary perspective, facilitate communication, and promote sustainable concept integration.

The circular economy has been seen as a valuable approach to reaching sustainability. When applied to the urban environment, a circular economy promotes the optimization of cities' metabolism, enhancing resource efficiency, while reducing environmental footprints. This could help in reducing resource consumption, allowing for the reuse of materials and components. It also helps to create local jobs and stimulate innovation, while promoting a more equitable and resilient society.

The purpose of this study is to discuss how City Information Modeling can be used to evaluate circular economy strategies in the urban environment. The findings have shown that the use of CIM to assess urban circularity is feasible. However, at first, it is necessary to relate the circularity concerns to indicators, in order to quantify them. The sustainability indicators used were based on a previous study. Using the Portuguese methodology SBTool^{PT} Urban as a baseline, the previous study determined which indicators could be integrated into the CIM concept for assessing urban sustainability. These prioritized indicators were related to urban circularity concerns and a correlation matrix was proposed. Then, for each circularity concern, it was evaluated whether the related indicators could be calculated through CIM. Overall, most of the concerns are related to indicators that can be calculated using CIM. Only 3 concerns presented negative results, concluding that the corresponding indicators cannot be determined. The research concluded that CIM can be used to assess circularity in the urban environment. However, further research is necessary to determine whether the proposal is applicable. The findings provide a foundation for further research into the application of CIM to assess circularity.

References

1. Dong L, Liu Z, Bian Y Match circular economy and urban sustainability: Re-investigating circular economy under sustainable development goals (SDGs). https://doi.org/10.1007/s43615-021-00032-1
2. Joensuu T, Edelman H, Saari A (2020) Circular economy practices in the built environment. J Clean Prod 276:124215. https://doi.org/10.1016/J.JCLEPRO.2020.124215
3. Sharifi A (2021) Urban sustainability assessment: An overview and bibliometric analysis. Ecol Indic 121:107102. https://doi.org/10.1016/j.ecolind.2020.107102
4. Morphet J, Morphet R (2019) New urban agenda: New urban analytics

5. Steiniger S, Wagemann E, de la Barrera F et al (2020) Localising urban sustainability indicators: The CEDEUS indicator set, and lessons from an expert-driven process. Cities 101:102683. https://doi.org/10.1016/J.CITIES.2020.102683
6. Salati M, Bragança L, Mateus R (2022) Sustainability assessment on an urban scale: Context, challenges, and most relevant indicators. Appl Syst Innov 5:41. https://doi.org/10.3390/asi5020041
7. Sharifi A, Murayama A (2013) A critical review of seven selected neighborhood sustainability assessment tools. Environ Impact Assess Rev 38:73–87. https://doi.org/10.1016/J.EIAR.2012.06.006
8. Kehmlani L (2016) City information modeling. In: AECbytes Mag. https://www.aecbytes.com/feature/2016/CityInformationModeling.html. Accessed 16 Dec 2021
9. Bi T, Yang X, Zhou F et al (2020) Research on CIM basic platform construction. ACM Int Conf Proceeding Ser Part F 16834:159–163. https://doi.org/10.1145/3446999.3447029
10. Bragança L, Mateus R (2018) Guia de Avaliação SBToolPT Urban V2018. Edições iiSBE Portugal
11. Ameen RFM, Mourshed M (2019) Urban sustainability assessment framework development: The ranking and weighting of sustainability indicators using analytic hierarchy process. Sustain Cities Soc 44:356–366. https://doi.org/10.1016/J.SCS.2018.10.020
12. Gil J (2020) City Information Modelling: Digital planning for sustainable cities. Built Environ 46:497–500. https://doi.org/10.2148/BENV.46.4.497
13. Dall'O' G, Zichi A, Torri M (2020) Green BIM and CIM: Sustainable planning using building information modelling. Res Dev 383–409. https://doi.org/10.1007/978-3-030-41072-8_17
14. Wang B, Tian Y (2021) Research on key technologies of city information modeling. IOP Conf Ser Earth Environ Sci 693. https://doi.org/10.1088/1755-1315/693/1/012129
15. Souza L, Bueno C (2022) City Information modelling as a support decision tool for planning and management of cities: A systematic literature review and bibliometric analysis. Build Environ 207:108403. https://doi.org/10.1016/j.buildenv.2021.108403
16. Xu Z, Qi M, Wu Y et al (2021) City information modeling: State of the art. Appl Sci 11:1–15. https://doi.org/10.3390/app11199333
17. Omrany H, Ghaffarianhoseini A, Ghaffarianhoseini A, Clements-Croome DJ (2022) The uptake of City Information Modelling (CIM): A comprehensive review of current implementations, challenges and future outlook. Smart Sustain Built Environ. https://doi.org/10.1108/SASBE-06-2022-0116
18. Xu X, Ding L, Luo H, Ma L (2014) From building information modeling to city information modeling. J Inf Technol Constr 19:292–307
19. Dantas HS, Sousa JMMS, Melo HC (2019) The importance of city information modeling (CIM) for cities' sustainability. IOP Conf Ser Earth Environ Sci 225. https://doi.org/10.1088/1755-1315/225/1/012074
20. Sabri S, Chen Y, Rajabifard A et al (2019) A multi-dimensional analytics platform to support planning and design for liveable and sustainable urban environment. Int Arch Photogramm Remote Sens Spat Inf Sci—ISPRS Arch 42:75–80. https://doi.org/10.5194/isprs-archives-XLII-4-W15-75-2019
21. Tao Z, Qian Z (2015) Study on the microenvironment evaluation of the architectural layout based on building information modeling: A case study of Chongqing, China. Int J Des Nat Ecodynamics 10:140–153. https://doi.org/10.2495/DNE-V10-N2-140-153
22. Beirão J (2014) City information modelling: Spatial planning and design with CAD and GIS—a workshop experiment
23. Salles A, Salati M, Bragança L (2023) Analyzing the feasibility of integrating urban sustainability assessment indicators with city information modelling (CIM). Appl Syst Innov 6(45)https://doi.org/10.3390/ASI6020045
24. Bellezoni RA, Adeogun AP, Paes MX, de Oliveira JAP (2022) Tackling climate change through circular economy in cities. J Clean Prod 381. https://doi.org/10.1016/j.jclepro.2022.135126
25. Pamucar D, Deveci M, Gokasar I, et al (2021) Circular economy concepts in urban mobility alternatives using integrated DIBR method and fuzzy Dombi CoCoSo model. J Clean Prod 323. https://doi.org/10.1016/j.jclepro.2021.129096

26. Schaubroeck S, Dewil R, Allacker K (2022) Circularity of building stocks: Modelling building joints and their disassembly in a 3D city model. Procedia CIRP 105:712–720. https://doi.org/10.1016/J.PROCIR.2022.02.119

Chapter 32
BIM-Based Sustainability Assessment: Insights for Building Circularity

José Pedro Carvalho, Luís Bragança, and Ricardo Mateus

Abstract Facing the increased pursuit for sustainable buildings, the construction industry has been adopting new project technologies such as Building Sustainability Assessment (BSA) and Building Information Modelling (BIM), providing new potentialities like continuous data storage, optimised building performance and integrated building design. Facing the opportunity to integrate BSA into BIM, a BIM-based application—SBTool^{BIM}—was developed to automate BSA during the project early stages. Visual Programming Language (VPL) was used to translate 19 criteria requirements from SBTool^{PT}-H and additional BIM connections were identified for the remaining 6 criteria. The applied procedure has established a novel framework to carry out BSA within the BIM environment in a faster reliable way. Such a method can provide valuable insights for a BIM-based application for building circularity assessment by using similar structures and procedures. This research aims to explore SBTool^{BIM} in order to identify possible insights which can be replicated for a BIM-based application to assess buildings circularity potential. Results show the potentialities of SBTool^{BIM} structure for circularity purposes, as different multidisciplinary data can be stored in the BIM model, novel Key Performance Indicators (KPIs) can be easily integrated and the evaluation structure can be adapted according to the analysis requirements.

Keywords Building information modelling (BIM) · Project tool · Buildings circularity

---

J. P. Carvalho (✉) · L. Bragança · R. Mateus
Institute for Sustainability and Innovation in Structural Engineering (ISISE), University of Minho, Guimarães, Portugal
e-mail: jpcarvalho@civil.uminho.pt

L. Bragança · R. Mateus
University of Minho, School of Engineering, Civil Engineering Department, Guimarães, Portugal

L. Bragança et al. (eds.), *Creating a Roadmap Towards Circularity in the Built Environment*, Springer Tracts in Civil Engineering,
https://doi.org/10.1007/978-3-031-45980-1_32

32.1 Introduction

General society and worldwide authorities are increasingly looking for more sustainable buildings [1]. To turn buildings more efficient, there is a need to act since the early project stages by introducing and analysing different sustainability measures. To support decision-making regarding effective sustainable measures and to monitor the building's sustainable performance, BSA methods have been developed all over the world and adapted to the specific conditions of each location and building [2]. However, carrying out a BSA is currently a time-consuming and complex process, requiring the knowledge of interdisciplinary data and a long manual calculation procedure. Thus, BSA is commonly used in the project later stages after finishing the project design, where modifications are unbearable or too expensive [3]. BSA is also a voluntary approach with an absence of mandatory regulation, establishing a scenario for its reduced adoption [4]. There is a need to develop a more expedited method to perform BSA, with a real-time and dynamic evaluation, able to support project improvements and modifications, instead of the traditional static approach [5]. With the deployment of digitalisation in the construction sector, building requirements are being translated into computer rules, to accelerate and guarantee building compliance in a faster and more reliable way [6]. BSA is based on a set of KPIs, which can also be represented by computer rules. Therefore, the opportunity arises to translate BSA methods into computable rules and effectively integrate them into existing computer software.

BIM has emerged to support designers in managing all the project data and improving project efficiency [7]. BIM creates an excellent opportunity to incorporate sustainable measures throughout the design process, especially during the project early stages, promoting the development of high-performance buildings. It allows the storage of multidisciplinary information and encourages real-time collaboration among stakeholders [3]. BIM potential for automated rule checking has been recognised and together with VPL, several compatible rules have been developed, creating better, faster, and more comprehensive assessment procedures [6].

Facing the absence of a proper connection of BIM with BSA, the authors have established a novel procedure for the automation of the Portuguese residential version of the BSA method SBTool^{PT}. By translating the BSA KPIs into a computer application integrated into the Open BIM environment, a fast and reliable sustainability analysis was achieved for the project early stages by using the SBTool^{BIM} application. Designers are provided with an intuitive tool to analyse their building sustainability performance, allowing them to introduce and compare improvement measures with few resources. VPL was used to translate the BSA requirements into computable rules and the application was hosted in Autodesk Revit to fit into the BIM environment. Taking advantage of the developed procedure and given the interconnection of Circular Economy (CE) and sustainability, the applied theory and methods can be easily replicated for building circularity purposes. Moreover, the implementation of CE principles will also promote the balance between the main dimensions of building sustainability—environment, society and economy. CE can act near local and regional competitiveness by improving resource allocation, utilisation and productivity. Primary resources exploitation is minimised by implementing a novel circular and ecological value chain, which will also create additional job opportunities and economic growth [8]. Thus, this research aims to explore and identify how SBTool^{BIM} can inspire future BIM-based applications to assess buildings circularity potential in the project early stages, to effectively enhance, implement and spread building circularity concepts.

32.2 Methodology

The aim of the enclosed paper is to explore an existing BIM approach to assess building sustainability—SBTool^{BIM}—in order to identify possible and valuable insights for a future circular building assessment. As SBTool^{PT}-H is based on KPIs, which are commonly discussed to be the basis of new circular assessment methods, a new circularity tool can benefit from the existing method to accelerate its development and practical implementation.

With the flexibility of SBTool^{BIM} input requirements and structure, new insights can be provided to inspire the development of new a generation of circular assessment methodologies. By analysing the most identified BIM uses for the circular economy, insights from SBTool^{BIM} will be framed under such applications and process replicability will be further explored to include new criteria and data, according to the circular building components and materials characteristics requirements.

32.3 SBTool^{BIM}

32.3.1 Objectives and Method

SBTool^{BIM} is a BIM-based application to automate sustainability assessment during the project early stages. It provides a simple and intuitive method to promote and encourage BSA implementation by designers, providing real-time feedback about the building sustainability level. The Open BIM concept, allows BSA to be effectively included in the common BIM project workflow. SBTool^{BIM} reflects the Portuguese residential version of SBTool (SBTool^{PT}-H), which is based on KPIs. A total of 19 criteria were fully automated in SBTool^{BIM} and an additional BIM linkage is proposed for the remaining 6 criteria. SBTool^{BIM} application is hosted in Autodesk Revit, as it has the capacity to import and export several file formats (also allowing to frame SBTool^{BIM} into the Open BIM environment) and, to develop personalised interfaces to perform specific tasks through Dynamo. The application has two parts, both based on VPL through Dynamo, which provides a convenient and automated data exchange procedure with the BIM model in Autodesk Revit.

The application algorithm follows an automated code checking procedure method [6]:

- Rule interpretation—SBTool^{PT}-H assessment requirements and calculation procedures are carefully analysed to identify all the necessary inputs and workflow. The collected data is later translated into a computer-processable language;
- Setup the BIM model—The model is created or imported in Autodesk Revit with all the required data for the assessment and a set of modelling guidelines must be followed. A template—the first part of SBTool^{BIM}—is used to host the new required shared parameters for the sustainability assessment;
- Rule execution—All the established "rules" during the interpretation phase and the identified research procedure were programmed in Dynamo. This phase is the core of the SBTool^{BIM} application (second part of SBTool^{BIM}), gathering and processing the required assessment data from the BIM model according to each criterion's needs;
- Rule report and validation—Building sustainability evaluation. The application displays the building results by presenting the individual score of each criterion, as well as the building global score. The analysis results can either be checked in Autodesk Revit environment or a Microsoft Excel spreadsheet can be extracted;
- BIM additional linkage—As Dynamo is not able to automate the assessment of all SBTool^{PT}-H criteria, especially the ones that require performance analysis data, an additional BIM linkage is proposed to collect the remaining assessment information.

32.3.2 Input Requirements and Assessment Model

As SBTool addresses a wide range of criteria categories, different type of data is required for the assessment. The assessment process of some criteria lies only in geometric quantitative data from the model. Other criteria require the knowledge of several specific regional data or even the results of performance simulations. Therefore, the different input types have been grouped into three sections:

- BIM Model—Data intrinsic to the model (identity and geometry data) or data that can be introduced into the model through shared parameters (mainly quantitative data);
- User interface—Specific regional data and qualitative data that the user must provide to SBTool^{BIM} through an interface (as local amenities, transports or recycling conditions);
- BIM tools—Data that requires the use of additional tools (performance simulation tools) for model simulation.

As SBTool^{BIM} is a Dynamo-based application, both the BIM model data and the user interface data are available inside the Autodesk Revit environment. The BIM tools data requires the exportation of the BIM model for additional software to carry

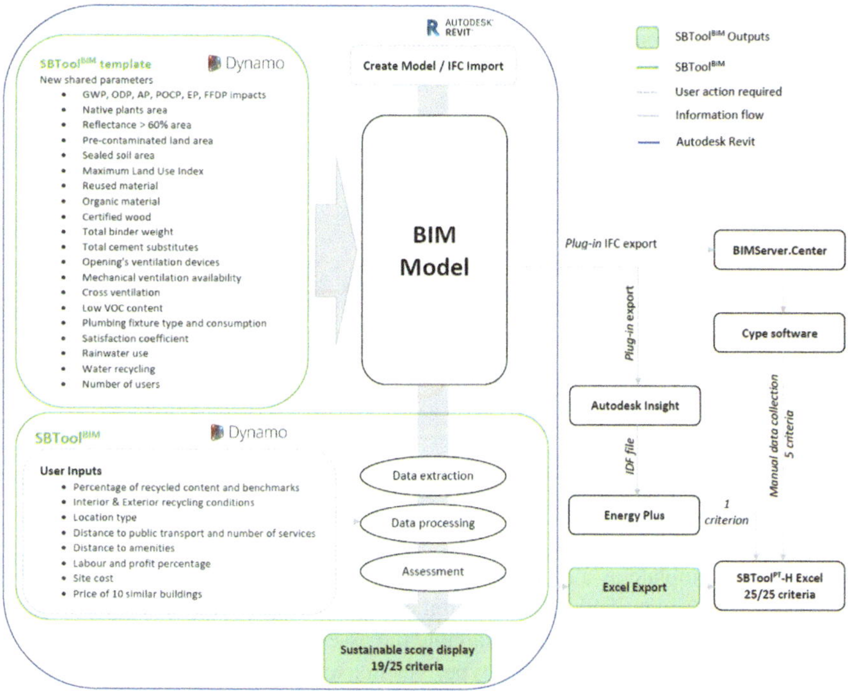

Fig. 32.1 SBTool^{BIM} process map

out performance simulations and data is not directly available in the Autodesk Revit environment.

Based on the adopted software capabilities, the identified input requirements, and the assessment method, the SBTool^{BIM} process map was established, and it is presented in Fig. 32.1.

32.3.3 Template and Model Characteristics

Autodesk Revit has a set of base templates with common transversal parameters to characterise the BIM model, such as:

- Elements dimensions;
- Element's cost;
- Material cost, density, class, and function.

This data is usually intrinsic to the model and is shared by all the templates. To introduce sustainability data, Autodesk Revit allows to create and edit templates according to the user's needs. SBTool^{BIM} template was developed to properly describe the BIM model for the sustainability evaluation. It uses Dynamo and the

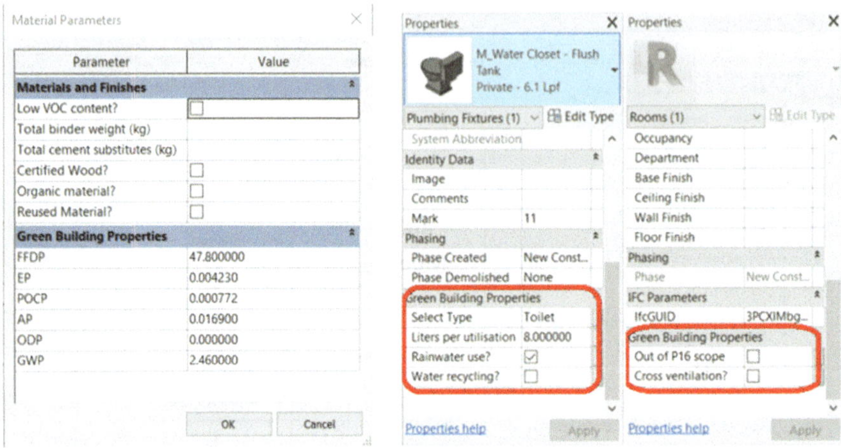

Fig. 32.2 SBTool^{BIM} shared parameters

architectural template as a basis. New shared parameters are automatically created according to the identified input requirements for SBTool^{PT}-H. In total, 28 shared parameters were created to make available the assessment of 13 criteria. Four types of parameters were used: number type, yes/no, text and area. Some examples of SBTool^{BIM} parameters are presented in Fig. 32.2. Most of these characteristics must be provided directly by the user. Although, a material database can also be used, where all the material characteristics are already defined, and it is just necessary to import them for the current model. The template can be saved as an Autodesk Revit template and marked as standard to create and develop models, avoiding the need to run the Dynamo file for every project.

32.3.4 Structure

To extract and process data from the BIM model, the SBTool^{BIM} application was developed through VPL in Dynamo. It automatically assesses the required building data for the sustainability evaluation and provides an interface for the user to complete the analysis information. SBTool^{BIM} code is based on the collection of the BIM model categories data. All elements of each category are assessed and filtered according to each criterion scope. However, SBTool^{BIM} also requires additional information which is not available in the model. Dynamo was used to create user interfaces for the analysis, where the designer can introduce the remaining required data for the assessment. After starting SBTool^{BIM}, a set of pop-up windows will consecutively appear on the Autodesk Revit environment. The aim is to collect external and local data for the analysis. Some examples of the popup windows are presented in Fig. 32.3.

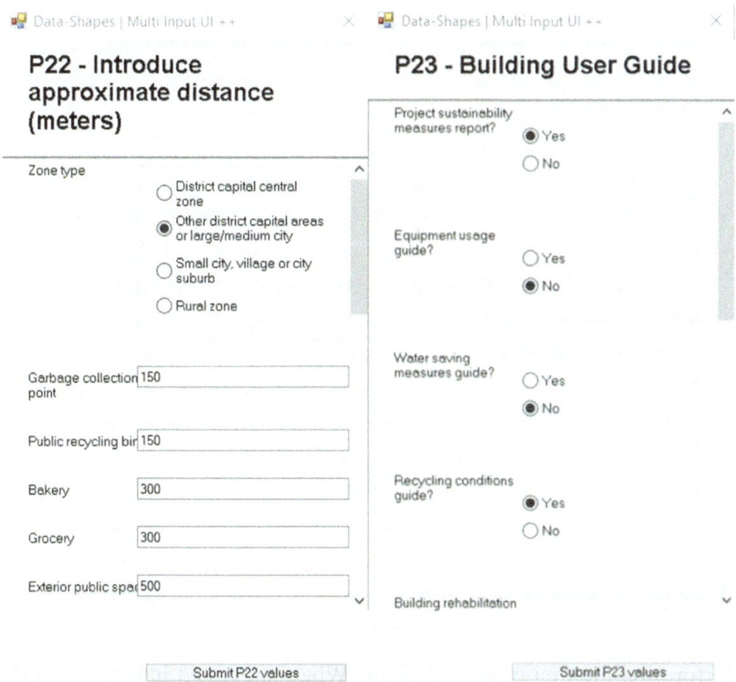

Fig. 32.3 Input forms through user interfaces in Autodesk Revit

After gathering the required data for each criterion, SBTool^{BIM} performs the assessment calculations, normalisations, and qualitative quantification. The building final score is displayed through a pop-up window in the Autodesk Revit (Fig. 32.4). Simultaneously, it produces an Excel spreadsheet with the building quantitative and qualitative scores.

32.4 Discussion

Estimating the performance of a complete building is a complex procedure and depends on several external factors. SBTool^{BIM} application reflects the KPIs of SBTool^{PT}-H and provides instant feedback for designers, making it a cutting-edge tool to support decision-making towards the practical implementation of the sustainable building concept. With this tool, project designs can be easily analysed and compared with fewer resources, creating the opportunity to integrate new sustainability measures and improve building efficiency. SBTool^{BIM} was found very reliable and can reduce the sustainability assessment time from a week to a couple of days

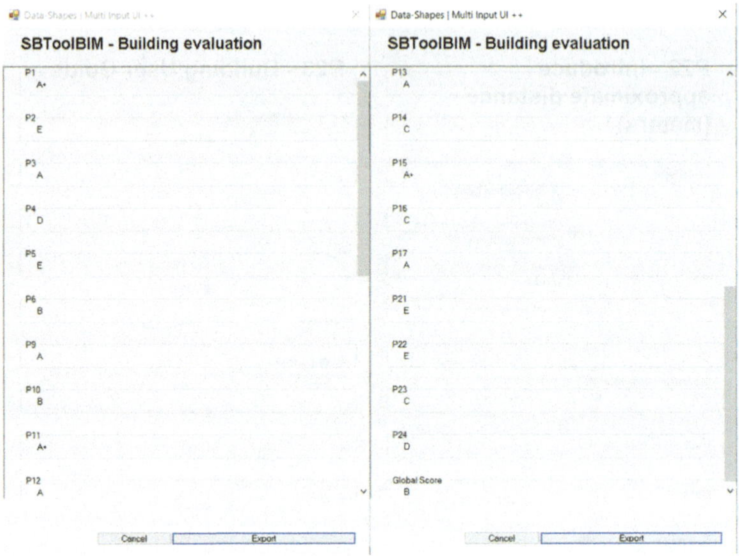

Fig. 32.4 SBtool^{BIM} results in Autodesk Revit

when performing the first project analysis. When integrated into the BIM environment, the prior existence of a BIM model can further reduce the assessment time and enhance the coordination of building sustainability with other project disciplines.

SBTool^{BIM} process replicability was found very accessible for other building types or locations, as Dynamo allows to adapt and/or introduced new criteria with similar established procedures. The replication for other BSA schemes is also feasible, as long as they are based on KPIs. Therefore, this approach can also be replicable for other kinds of building assessments, such as circular building assessments, pre-demolition audits or building deconstruction assessments, making SBTool^{BIM} a valuable bank of contributions for cutting-edge assessment models.

Within the frame of CE, BIM has been recognised as a key feature to enhance the building circularity potential. The BIM method simplifies and automates the integration of CE strategies during the design process, as well as integrates them with traditional assessment procedures, such as BSA or Life Cycle Analysis (LCA). Given BIM applicability throughout the building life cycle, it offers three main features that make it suitable for CE assessment [9]:

- Parametric modelling which regulates building design form and functionalities using established parameters and rules;
- Bi-directional associativity which provides support and real-time updates;
- Intelligent modelling ensures that supplementary data is provided in addition to 3D geometric data, such as building local data, cost information, energy analysis details, waste management plan, etc. This data can also be stored in the BIM

model during the building lifecycle, making it a data storage, which is available for all the stakeholders in any life cycle stage.

Although, BIM can also be applied according to the building life cycle stage [10]. During the project early stages, it can be used for automatic data extraction and can be coupled with material passports, BSA and LCA to provide the environmental, sustainability and recyclability potential of building materials and components. In the operation phase, BIM can act as a management tool to aid in maintenance planning and provide more efficient management of the applied resources. At the end of life, it can support waste management, especially in quantifying waste and in the evaluation of the building and materials recycling, deconstruction or disassembly potential.

Moreover, Charef and Emmitt [11] have identified seven BIM uses for supporting CE: (1) digital model for building end-of-life evaluation, (2) material passport development, (3) project database, (4) data checking, (5) circularity assessment, (6) materials' recovery processes and (7) materials' bank.

Facing the common application of BIM for CE purposes and the need for establishing a new assessment framework to assess buildings circularity potential, SBTool^{BIM} method replicability can provide valuable insights which can be framed under the majority of BIM uses for CE.

BIM models have enough flexibility for being used for different purposes. In SBTool^{BIM}, a properly characterised BIM model is required. Such a model can also be used to extend the building analysis, as it was already developed through parametric and intelligent modelling, acting as a data storage tool. To proceed with further CE analysis, new parameters can be integrated into the BIM model to reflect the circularity assessment, by using a similar procedure as SBTool^{BIM}. The personalised BIM model characterisation made in SBTool^{BIM}—by creating new shared parameters—can be replicated to new assessment methodologies such as circularity or deconstruction assessment of buildings. Different types of data are allowed to be used, in order to identify key building and material features, such as specifications of reusable/recyclable materials, use of nut/bolt joints, use of prefabricated assemblies, minimisation of building components, materials toxic content and/or material maintenance/deconstruction data. According to the type of data, Autodesk Revit and Dynamo flexibility allows to create and adapt new parameters according to those characteristics for building components and/or materials. Quantitative data can be easily introduced by creating new shared parameters for the BIM model components or materials, while other specific data such as analysis results, landfill locations, transportation requirements or recyclable procedures can be added as supplementary information through pop-up windows, by using similar procedures as SBTool^{BIM}. Overall, a template for circularity assessment can also be developed, where all the new shared parameters are already defined and all the supplementary data is listed to be integrated into the BIM model. Such a template can reduce the required time to carry out the circularity assessment, minimise model characterisation errors and enhance the proper development of the BIM model according to the life cycle circular requirements. Moreover, to support model characterisation, a circular material and components database can be developed, where all the materials can already have pre-defined circular characteristics, as well as other building components, such as walls, windows or roofs (for example, the type of joints and connections between element layers).

For a KPIs-based circularity scheme, each indicator assessment procedure can be coded into Dynamo—as made in SBTool^{BIM}—and all the new circularity data, as well as the intrinsic model data, can be collected and processed according to the circular indicator requirements. The building circular performance is provided in real-time during early project stages, allowing designers to compare different strategies and enhance the building circular performance.

Furthermore, all this data can either be used to define material passports or stored as a project database, where the building materials and components characteristics can be listed during any life cycle stage, making it an important tool for building operation and maintenance. Besides the BIM model usability as a life cycle storage, if proper characterised, it can also be used to predict the end-of-life scenarios, from construction and demolition waste management to the building deconstruction and recyclable potential.

32.5 Conclusions

The circular design has a lot to gain by integrating BIM-based workflows in the process. SBTool^{BIM} has proved its applicability and functionality in improving buildings sustainability performance during the project early stages. Its replicability potential allows gathering valuable insights for future researchers to increase investigation on new BIM-based procedures, such as building circularity assessment, providing a broader and detailed assessment of buildings and, consequently, developing high-performance constructions. A new framework for a circular assessment can be developed by using the same host platforms as SBTool^{BIM}—Autodesk Revit and Dynamo—to automate circular assessment data collection and calculations. The BIM model can store different types of building circular data during its life cycle, making it a support tool for building maintenance and operation. Local complementary data can also be introduced into the analysis through specific popup windows. SBTool^{BIM} method still has a set of features which can be improved, such as the minimisation of user interference (there is still a need to manually collect specific regional data), optimisation of the dynamo script to minimise the analysis time and software requirements, as well as the development of new user-friendly interfaces to extend the application usability. Regarding the insights for CE, future research should include a detailed definition and structure of those insights, together with the development of a process map for future BIM-based applications for CE. Overall, the SBTool^{BIM} framework has enough flexibility to be adapted for circularity purposes, only by modifying the KPIs and the BIM model requirements according to the assessment scheme. This framework marks a new era of BIM-based assessments for decision support in the project early stages, allowing to achieve more sustainable buildings and meeting decarbonisation and climate targets. Moreover, the corresponding fully detailed BIM model can be integrated into local or national databases, creating an effective national buildings database, which can enhance the concept of buildings as material banks.

References

1. Araújo C, Almeida M, Bragança L (2013) Analysis of some Portuguese thermal regulation parameters. Energy Build 58:141–150. https://doi.org/10.1016/J.ENBUILD.2012.11.024
2. Mateus R, Bragança L (2011) Sustainability assessment and rating of buildings: Developing the methodology SBToolPT-H. Build Environ 46:1962–1971. https://doi.org/10.1016/J.BUILDENV.2011.04.023
3. Carvalho JP, Bragança L, Mateus R (2021) Sustainable building design: Analysing the feasibility of BIM platforms to support practical building sustainability assessment. Comput Ind 127:103400. https://doi.org/10.1016/j.compind.2021.103400
4. Ade R, Rehm M (2020) The unwritten history of green building rating tools: A personal view from some of the 'founding fathers.' Build Res & Inf 48:1–17. https://doi.org/10.1080/09613218.2019.1627179
5. Tagliabue LC, Cecconi FR, Maltese S, Rinaldi S, Ciribini ALC, Flammini A (2021) Leveraging digital twin for sustainability assessment of an educational building. Sustainability 13:480. https://doi.org/10.3390/SU13020480
6. Burggräf P, Dannapfel M, Ebade-Esfahani M, Scheidler F (2021) Creation of an expert system for design validation in BIM-based factory design through automatic checking of semantic information. Procedia CIRP 99:3–8. https://doi.org/10.1016/J.PROCIR.2021.03.012
7. Santos R, Costa AA, Silvestre JD, Pyl L (2019) Integration of LCA and LCC analysis within a BIM-based environment. Autom Constr 103:127–149. https://doi.org/10.1016/j.autcon.2019.02.011
8. Xue K, Hossain MU, Liu M, Ma M, Zhang Y, Hu M, Chen X, Cao G (2021) BIM integrated LCA for promoting circular economy towards sustainable construction: An analytical review. Sustainability 13(3):1310. https://doi.org/10.3390/su13031310
9. Akanbi L, Oyedele L, Akinade O, Ajayi A, Delgado M, Bilal M, Bello S (2018) Salvaging building materials in a circular economy: A BIM-based whole-life performance estimator. Resour Conserv Recycl 129:175–186. https://doi.org/10.1016/j.resconrec.2017.10.026
10. Caldas LR, Silva MV, Silva VP, Carvalho MTM, Toledo Filho RD (2022) How different tools contribute to climate change mitigation in a circular building environment?—A systematic literature review. Sustainability 14(7):3759. https://doi.org/10.3390/su14073759
11. Charef R, Emmitt S (2021) Uses of building information modelling for overcoming barriers to a circular economy. J Clean Prod 285:124854. https://doi.org/10.1016/j.jclepro.2020.124854

Chapter 33
Enhancing Life Cycle Costing (LCC) in Circular Construction of Buildings by Applying BIM: A Literature Review

Esam Alasmari, Abdulaziz AlJaber, Pedro Martinez-Vazquez, and Charalampos Baniotopoulos

Abstract Several challenges are associated with implementing Circular Economy (CE) in buildings. These include legal, technical, social, behavioural, and economic barriers. As a result of these challenges, Building Information Modeling (BIM) has emerged as a tool to address them, supporting the development of digital models for sustainable end-of-life and offering material passports for efficient recovery of materials. This paper aims to review recent publications on the topic to explore strategies, material selection criteria and the role of circular components at various stages of building construction. This literature review is based on a review of 50 articles that contributes to the understanding of how BIM can enhance Life Cycle Costing (LCC) in the circular construction of buildings. This review identifies the barriers to implementing CE in buildings by examining recent publications in CE and highlights BIM potential to address these challenges. In this paper, the role of BIM is discussed in relation to sustainable design, material recovery, and components selection for buildings in circular construction. In addition, the review examines whether BIM can be used in circular construction to reduce LCC and promote sustainability. In constructing buildings in circular construction, BIM can be instrumental in enabling decision-makers to conduct comprehensive economic studies, leading to more holistic decision-making.

Keywords Circular Economy · BIM · LCC · Barriers · Buildings

E. Alasmari · A. AlJaber (✉) · P. Martinez-Vazquez · C. Baniotopoulos
School of Engineering, Department of Civil Engineering, University of Birmingham, Edgbaston, Birmingham B15 2TT, UK
e-mail: asa050@student.bham.ac.uk

E. Alasmari
e-mail: exa855@student.bham.ac.uk

P. Martinez-Vazquez
e-mail: p.vazquez@bham.ac.uk

C. Baniotopoulos
e-mail: c.baniotopoulos@bham.ac.uk

L. Bragança et al. (eds.), *Creating a Roadmap Towards Circularity in the Built Environment*, Springer Tracts in Civil Engineering,
https://doi.org/10.1007/978-3-031-45980-1_33

33.1 Introduction

Most of the world's population live in cities [1], whose most prominent component is buildings. In the UK alone, up to 50% of carbon emissions are attributable to the built environment [2], which globally produce around 40% of city carbon emissions [1]. Therefore, the need to scale down these levels of environmental impact is inevitable. There are multiple strategies to attempt this goal, like using sustainable materials with a low carbon footprint, identifying more resilient yet energy-efficient materials that could reduce the volume of constructions, and refurbishing or maintaining old building stock. Researchers have come to realise that all those strategies converge to the broader concept known as circular economy.

Circular Economy (CE) is "An industrial system that is restorative or regenerative by intention and design. It replaces the 'end-of-life' concept with 'restoration', shifts towards the use of renewable energy, eliminates the use of toxic chemicals, which impair reuse, and aims for the elimination of waste through the superior design of materials, products, systems, and, within this, business models" [3]. Similarly, circular construction in buildings is a strategy that prioritises energy consumption and recycling opportunities when selecting materials and production methods.

There are multiple strategies for implementing circular construction in buildings at the planning, design, and construction level. One of these is the Design for Adaptability and Disassembly (DfAD), which combines the benefits of DfD and DfA to allow building components be dismantled and replaced or repaired for any required layout change [4]. Notwithstanding, CE strategies are often hampered as buildings are traditionally built to be demolished at the end of their useful life rather than deconstructed or adapted. Further barriers to implementing CE in buildings derive from a lack of understanding of the new CE concepts among urban planners and designers [5, 6], which add to the lack of incentives offered by the governing bodies [5, 7–9]. Technical challenges are abundant like the mismatch between supply and demand of reused materials and the uncollaborative nature of the building industry stakeholders [9, 10]. According to researchers, these challenges could be overcome through the digitisation of building materials and components.

On the other hand, tools have been developed to aid with implementing CE in buildings. Most of these tools judge the level of achieved circularity by scrutinising material and components during the design stage [11, 12] or by enforcing collaboration between different parts of the construction supply chain [13]. For example, Akanbi et al. [14] developed a BIM-based Whole-life Performance Estimator (BWPE), to assess the performance of structural components during design and interact with Material Passports to manage building performance and restoration when needed.

This paper addresses the need to define circular building construction and identify barriers to its adoption. It also explores the role of BIM in removing such barriers while helping to reduce LCC in constructing buildings applying a circular economy concept.

33.2 Methodology

This study analyses existing literature regarding the enhancement of LCC through BIM in the circular construction of buildings. An in-depth search was conducted in academic databases from 2013 to 2023, concentrating on publications published within the past decade, utilising academic databases such as Scopus, Web of Science, and Google Scholar. Specifically, the search strategy employed keywords such as "Building Information Modelling", "BIM", "Life Cycle Cost", "LCC", "Circular Economy", "CE", and combinations of phrases such as "BIM enhances LCC in the circular economy of buildings", specifically in the context of civil engineering.

Developing appropriate keywords was informed by preliminary research, which identified relevant synonyms and keyword combinations. Using keyword search, many results were generated, which were filtered according to the relevance of the research topic. Further criteria were used to refine the selection, including the exclusion of duplicate articles, non-English publications, and articles that did not directly address the research question. A full-text review of the remaining articles was conducted to assess their relevance and quality. An analysis of the abstracts of 50 articles was conducted in order to ensure alignment with the research topic. Following the identification of the key articles, the study extracted vital information from each paper, which was compiled during the metadata extraction phase. This process involved evaluating the articles in relation to the keywords, as these keywords assist in guiding citations and article recommendations. Figure 33.1 shows the data mining implemented for processing for this review. We expect the results to provide valuable information for architects, engineers, and building owners, to improve the sustainability and efficiency of circular buildings.

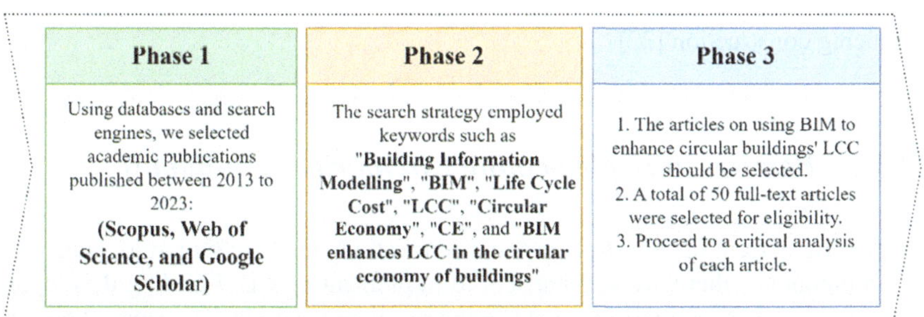

Fig. 33.1 Phases in processing published papers

33.3 Circular Economy in Building Construction

In a CE approach, building components are kept in a cycle involving use, reuse, repair, and recycling, allowing resources to retain their highest intrinsic value for as long as possible. Furthermore, the lifetime of a circular building should be a closed-loop system in which components and materials are used and retained appropriately. Durability and resilience must be considered when selecting materials to prevent the loss of quality. Materials selection, contextual building design, and layout should be examined to decrease energy usage over the lifespan of the building [6].

In addition to design strategies, material and component selection are critical for ensuring optimal CE implementation [15]. Rahla et al. [16] identified nine material selection criteria that are being used to promote the adoption of CE principles in the construction sector. These criteria relate to the type of material used, how it is used, and what happens to it at the end of its life. The nine criteria are: (Recycled and recovered content, Recyclability, Reusability, Ease of deconstruction, Maintainability, Durability, Energy recoverability, Upcycling potential, and Biodegradability).

Several types of circular systems are being developed, including methodologies to evaluate their circularity [17]. The building skin is predominantly façade. Façades are commonly built with steel and aluminium [18] or wood [19], which have a relatively high potential for reuse compared to traditional masonry and concrete. It could be better verified through a Morphological Design and Evaluation Model (MDEM), a tool that was specifically created to guide the designer in creating and assessing circular facades [20]. A study by Buyle et al. [21] estimated the environmental impact of introducing circular design options for internal wall assemblies. Moreover, demountable and reusable wall assemblies with metal substructures were demonstrated to reduce or match conventional wall life cycle effects. It was shown that removable interior linings and dismountable connections are crucial for advancing construction [22].

33.3.1 Barriers in Implementing Circularity in Buildings

Despite the current design strategies and innovations made in buildings' materials and components, there are still barriers to implementing CE. Building developers are progressively becoming aware of CE and regenerative design concepts, although, they have not a unified definition or understanding of such concepts. Munaro et al. [7] highlighted the scarcity of academic literature covering ways for implementing CE throughout the supply chain, which somehow explains the existing knowledge gaps, but it fails to clarify the extent to which these concepts permeate professional practice [5, 23]. CE plans are therefore at their early stages, concentrating on optimising the reuse and recycling of construction and demolition waste in the construction industry while excluding some of the industry's players, goods, services, and systems [24]. The implementation of CE design strategies in buildings remains challenging due to a lack

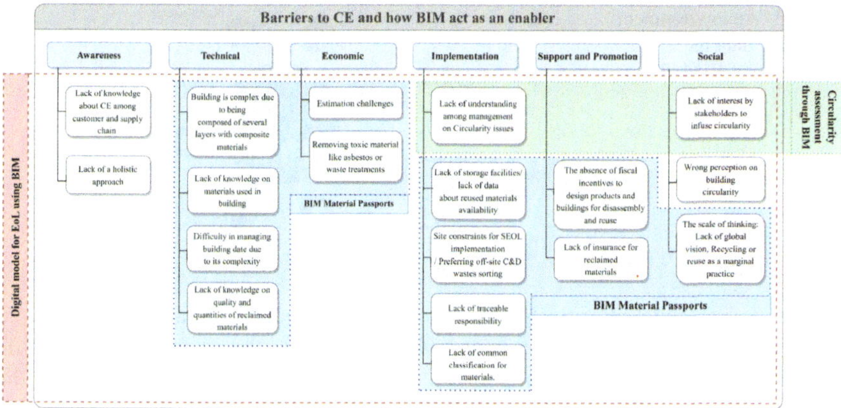

Fig. 33.2 BIM as an enabler to CE barriers

of practical guidelines and design-support tools that facilitate their implementation. These would ideally enable performance assessment and simulation of added value throughout full life-cycle, particularly when their service life comes to an end [25].

It has been suggested to divide the challenges of transitioning from a linear economy to a CE in the construction and demolition waste sector into five main categories: legal, technical, social, behavioural, and economic [26]. Within each category, there are specific challenges, such as policy and regulations, permits and specifications, technological limitations, quality and performance, knowledge and information, and the costs associated with implementing the CE model. The scale of the challenges can also vary depending on the size of the project and the country. Several tools and innovations could speed up the implementation of CE in the building sector. BIM is one of the most prominent and flexible tools to allow that change (see Fig. 33.2).

33.3.2 BIM as an Enabler for Promoting Circular Economy

BIM is a digital representation of a building's physical and functional characteristics. It can be used for analysis and optimisation, reducing costs and improving efficiency. Recent studies have demonstrated that BIM enhances design and construction practices by effectively integrating information into various phases [27, 28].

Charef and Emmitt [29] conducted interviews with experts in sustainable buildings and BIM to explore ways in which BIM could be adopted to assist practitioners in embracing a CE approach. The findings showed seven new ways to use BIM to overcome CE barriers. The seven new BIM uses were: a digital model for Sustainable End of Life (SEOL), a material passport development that stores information about building materials, a project database that stores information about the entire building

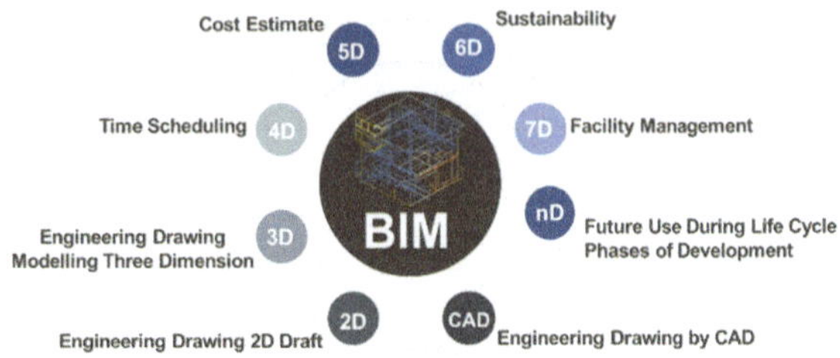

Fig. 33.3 An environment for sharing data in dimensions of BIM [28]

process, a data checking process which makes sure that the data meet the desired requirements, a circularity assessment which evaluates the sustainability and life cycle of building materials, materials recovery processes which help facilities the reclaiming and recycling of materials, and materials bank which stores information about material types.

BIM virtual environment provides spatial information, building properties, cost estimations, geometric and geographical information, inventory management, and schedules [30–32]. Furthermore, Juan and Hsing [33] added that BIM has applications in renovations and building maintenance. A BIM system with a high level of integration and add-ins can support interoperability [34]. The BIM process incorporates risk management techniques as well as opening communication channels for all stakeholders. Essentially, it involves more than just 3D modelling, but also covers 4D (time), 5D (cost), 6D (operation), 7D (sustainability), and 8D (safety) [35]. Ghaffarianhoseini et al. [36] described BIM as a seven-dimensional, global information system (see Fig. 33.3).

33.4 Role of BIM in Enhancing LCC

Building projects use LCC to estimate long-term costs. This process analyses all costs associated with a construction project, operation, maintenance, demolition, and decommissioning. According to Lee et al. [37], building lifecycle cost is the sum of all fundamental processes that take place over the lifecycle of a building. LCC includes planning, obtaining paperwork, implementing environmental management measures before construction starts, dismantling maintenance, and disposal [37]. LCC can switch to CE by integrating BIM into LCC for economic and environmental sustainability. Studies by Ghaffarianhoseini et al. [36], Marzouk et al. [38], Ullah et al. [39], Alasmari et al. [40], have shown that BIM saves time, estimates costs, minimises changes, analyses sustainability, eliminates omissions, manages quality and logistics, establishes life cycles, manages LCC, ensures energy efficiency, facility management, daylight analysis, thermal design, transparency costs, quantity surveys, quantity takeoff, among other benefits. By using BIM, decision-makers can gain critical information to use in their decision-making process.

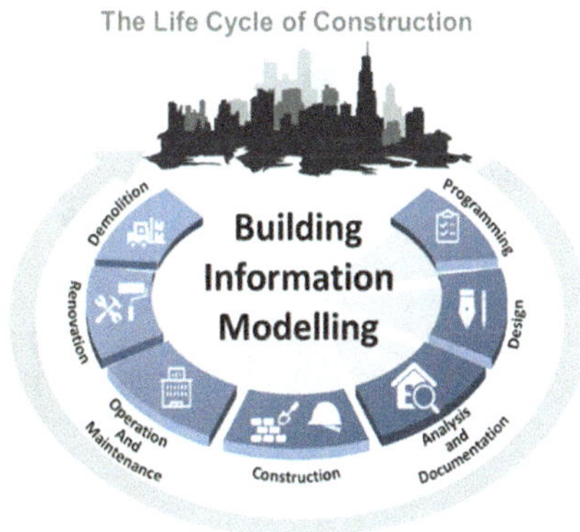

Fig. 33.4 The life cycle of construction processing and BIM

In addition, using BIM to LCC in circular construction can reduce costs and improve sustainability by optimising the design. In BIM, for example, the LCC of different materials and construction methods can be analysed. Furthermore, BIM can be used to analyse a project energy and resource efficiency over its life cycle and identify ways to reduce its environmental impact (see Fig. 33.4). Furthermore, BIM can be applied to project review, standardisation, certification, project design, engineering analyses, site design, coordination, and planning, in addition to documentation and quantity take-off [35, 41, 42]. It has been demonstrated that when using the BIM platform, the results of lifecycle analysis are more accurate and consistent [43]. For example, Saudi Arabia's King Abdulaziz Cultural Center (Ithra) received LEED gold certification for utlising reused materials. Seele [44] reports that in the Ithra project, stainless steel tubes are carefully woven around the building's exterior cladding, which stretches 350 kms.

Santos et al. [45] have published substantial publications on using BIM to simulate a variety of domains. These limitations can be addressed by integrating life cycle assessment (LCA), LCC, and BIM. By improving our understanding of BIM-LCA/LCC, this study examined how BIM technology can improve LCA and LCC results. Love et al. [46] argued that BIM can contribute to quality decisions regarding public infrastructure, resulting in a better return on investment. Furthermore, Santos et al. [45], Dawood [47], Di Biccari et al. [48] suggested that BIM promotes accountability and transparency in construction projects by improving communication and collaboration. Based on the findings of the studies, it has consistently been shown that the use of BIM can improve the efficiency of LCC processes. BIM has been found to resolve non-standardisation issues, assist with document preparation, correct and obscure data, and provide intelligent decision-making pathways that contribute to building sustainability [49, 50]. The advantage of using BIM on LCC is that the materials used in construction can be easily identified and analysed. BIM can also reduce repair and replacement costs by utilising more durable and sustainable materials (see Fig. 33.5).

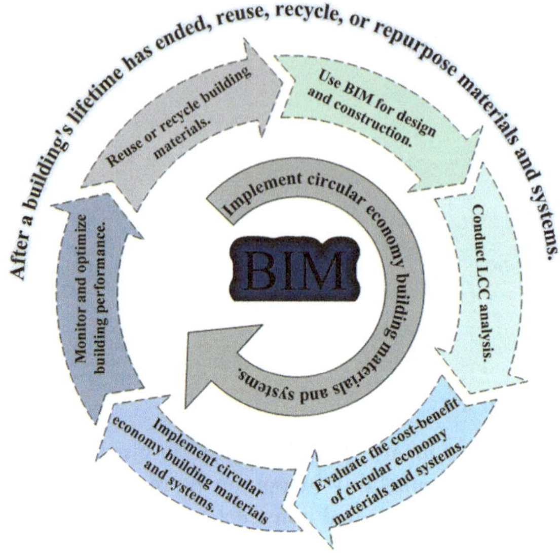

Fig. 33.5 On the incorporation of the principles of circular economy into the LCC building by BIM

33.5 Concluding Remarks

Various tools and innovations help in applying CE in the building industry at different building tiers. However, widespread implementation is hampered by several factors, including a lack of government support in the form of regulations and incentives, as well as a lack of understanding among construction experts about the circularity components of a building. A detailed understanding of the significance of the various barriers, as well as a plan to overcome them, are required. There are several theoretical guidelines and tools available that show the essential concepts of circular construction. Most of the tools, however, serve the same goal. There is a need for practical evidence concerning their usability and influence on the design process to highlight best practices and evaluate numerous options. Several research studies have examined the use of BIM on the LCC. Nevertheless, BIM on LCC is a relatively new construction trend. Therefore, building and construction management offer potential areas for exploration. Overall, the integration of BIM into LCC can significantly impact a project economic and environmental sustainability and can support the transition to a circular economy by improving economic and environmental sustainability.

This study reviewed the current state of circular building design and LCC analyses using BIM. A literature review analysis showed that BIM could be a valuable tool for enhancing LCC in circular buildings by providing a holistic view of the building's life cycle and enabling efficient decision-making. Circular buildings can be designed, constructed, operated, and decommissioned more sustainably by using BIM in LCC analysis. As a result, architects, engineers, and building owners can gain valuable information about improving the efficiency of circular buildings.

However, the study has several limitations, including a lack of practical case studies on the implementation of BIM in circular construction projects and the need to gain a thorough understanding of the integration of BIM with LCC. Future studies should address the feasibility of some significant tools like BIM applications in real-life, and case studies of circular building construction. Additionally, a comprehensive priority ranking of the barriers and BIM as an enabler should be outlined with built environment professionals to streamline further efforts to implement circular building.

References

1. Zarco-Periñán PJ, Zarco-Soto FJ, Zarco-Soto IM, Martínez-Ramos JL, Sánchez-Durán R (2022) CO_2 emissions in buildings: a synopsis of current studies. Energies 15(18):6635
2. Hammond GP, Jones CI (2008) Embodied energy and carbon in construction materials. In: Proceedings of the institution of civil engineers-energy, vol 161, no 2, pp 87–98. Thomas Telford Ltd., Bath
3. MacArthur E (2013) Towards the circular economy. J Ind Ecol 2(1):23–44
4. Munaro MR, Tavares SF, Braganca L (2022) The ecodesign methodologies to achieve buildings' deconstruction: a review and framework. Sustain Prod Consum 30:566–583
5. Adams KT, Osmani M, Thorpe T, Thornback J (2017) Circular economy in construction: current awareness, challenges and enablers. In: Proceedings of the institution of civil engineers-waste and resource management, vol 170, no 1, pp 15–24. Thomas Telford Ltd, Loughborough
6. Rahla KM, Bragança L, Mateus R (2021) Practical strategies for embodying the circular economy concept in the construction sector. In: Sustainable urban development: topics, trends and solutions. IOP Publishing, Bristol
7. Munaro MR, Tavares SF, Bragança L (2020) Towards circular and more sustainable buildings: a systematic literature review on the circular economy in the built environment. J Clean Prod 260:121134
8. Bilal M, Khan KIA, Thaheem MJ, Nasir AR (2020) Current state and barriers to the circular economy in the building sector: towards a mitigation framework. J Clean Prod 276:123250
9. Hart J, Adams K, Giesekam J, Tingley DD, Pomponi F (2019) Barriers and drivers in a circular economy: the case of the built environment. Procedia Cirp 80:619–624
10. Kanters J (2020) Circular building design: an analysis of barriers and drivers for a circular building sector. Buildings 10(4):77
11. Hradil P, Fülöp L, Ungureanu V (2019) Reusability of components from single-storey steel-framed buildings. Steel Constr 12(2):91–97
12. Dams B, Maskell D, Shea A, Allen S, Driesser M, Kretschmann T, Walker P, Emmitt S (2021) A circular construction evaluation framework to promote designing for disassembly and adaptability. J Clean Prod 316:128122
13. Leising E, Quist J, Bocken N (2018) Circular Economy in the building sector: three cases and a collaboration tool. J Clean Prod 176:976–989
14. Akanbi LA, Oyedele LO, Akinade OO, Ajayi AO, Delgado MD, Bilal M, Bello SA (2018) Salvaging building materials in a circular economy: a BIM-based whole-life performance estimator. Resour Conserv Recycl 129:175–186
15. Rahla KM, Mateus R, Bragança L (2021) Implementing circular economy strategies in buildings—from theory to practice. Appl Syst Innov 4(2):26
16. Rahla KM, Mateus R, Bragança L (2021) Selection criteria for building materials and components in line with the circular economy principles in the built environment—a review of current trends. Infrastructures 6(4):49

17. Overend M, Hartwell R (2020) End-of-life challenges in Façade design: a disassembly framework for assessing the environmental reclamation potential of façade systems. In: Better buildings through better skin. Facade Tectonics, Los Angeles
18. Hartwell R, Macmillan S, Overend M (2021) Circular economy of façades: real-world challenges and opportunities. Resour Conserv Recycl 175:105827
19. Androsevic R, Durmisevic E, Brocato M (2019) Measuring reuse potential and waste creation of wooden façades. In: IOP conference series: earth and environmental science, vol 225, no 1, pp 012017. IOP Publishing, Brussels
20. Beurskens PR, Bakx MJM, Ritzen MJ, Durmisevic E, Lichtenberg JJN (2016) A morphological design and evaluation model for the development of circular facades. In: Proceedings of the Utrecht SBE16 conference, pp 252–268. Sustainable Built Environment, Transition Zero, Utrecht
21. Buyle M, Galle W, Debacker W, Audenaert A (2019) Consequential LCA of demountable and reusable internal wall assemblies: a case study in a Belgian context. In: IOP conference series: earth and environmental science, vol 323, no 1, pp 012057. IOP Publishing, Graz
22. O'Grady TM, Minunno R, Chong HY, Morrison GM (2021) Interconnections: an analysis of disassemblable building connection systems towards a circular economy. Buildings 11(11):535
23. Sala Benites H, Osmond P, Prasad D (2023) Inquiry on perceptions and practices of built environment professionals regarding regenerative and circular approaches. Buildings 13(1):63
24. Munaro MR, Freitas MDCD, Tavares SF, Bragança L (2021) Circular business models: current state and framework to achieve sustainable buildings. J Constr Eng Manag 147(12):04021164
25. Askar R, Bragança L, Gervásio H (2022) Design for adaptability (DfA)—frameworks and assessment models for enhanced circularity in buildings. Appl Syst Innov 5(1):24
26. Purchase CK, Al Zulayq DM, O'Brien BT, Kowalewski MJ, Berenjian A, Tarighaleslami AH, Seifan M (2022) Circular economy of construction and demolition waste: a literature review on lessons, challenges, and benefits. Materials 15(1):76
27. Liu Z, Chi Z, Osmani M, Demian P (2021) Blockchain and building information management (BIM) for sustainable building development within the context of smart cities. Sustainability 13(4):1–17
28. Alasmari E, Martinez Vazquez P, Baniotopoulos C (2022) Building information modeling (BIM) towards a sustainable building design: a survey. In: Proceedings of the CESARE22, 3rd coordinating engineering for sustainability and resilience, pp 6–9. CESARE22, Irbid
29. Charef R, Emmitt S (2021) Uses of building information modelling for overcoming barriers to a circular economy. J Clean Prod 285:124854
30. Olawumi TO, Chan DW (2019) Building information modelling and project information management framework for construction projects. J Civ Eng Manag 25(1):53–75
31. Kocakaya MN, Namlı E, Işıkdağ Ü (2019) Building information management (BIM), a new approach to project management. J Sustain Constr Mater Technol 4(1):323–332
32. Elshafey A, Saar CC, Aminudin EB, Gheisari M, Usmani A (2020) Technology acceptance model for augmented reality and building information modeling integration in the construction industry. J Inf Technol Constr 25:161–172
33. Juan YK, Hsing NP (2017) BIM-based approach to simulate building adaptive performance and life cycle costs for an open building design. Appl Sci 7(8):837
34. Sandberg M, Mukkavaara J, Shadram F, Olofsson T (2019) Multidisciplinary optimization of life-cycle energy and cost using a BIM-based master model. Sustainability 11(1):286
35. Badran D, AlZubaidi R, Venkatachalam S (2020) BIM based risk management for design bid build (DBB) design process in the United Arab Emirates: a conceptual framework. Int J Syst Assur Eng Manag 11:1339–1361
36. GhaffarianHoseini A, Zhang T, Nwadigo O, GhaffarianHoseini A, Naismith N, Tookey J, Raahemifar K (2017) Application of nD BIM integrated knowledge-based building management system (BIM-IKBMS) for inspecting post-construction energy efficiency. Renew Sustain Energy Rev 72:935–949
37. Lee J, Yang H, Lim J, Hong T, Kim J, Jeong K (2020) BIM-based preliminary estimation method considering the life cycle cost for decision-making in the early design phase. J Asian Arch Build Eng 19(4):384–399

38. Marzouk M, Azab S, Metawie M (2018) BIM-based approach for optimizing life cycle costs of sustainable buildings. J Clean Prod 188:217–226
39. Ullah K, Lill I, Witt E (2019) An overview of BIM adoption in the construction industry: Benefits and barriers. In: 10th Nordic conference on construction economics and organization, vol 2, pp 297–303. Emerald Publishing Limited, Bingley, Nordic
40. Alasmari E, Martinez-Vazquez P, Baniotopoulos C (2022) A systematic literature review of the adoption of building information modelling (BIM) on life cycle cost (LCC). Buildings 12(11):1829
41. Tomek A, Matějka P (2014) The impact of BIM on risk management as an argument for its implementation in a construction company. Procedia Eng 85:501–509
42. Wetzel EM, Thabet WY (2015) The use of a BIM-based framework to support safe facility management processes. Autom Constr 60:12–24
43. Zoghi M, Kim S (2020) Dynamic modeling for life cycle cost analysis of BIM-based construction waste management. Sustainability 12(6):2483
44. SEELE, https://seele.com/references/king-abdulaziz-center-for-world-culture/. Last Accessed 22 Nov 2022
45. Santos R, Costa AA, Silvestre JD, Vandenbergh T, Pyl L (2020) BIM-based life cycle assessment and life cycle costing of an office building in Western Europe. Build Environ 169:106568
46. Love PE, Liu J, Matthews J, Sing CP, Smith J (2015) Future proofing PPPs: life-cycle performance measurement and building information modelling. Autom Constr 56:26–35
47. Dawood MH (2016) BIM based optimal life cycle cost of sustainable house framework. In: 3rd MEC international conference on big data and smart city, pp 1–5. IEEE, Muscat
48. Di Biccari C, Abualdenien J, Borrmann A, Corallo A (2019) A BIM-based framework to visually evaluate circularity and life cycle cost of buildings. In: IOP conference series: earth and environmental science, vol 290, no 1, pp 012043. IOP Publishing, Prague
49. Kehily D (2016) Leveraging building information modelling to address the barriers that prevent the widespread adoption of life cycle costing by quantity surveyors. University of Salford
50. Nwodo MN, Anumba, CJ, Asadi S (2017) BIM-based life cycle assessment and costing of buildings: current trends and opportunities. In: Computing in civil engineering, pp 51–59. ASCE, Seattle

Chapter 34
BIM for Enhancing the Energy Efficiency and Sustainability of Existing Buildings

Liljana Dimevska Sofronievska, Meri Cvetkovska, Ana Trombeva Gavriloska, and Teodora Mihajlovska

Abstract Most of the buildings of the post-earthquake period in Skopje are built in "beton-brut" architectural style and nowadays, they are treated as an important cultural heritage of the city. Unfortunately, the envelope of those buildings is built without thermal insulation materials and consequently, they are large energy consumers responsible for tons of CO_2 emission per year. They don't meet today's criteria for energy efficiency and the circular economy (CE) practices can hardly be implemented. But, if circularity is understood from the aspect of prolonging their use and improving comfort instead of destroying them, as well as reducing the negative effects caused by their use, then it can be said that the principles of CE have been achieved. In order to improve the energy performance and the building sustainability, as well as to prolong the lifespan with minimal impact on the authentic façade appearance, a renovation with a new façade nanomaterial with high thermal properties is proposed. The simulations of the existing building and the renovated scenario using Building Information Modeling (BIM) is conducted and comparative analyses between the actual condition and the improved scenario of the building are carried out. According to the simulation results, by adding the new material on the façade a significant improvement in the building's energy performance is achieved. It can be concluded that BIM has a great potential in assessing the improvement of energy efficiency of buildings and the possible implementation of CE for existing buildings.

Keywords CE · BIM · Energy efficiency · CO_2 emissions · Renovation

L. D. Sofronievska (✉) · A. T. Gavriloska · T. Mihajlovska
Faculty of Architecture, Ss. Cyril and Methodium University, Skopje, North Macedonia
e-mail: dimevska.liljana@arh.ukim.edu.mk

M. Cvetkovska
Faculty of Civil Engineering, Ss. Cyril and Methodium University, Skopje, North Macedonia

L. Bragança et al. (eds.), *Creating a Roadmap Towards Circularity in the Built Environment*, Springer Tracts in Civil Engineering,
https://doi.org/10.1007/978-3-031-45980-1_34

34.1 Introduction

The post-earthquake architecture, also known as architecture of Modernism, is built after the catastrophic earthquake in 1963 and represents an important cultural heritage of Skopje. According to the construction standards of that time, those buildings are built with lack of thermal insulation materials, which results in a poor thermal comfort, high heating, cooling and maintenance costs, degradation and deterioration. These buildings need to be properly renovated according to current energy efficiency standards. On the other hand, the authentic appearance of the architecture that is considered as a cultural heritage should not be compromised in the renovation process. Since the aim of the circular economy (CE) is to eliminate waste and pollution, circulate products and materials and regenerate the nature, the building sector has one of the biggest potentials for CE implementation. Improved design and new construction techniques based on CE will produce highly efficient new buildings, but most of the existing construction sector does not meet CE criteria [1]. Most of these buildings are historical buildings with cultural values and no change in appearance is allowed. The real challenge is to implement the CE strategies in existing buildings, by improving their sustainability, energy efficiency and durability, which will lead to cutting the CO_2 emissions, reduction in energy consumption and financial costs. The adoption of CE principles in the renovation of buildings can reduce the amount of materials used for renovation and thus minimize the emissions embedded in building materials [1]. Building information modeling (BIM) is one of the most promising developments in construction process, which encourages collaborative working between all the disciplines involved in design, construction, maintenance and use of buildings.

This paper deals with the potential renovation activities that could improve the sustainability and energy efficiency of an existing building and reduce the embedded emissions, by using BIM. A simulation of a renovated scenario of a post—earthquake building is made by using EnergyPlus. In order to enhance the energy efficiency and in the same time to keep the authenticity of the building, a sustainable insulation façade material is used in the scenario. The results showed positive impact of the new material in terms of energy savings, emissions, sustainability and durability of the building, which indicates that BIM has great potential to facilitate CE adoption.

34.2 Case Study Building

34.2.1 Building Description

The Hydrometeorological Service building in Skopje is built in the post- earthquake period in a specific architectural style, known as "brutalist" architecture, which means that the whole building is constructed in raw exposed natural concrete, known as "beton—brut", (see Fig. 34.1). The selected case study building has and architectural

Fig. 34.1 Different views of Hydrometeorological Service building

and cultural value, not only for its brutalist appearance, but also for its unique plan and design. Since the building facade is made only of natural concrete and doesn't have any exterior finishing layers or insulation, the energy efficiency and thermal comfort are quite poor, which results with large energy consumption for its maintenance and also carbonization and decay of the unprotected concrete. The proper functioning of the building is compromised and her authentic appearance and lifespan is endangered. In order to extend the building lifespan by improving it's durability, sustainability and energy efficiency, and to ensure it's better functioning, the building needs to be properly renovated, by implementing the CE principles such as adaptive reuse of the building and by using sustainable materials on the façade which won't make a significant imapct on the original appearance of the building.

34.2.2 In Situ Measurements

In order to determine the real U values of the concrete facade walls of the building, ''in situ'' measurements of the heat flux and the surface temperatures of the facade walls were performed. The measurements were conducted in the period from 6 to 13th of March, 2020. The measurements were performed by TRSYS01 [2] which is a high-accuracy building thermal resistance measuring system with two measurement locations (see Fig. 34.2).

The sensors were placed on the both sides of the façade wall (a—inside sensor position; b—outside sensor position), shown in Fig. 34.3. Two of the internal sensors measure the heat flux trough the wall, and the other two sensors measure the surface indoor surface temperature of the wall. The other pair of external sensor measure the outdoor surface temperature of the wall. After the eight days measurement the data were processed by LoggerNet software and U values calculations were made.

Figure 34.4 shows the graph of the measured U values of the façade wall and the average value is 1.5 W/m^{2}K. The maximum prescribed U value for external façade walls is is 0.35 W/m^{2}K, which indicates that the building has very poor thermal properties. Figure 34.5 shows the graph of the surface temperatures of the wall,

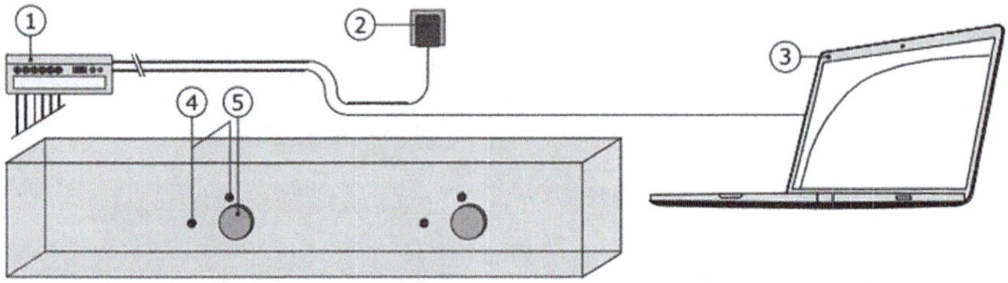

Fig. 34.2 Method of connecting the measuring equipment with the structural element: (1) measurement and control unit; (2) power adapter; (3) computer data reading; (4) two pairs of temperature sensors; (5) two heat flow plates (fluxometers)

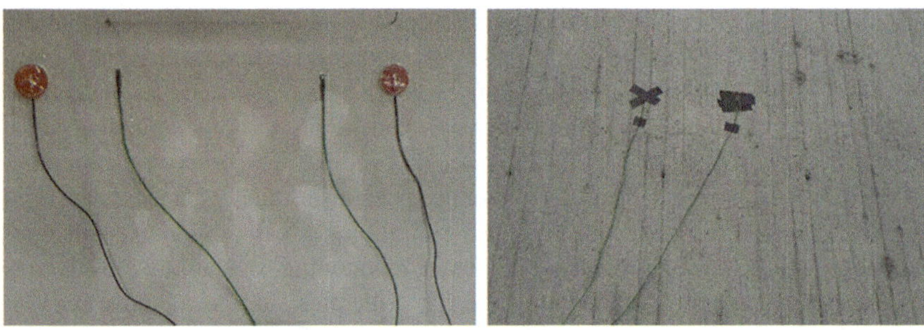

Fig. 34.3 TRSYS01—a building thermal resistance measuring system

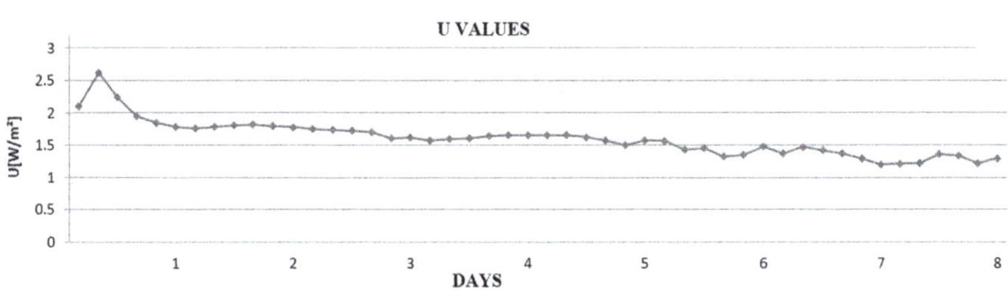

Fig. 34.4 Average U values during the measurement period

where the big difference between the indoor and outdoor surface temperature can be seen, which leads to the conclusion that a lot of energy is needed for heating the indoor spaces.

34.2.3 Nano Materials Selection and Properties

Recently, the use of nanomaterials as thermal insulation building materials has been increased. A façade nanomaterial such as silica aerogel based plaster with good

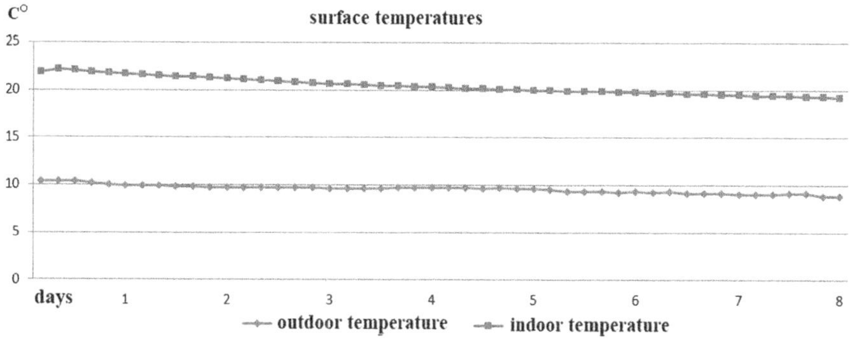

Fig. 34.5 Average values of surface wall temperatures during the measurement period

thermal insulation properties, which has a minimal impact on the original architectural appearance of the building is used for the analysis. The aerogel plaster has a low thermal conductivity λ, ranging from 0.028–0.014 W/mK, thanks to its porous nano structure [3]. Silica aerogel material has many applications and it can be modified to meet a number of specific purposes required by CE, since they have low embodied energy, lower than traditional insulation products [4]. Aerogel can be mixed to develop a green building material with unique characteristics and have a great potential for an application in green and sustainable buildings [5]. According to the criteria for protection of historical buildings, aerogel plasters have a mild impact on their authenticity, but it is important that they are compatible with the chemical composition of the original materials, and can be easily removed without damaging them and there is no need for additional fastening that would damage the original material [6]. They have great flexibility in applying on uneven surfaces and complex architectural details [7]. The application of the new material will not only improve the energy efficiency and sustainability of the building but also it will protect it from climate conditions and expand its lifespan. Due to the composition and method of application, aerogel plasters perfectly mimic the texture of natural concrete and it's difficult to distinguish it from the original (see Fig. 34.6), while the original material remains preserved. The thermal properties of the material are shown in Table 34.1.

Fig. 34.6 Façade material—concrete texture

Table 34.1 Facade nanomaterial thermal properties

Aerogel based thermal insulation plaster thermal properties	
λ (W/mK)	0.028
c (J/kgK)	990
ρ (kg/m^3)	220
d (m)	0.06
μ [−]	04-May

Two dynamic energy simulations have been conducted: a simulation of the existing condition and of an improved scenario by adding aerogel plaster to the facade walls. Conventional thermal insulation materials such as rock wool and EPS are added to the rest of the envelope (floor slab and roof). The existing windows have been replaced by aluminum framed double glazed low emission glass with U value of 1.06 W/m^{2}K.

34.3 Energy Modeling and Simulation

34.3.1 Modeling, Zoning, Geometry and Materials

The detailed building modeling is obtained by using BIM software and the dynamic energy simulation is carried out by using EnergyPlus together with Design Builder software tool. The modeled capacity of the building is 150 employees, organized into several types of offices and departments. Figure 34.7 shows the ground floor position of thermal zones. All four facades of the building are exposed to wind, its entrance part is north oriented, surrounded by vegetation.

The building is divided into a total of 140 thermal zones. Each of the zones is defined with its own design temperature, orientation, number of people, lighting, electrical equipment and appliances, type of heating, cooling, ventilation, glazing area, etc. Figures 34.7 and 34.8 show the ground floor thermal zones divided in the following groups: occupied zones (zones with people, conditioned during working hours); corridors (no people, but conditioned during working hours); unconditioned zones (stairs and toilets); 24 h occupied zones (conditioned zones with permanent stay of people).

Table 34.2 shows some of the most important parameters of the building geometry, summarized for all zones (conditioned and unconditioned). The building envelope is mainly composed of reinforced concrete solid walls without external finishing layer, except in certain parts, where a brick masonry wall appears. The facade walls, the roof and floor slabs are not insulated. The façade fenestration is made of aluminum framed single glazed windows with U value 5.7 W/m^{2}K. All of the parameters indicate poor energy performance of the building and bad thermal comfort.

Fig. 34.7 Thermal zones—ground floor

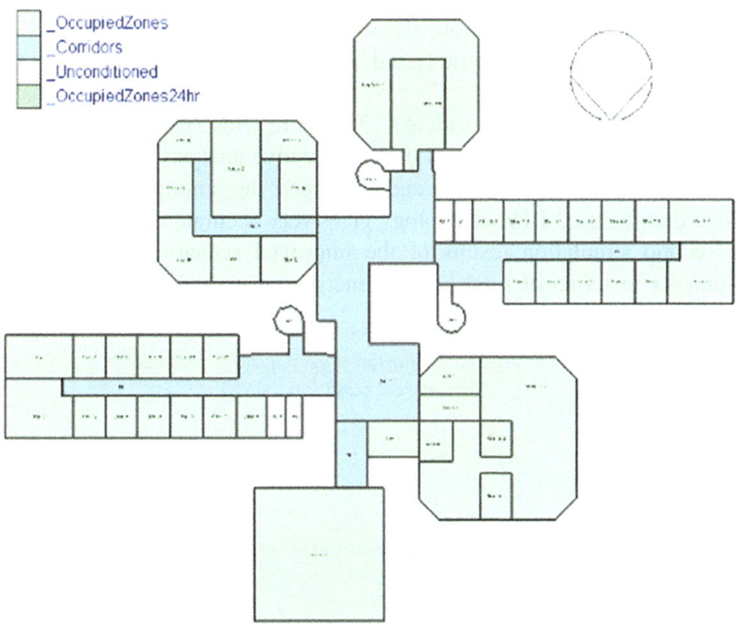

Fig. 34.8 Thermal zones—first floor

Table 34.2 Building parameters

Zone summary	Area [m^2]	Volume [m^3]	Above ground gross wall area [m^2]	Underground gross wall area [m^2]	Window glass area [m^2]	Total openings area [m^2]
Total conditioned area (m^2)	2647.7	8085.9	2468.9	9.69	658.8	712.3
Total unconditioned area (m^2)	694.5	2103.8	694.7	325.57	43.3	49.3

34.3.2 Simulation Results

First, a simulation of the existing condition of the building is conducted, and the results are compared with the actual heating and electricity bills. Comparisons between the U values obtained from the software simulation and the results obtained from the "in situ" measurements are also discussed. The bills' coincidences with the simulated scenario for both electricity and heating energy consumption are over 90% (see Figs. 34.9 and 34.10).

The simulated U value of the wall is 1.73 W/m^2K, which is slightly higher than the measured one of 1.5 W/m^2K. From the comparative analysis between the existing condition ("in situ" measurements and bills) and the simulated scenario, can be concluded that the used BIM technology gives very accurate results.

The second simulation results of the improved scenario, showed significant improvements in terms of total heating energy consumption and total electricity

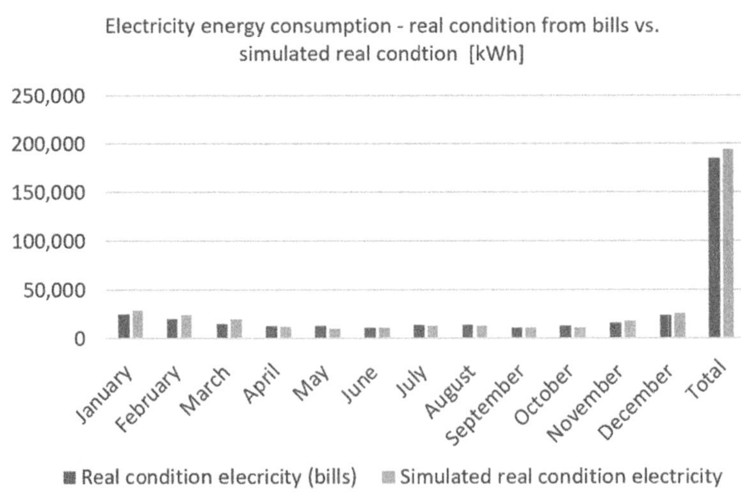

Fig. 34.9 Electricity consumption (bills and simulation)

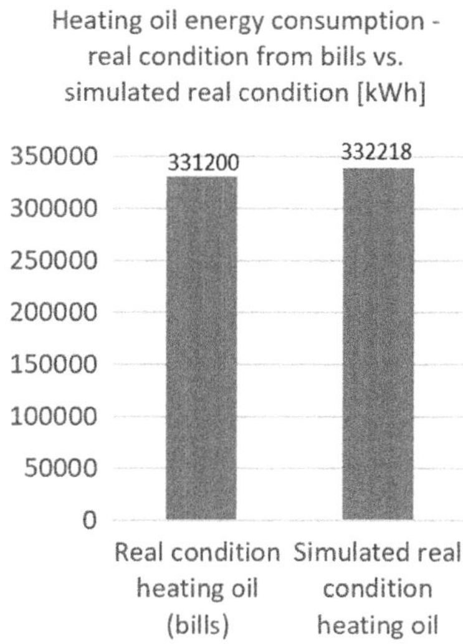

Fig. 34.10 Heating consumption (bills and simulation)

energy consumption (electricity for heating, cooling, lighting and appliances), as well as a drastic reduction in CO_2 emissions. The comparative analysis between the heating energy consumption of the existing condition and the improved scenario showed a big reduction in the improved scenario by 68%. The total heating energy consumption in the existing condition is 332 218.5 kWh annually, i.e. an average of 27 684.8 kWh monthly. The total heating energy consumption in the improved scenario is 105 188.4 kWh annually, i.e. an average of 8 765.7 kWh monthly. The graphs of the heating energy consumption for both, baseline and improved scenario are shown in Fig. 34.11.

The comparative analysis between the electricity energy consumption of the existing condition and the improved scenario showed a reduction in the improved

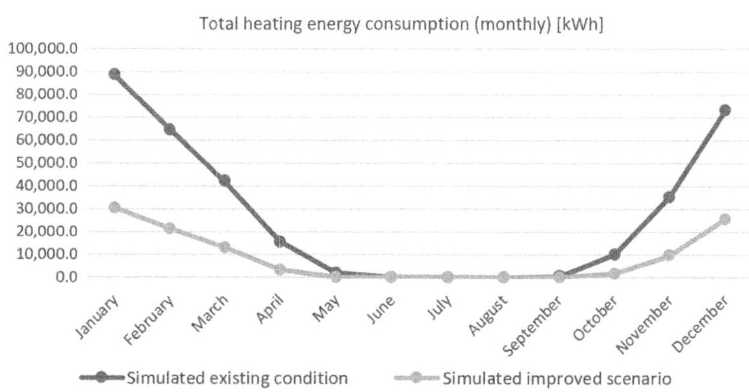

Fig. 34.11 Heating energy consumption (existing condition and improved scenario)

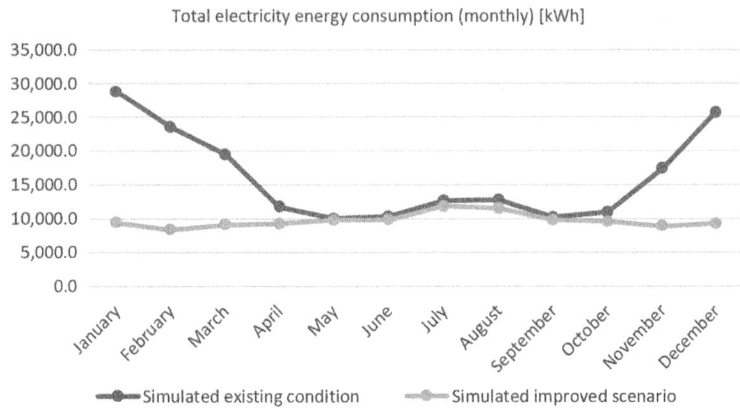

Fig. 34.12 Electricity energy consumption (existing condition and improved scenario)

scenario by 40%. The total energy consumption for electricity in the existing condition is 193.888 kWh annually, i.e. an average of 16 157 kWh monthly. The total electricity energy consumption in the improved scenario is 116 833.1 kWh annually, i.e. an average of 9 736.1 kWh monthly. The graphs of the electricity energy consumption for both, existing condition and improved scenario are shown in Fig. 34.12.

The CO_2 emission for the existing condition of the building is 168.269 kg annually, i.e. an average of 14 022 kg monthly, while the CO_2 emission in case of the improved scenario is 84 215 kg annually, i.e. an average of 7 018 kg monthly. The reduction of CO_2 emission in the improved scenario is 50%.

34.4 Conclusions

The case study building is a significant cultural heritage built only in natural concrete with no protective façade layers or thermal insulation. That leads to a process of decay, lack of thermal comfort, huge energy consumption, pollution and endangered lifespan of the building. In order to improve the building's function, sustainability, thermal comfort and energy efficiency, an urgent renovation needs to be done. Since the building has an important architectural and cultural value, its authentic appearance should not be compromised by the renovation process. From the investigated nanomaterials, the aerogel plaster has a minimal impact on the authenticity of building façade, and it is a very promising sustainable material and has a great potential for its implementation in the CE practices according to the new state of the art literature [4, 5, 8]. On the other hand BIM technology allows us to see how the building will behave if it is renovated with the proposed material and what will be the benefits (saving energy, reducing emissions, financial costs and even predicting the life cycle of the building). The purpose of the paper is to connect these two things together, to show the potential that BIM has in CE, and in same time TO show the opportunities and improvements that aerogel plaster brings to the existing buildings with cultural values. Based on the "in situ" measurements and on dynamic energy simulations of existing condition and the improved scenario, comparative analysis between the two scenarios were obtained. The analysis showed a significant reduction of the heating energy consumption in the improved scenario by 68% compared to the existing condition of the building. Also, the electricity energy consumption is reduced by 40% and CO_2 emissions by 50%.

Finally, it can be concluded that the application of the aerogel based thermal plaster on the facade walls, not only improves its energy efficiency, comfort and CO_2 emissivity, but also has a minimal impact on the authentic appearance of the building. BIM technology has proven to be an excellent method for predicting the benefits of the renovated scenario of existing building, giving accurate results and it has a great potential in enabling the design team to simulate renovations and the related cost. BIM provides support to decision-making for component selection; assess the sustainability and even the circularity of the asset during the design process.

References

1. EEA (2022) Building renovation: where circular economy and climate meet. European Environment Agency
2. High-accuracy building thermal resistance measuring system with two measurement locations. User manual TRSYS01, Hukseflux thermal sensors. www.hukseflux.com
3. Stahl TH, Brunner S, Zimmermann M, Ghazi WK (2016) Thermo-hygric properties of a newly developed aerogel based insulation rendering for both exterior and interior applications. Energy Build 44:114–117
4. Curto DD, Cinieri V (2020) Aerogel-based plasters and energy efficiency of historic buildings. Literature review and guidelines for manufacturing specimens destined for thermal tests. Sustainability 12:9457. https://doi.org/10.3390/su12229457
5. Djati Utomo H, Li X, Ng ET (2022) Sustainable production in circular economy: aerogel upscaling production. Environ Sci Pollut Res 29:20078–20084. (Springer)
6. Carty L (2017) Analysis of the effects of aerogel insulation on the thermal performance of existing building envelopes. Edinburgh Napier University, p 09013398
7. Ganobjak M (2019) Aerogel materials for heritage buildings: materials, properties and case studies. J Cult Herit Elsevier
8. Castro-Díaz M, Osmani M, Cavalaro S, Parker B, Lovato T, Needham P, Thompson J, Philippe K, Ruiz F (2022) Impact of circular silica aerogel on plasterboard recycling. In: Conference contribution. Loughborough University. https://hdl.handle.net/2134/21435765.v1

Part XVII
Standardization of CE Definitions in Buildings

Chapter 35
Standards for Recycling Oriented Deconstruction and Classification of Waste

Ruben Paul Borg

Abstract Construction and Demolition (C&D) Waste is a priority waste stream due to the substantial amounts of such waste generated. There is a high potential for reuse and recycling of waste and byproducts. The Constuction and Demolition Waste Strategy for Malta supports the transition towards a more circular economy and closes the loop of construction products life cycle. The quality of recycling and recovery of this waste stream needs to be improved, for market conditions to be developed to increase the demand for secondary raw materials. The paper reviews the strategic framework leading to the design, development and eventual implementation of National Standards for Excavation, Construction and Demolition Waste in Malta. The Standards presented, consist of two complimentary documents, developed to cover deconstruction, excavation and classification of waste and recycled aggregate. Standard SM810—Recycling-oriented Deconstruction, Controlled Excavation Works and Classification of Waste: Requirements for planning and execution, sets to prioritise the reduction of waste generation and highlights the importance of saving raw material resources. Standard SM820—Classification of Recycled Aggregate, sets out technical engineering attributes for the classification of waste aggregate, enabling its exploitation as a resource. A regulatory and legislative framework was developed for the implementation of the Standards. This important development sets the scene for the effective implementation of Circular Economy in the Construction Industry in the Maltese Islands.

Keywords Excavation · Construction and demolition waste standards · Recycling · Deconstruction

R. P. Borg (✉)
Faculty for the Built Environment, University of Malta, Msida MSD 2080, Malta
e-mail: ruben.p.borg@um.edu.mt

R&I, Building Industry Consultative Council, Valletta, Malta

L. Bragança et al. (eds.), *Creating a Roadmap Towards Circularity in the Built Environment*, Springer Tracts in Civil Engineering,
https://doi.org/10.1007/978-3-031-45980-1_35

35.1 Construction and Demolition Waste

35.1.1 Circular Economy and C&D Waste

The European Union has identified Construction and Demolition (C&D) Waste as a priority waste stream due to the substantial amounts of such waste generated. Given that several components within the waste stream have a high resource value, there is a high potential for the re-use and also recycling of such waste. To this aim, the quality of recycling and recovery of this waste stream needs to be increased, in order for market conditions to be developed to increase the demand for secondary raw materials. Therefore, construction materials and resources are maintained in the economy for as long as possible, supporting a proper transition towards a circular economy [1]. A circular economy, from construction through the demolition at the end of life, or to recycling of waste in new construction work, supports growth in the construction industry. This paper reviews the strategic framework developed, leading to the design, development and eventual implementation of National Standards for Excavation, Construction and Demolition Waste in Malta. The Standards presented in this paper, consist of two complimentary documents developed to cover deconstruction, excavation and classification of waste (SM810) and the classification of recycled aggregate (SM820).

35.1.2 Resources and Waste

The construction sector is responsible for contributing, through innovation-led growth, to build up a carbon–neutral society and economy, as per intended goals of the European Commission. The construction industry is currently the largest consumer of raw materials and resources and as a consequence one of the largest contributors to carbon dioxide emissions, mainly coming from the operational life of buildings and structures. It is also one of the largest producers of waste, with construction and demolition waste accounting for 30–40% of the solid waste produced worldwide [2, 3]. One of the main reasons for the high share in both CO_2 emissions and waste production can be identified in the need to renovate the building stock in developed countries, as well as population growth and urbanization in developing countries [4]. The largest share in the environmental footprint of construction interventions is represented by construction materials [4], with cement and concrete contributing to the largest share, since concrete, and cement as its main constituent, is the mostly used construction material in the world [5]. The large production of solid waste from the construction industry also results in problems related to transportation and storage of C&D Waste before processing and disposal to landfill sites. In the European Union (EU), construction and demolition activities generate 850 million tonnes of construction and demolition waste per year, and C&D Waste generation across the EU member states is correlated with their respective national construction sector turnover, population, and gross domestic product [6].

35.1.3 The Waste Framework Directive

There is an increased awareness about the need to effectively address the high resource consumption and waste generation through construction and demolition activity. The Waste Framework Directive (WFD) 2008/98/EC provides a general framework for waste management, setting definitions for EU member states [7]. The Waste Framework Directive refers the stage when waste ceases to be waste and lays down end-of-waste criteria that provide a high level of environmental protection and an environmental and economic benefit. The Directive identifies possible categories of waste for which ‘end-of-waste’ specifications and criteria should be developed including construction and demolition waste, some ashes and slags, scrap metals, aggregates, tyres, textiles, compost, waste paper and glass. The Directive refers to recovery operations for the purposes of reaching end-of-waste status [7]. The WFD introduces a detailed clarification of the “waste hierarchy,” which makes prevention the top waste management priority, followed by preparation for reuse, recycling, and recovery, with landfilling being the least desirable option. It also requires that responsible authorities in all European member states establish one or more waste management plans covering their entire geographical territory. Member states are required to implement measures to recover a minimum of 70% (by weight) of non-hazardous C&D Waste (excluding soils) by 2020. Europe as a whole has achieved the target of recovery for the year 2020, including backfilling [6]. However, 15 member States have already reached the 70% recovery target set in the WFD; while 11 member States still need to improve their recovery performance in order to achieve the EU target, with the recovery rate varying from 10 to 95% among the member States. In this context, the construction sector is actively promoting initiatives to address resource consumption and waste generation through the exploitation of C&D Waste in recycling, which could reduce both waste disposal and also demands for natural resources [8, 9].

35.1.4 C&D Waste in Malta

Construction industry activities including construction of buildings and civil infrastructure, demolition of buildings and excavation generate large amounts of waste. C&D waste is the heaviest and most voluminous waste stream generated in Malta (over 1.5 million tonnes of waste generated annually) [1]. This constitutes the largest share of waste generated in Malta, accounting to roughly 80% of the total waste arising each year. This percentage share is considered as significantly high, particularly when compared to the EU average, which accounts to approximately 25–30%

of all waste generated in the EU [1]. Mineral waste from excavation and construction and demolition constitutes the largest share amounting to approximately 74% of all C&D waste generated annually (Fig. 35.1). Prior to 2012, Malta relied heavily on the disposal of C&D waste, both on land and at sea; between 2007 and 2012 circa 65% of all C&D waste treated was disposed in inert landfills; another 25% were disposed at sea, and only around 10% were recycled. During this period, although spent quarries were accepting C&D waste for rehabilitation purposes, they were permitted as inert landfills and thus considered as disposal [1]. In 2012, the reviewing of existing permits for spent quarries enabled backfilling operations, considered as a recovery operation. This resulted in a drastic shift from disposal to recovery of inert waste with backfilling of excavation voids for the management of such waste (Fig. 35.2). In 2016 Malta recorded the highest percentage rate of inert waste directed for backfilling; thereby 77% of the total C&D waste treated was backfilled. This backfilling activity has significantly contributed towards Malta's achievement in attaining the target to prepare for re-use, recycle and recover a minimum of 70% by weight of the C&D waste generated (WFD). Malta has to put greater effort by promoting reuse and recycling of construction, demolition and excavated waste, and therefore reducing reliance on backfilling.

Recycling of C&D waste remained constant over the years, with an annual average of 16% being recycled, reaching an all-time peak of 26% in 2015. Approximately 90% of the total C&D waste recycled annually, is recycled locally. Recycled material in Malta is primarily used for the following: aggregates for concrete and roadworks; crushed material as 'torba'; other material used for renovation works. Other waste

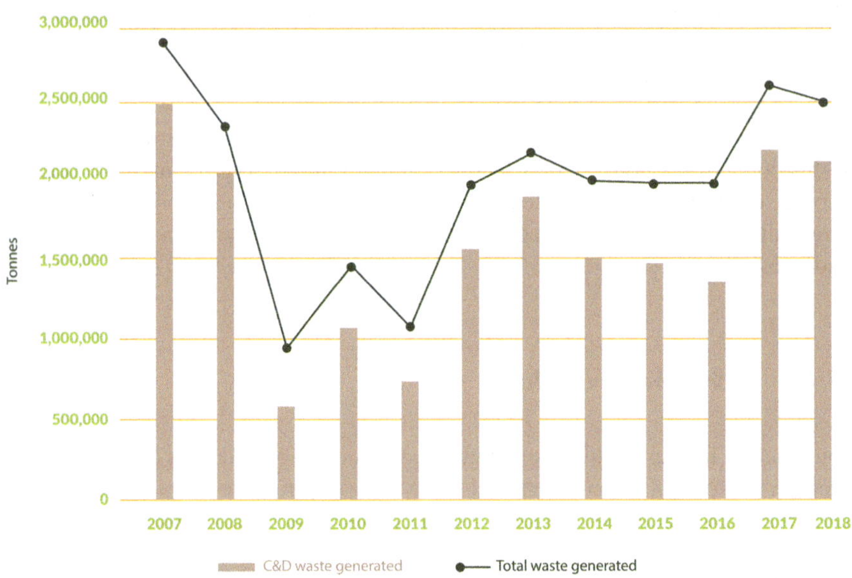

Fig. 35.1 The Construction and Demolition Waste generated, compared to the total waste generated in the period 2007–2018. *Source* ERA [1]

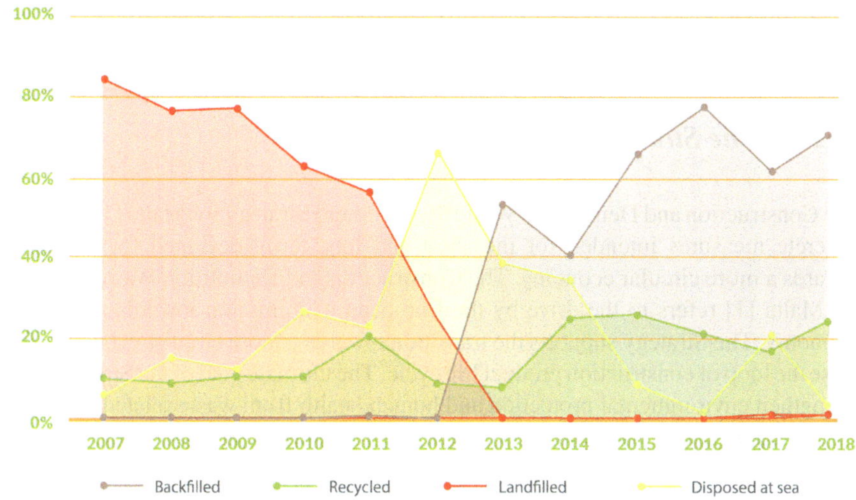

Fig. 35.2 Treatment of Construction and Demolition Waste in Malta (2007–2018). *Source* ERA [1]

deriving from construction and demolition activities, namely iron, steel and other mixed metals are exported for recycling. Given the lack of local facilities for the treatment of hazardous waste, such waste arising from the construction and demolition industry, including construction waste containing asbestos is also exported for further treatment. Intensive development and the subsequent large volumes of waste arising from excavations and construction and demolition activities, coupled with the high reliance on backfilling of inert waste, is causing a major problem as a result of the lack of void space for disposal available in Malta. In the coming years the volume of authorised void space might not meet the increasing demand for the backfilling of C&D waste [1]. Through the Tax Credit (Construction Waste Recycling) Rules, S.L. 123.186 [10], the Government announced a tax incentive, aiming to increase backfilling space, whereby operators of quarries authorised to accept C&D waste may claim a tax credit equivalent to 25% of the gross fees paid to them where the price of C&D waste accepted does not exceed €8 per tonne [1].

35.2 Construction and Demolition Waste Management Strategy

35.2.1 The Strategy

The Construction and Demolition Waste Management Strategy for Malta [1] presents concrete measures intended for the short and long-term, designed for a change towards a more circular economy. The Construction and Demolition Waste Strategy for Malta [1] refers to the drive by the European Commission towards a Circular Economy. The Strategy supports the transition towards a more circular economy and close the loop of construction product life cycle. The ultimate goal of the Strategy is to strengthen environmental protection and human health from waste related pollution while reducing the consumption of raw materials. In addition, the goal of the Strategy is to increase the quality and quantities of C&D waste recycled in the Maltese Islands as an archipelagic State. The Strategy consists of a framework acting "as a driver intended to bring about a cultural and behavioural shift within the sector in terms of its attitude towards excavation, demolition and construction methods" [1].

35.2.2 The Implementation Programme and Proposed Measures

The implementation programme consists of four main priority areas (Fig. 35.3), intended to identify options for C&D Waste Management. Each priority area is assigned specific measures to be implemented in the short and long-term, fulfilling the respective objectives. The Strategy refers to periodic review of implementation timeframes to adapt to changing circumstances and operational needs. The measures proposed in the Construction and Demolition Waste Strategy for Malta (2021–2030) are presented in Table 35.1.

35.3 The Strategic Framework: C&D Waste Standards

35.3.1 Waste Management Strategy and Standards

The Waste Management Strategy presents the required development of National Standards as a key measure [1]. The Strategy refers to the Standards development through the Building Industry Consultative Council and referred to in Sects. 35.3.2 and 35.3.3 below. The Strategy aims to build upon the measure highlighted in the Waste Management Plan for the Maltese Islands, by putting in place a set of standards, presented as follows:

Fig. 35.3 Construction and demolition waste strategy—implementation plan. *Source* ERA [1]

Table 35.1 The measures proposed: C&D waste strategy for Malta 2021–2030 ERA [1]

References	Proposed measure
1	Establish Standards for the Construction Industry
2	Promote Innovation through Research and Development
3	Introduce a New Regulatory Framework
4	Allow for the Provision of Training
5	Improve Waste Classification and Source Separation
6	Encourage Home Restoration Projects
7	Recognise the need for Resource Recovery and Storage Depots
8	Explore ways of applying the Polluter-Pays Principle
9	Extraction of Resources at Development Sites
10	Promoting markets for secondary raw materials
11	Set Re-use and Recycling Targets
12	Enforce Recovery through Restoration of Void Spaces
13	Discourage Landfilling
14	Explore the viability for Land Reclamation
15	Set Standards in Tenders Published by Government Entities

1. Best practices for (de)construction, aimed at reducing the C&D waste generated and purifying the resulting waste streams;
2. The classification of C&D waste by type, material, composition and weight, aimed to encourage on-site separation as well as improve the quality of the waste streams for subsequent re-use or recycling;
3. Appropriate excavation works, with the aim to re-use excavated rock for the purposes of construction;

The Strategy refers to the incorporation of the Standards within the regulatory framework, as an essential requirement prior to the issuance of an executable Development Permit [1].

35.3.2 The Strategic Framework

The Strategic framework for Excavation, C&D Waste in Malta was developed by the Building Industry Consultative Council (BICC) Research and Innovation Committee [11] and presented to BICC members and other stakeholders, including the Environment and Resources Authority in 2018. The Waste Strategy for Malta [1] refers to this action in Measure No. 1 (Table 35.1). The Framework was developed through stakeholder consultation and with reference to key strategic action at the following levels: (A) Policy and Legislation; (B) Waste Management, Technical Development and Research; (C). Waste Standards for Malta. The document presents action organised in different, yet complimentary components. The policy and legislation component refers to different policy recommendations which are intended to support the implementation of Circular Economy in Malta. The Waste Management, technical development and research component, presents recommendations for storage depots where the classification of waste can be implemented, recommendations for the research agenda and proposed technical developments of different high-performance products based on industrial byproducts and waste. The National Standards component is key in supporting the Strategic framework.

35.3.3 The C&D Waste Standards

The Standards were developed by the University of Malta, Construction Material Engineering Research Group through research-based approaches led by the Research Group including: (i.) the C&D Waste Research Standards activity developed in the period 2007–2009; (ii.) the C&D Waste Research Project (2019–2021). The first activity was based on the results of the European Union funded twinning research project: Recycling of Construction and Demolition Waste in Malta [12] and on further development of new draft standards in 2009 by the University of Malta Construction Materials Research Group together with the Malta Standards Authority. The second activity started in 2013 by the Building Industry Consultative Council, Research and Innovation Work Group leading to the setting up of a new National Technical Committee (TC800) in 2017 within the Malta Competition and Consumer Affairs Authority (MCCAA) on the request of the University of Malta, Construction Materials Research Group. These initiatives led to a research project developed by the Group [13] with the presentation of two draft National Standards at the BICC National Conference in December 2021. The Standards Framework was proposed based on two new key Standards for Malta: (1) Recycling Oriented Deconstruction and Classification of Waste [14]; (2) The Classification of Recycled Aggregate [15]. The National Technical Committee (TC800) was tasked with the final presentation and development leading to the public consultation and eventual publication of the C&D Waste Standards [14, 15]. The Standard SM810 was published in 2022 addressing the National Strategy [1], Measure No. 1: Standards for the Construction Industry. The Standard was published for public consultation and eventually as a New National Standard: SM 810:2022—*Recycling-oriented Deconstruction, Controlled Excavation Works and Classification of Waste—Requirements for planning and execution* [16]. The new Standard was rendered mandatory in the Maltese islands through its inclusion in the Construction Management Site Regulations (S.L. 623.08) [17], Legal Notice 340 of 2022 of the Government of Malta. The standard SM820—*Classification of Recycled Aggregate,* has been presented to the MCCAA committee TC800 for final development and review [18].

35.4 Recycling Oriented Deconstruction, Controlled Excavation and Classification of Waste

The SM810 Standard [16] is intended as a guide for good practice and a reference for building owners, developers, designers, and contractors. The document is set to prioritise the reduction of waste generation and highlights the importance of saving raw material resources. The Standard is intended to serve as an aid for the construction industry stakeholders to facilitate planning, classification of waste and conducting demolition operations, through deconstruction, and excavation operations with a view to reuse and recycling: the standard refers to recycling oriented deconstruction and controlled excavation works to reduce/eliminate waste disposal. This standard also makes reference to the code of practice for demolition operations as presented in BS 6187:2011 [19] and also to SM 820 Classification of Recycled Aggregates [18]. The standard applies to all demolition, deconstruction and excavation work in all projects in the Maltese Islands. It is noted that the applicability of this standard should take into consideration the type and the scale of deconstruction and excavation being undertaken and work practices on different sites should be adapted accordingly. The framework as presented in SM810, applies also to waste generated during construction operations [16]. The deconstruction process at end of life of a building needs to be supported through the principle of design for deconstruction. The ultimate goal of the Design for Deconstruction is to responsibly manage end-of-life building materials to minimise consumption of raw materials. SM810 refers to Deconstruction and Controlled Excavation, with the objective of reducing waste generated, and reusing/recycling building elements and components [18]. The Standard also promotes the reduction of waste primarily through the adaptation and retrofit of existing building assets. Deconstruction is encouraged, instead of demolition for the following reasons: (a) Deconstruction allows for reduced generation of waste; (b) Preservation of primary raw material resources; (c) Reducing reliance on backfilling; (d) Better management of waste at end of life, and facilitation of reusing and recycling. The Standard presents Controlled Excavation as an important principle. Waste classification is also presented as an essential step in the proper management of waste generated during deconstruction and controlled excavation processes; any waste generated during such activities shall be managed according to the waste hierarchy as laid down in Regulation 4A. of S.L. 549.63 Waste Regulations. The Standard foresees the compilation of a waste catalogue. The Standard presents key considerations for developers and contractors to guide through best practices during the planning and the execution of deconstruction and/or controlled excavation operations. Deconstruction operations shall be planned through a structured approach, supported with an engineering appraisal. Deconstruction shall be oriented towards recycling to enable and facilitate the recyclability of materials originating from building/excavation sites. For this to be possible, the Standard presents the classification of waste arising prior, during and after deconstruction. Controlled excavation works are proposed as a first consideration in excavation sites. Controlled excavation shall give due consideration to the potential resources that may be extracted from the site in the form of blocks instead of traditional excavation, to minimise the generation of waste. It is noted that excavation waste, primarily limestone, accounts for the largest volume of waste generated in Malta. The Standard refers to storage requirements, as the deconstruction and controlled excavation processes yield waste streams which may require temporary storage prior to recycling. SM810 refers to Skills and Training; operators in deconstruction, demolition, and excavation works are required to follow the National Occupational Standards including NOS for Excavation and Demolition, the training regime and respective skill cards.

35.5 Classification of Recycled Aggregate

The Standard SM820 for the Classification of Recycled Aggregate [18]; sets out technical engineering attributes for the classification of waste aggregate, enabling its exploitation as a resource. SM820 provides a first classification of Construction, Demolition and Excavation waste in the Maltese Islands. The purpose of the standard is to enable the classification of waste in order to transform it into a resource for construction which is of adequate quality and safe to consumers and the environment. The standard gives the opportunity to different stakeholders, to exploit waste, as a key resource in construction, enabling the production of construction materials for intended applications. The classification is designed to respect the possible current and future applications of materials in construction, in the context of the construction products regulations, and to provide the designer and producer with the necessary tools to enable wide exploitation of materials with respect to the principles of sustainable use [18].

35.6 Conclusion

The Strategy [1] proposes that the measures highlighted therein should not only be introduced as legislation but should be followed up by further discussions with the relevant stakeholders, and accompanied by standards and adequate training, in order to ensure effective implementation and positive change. The Construction and Demolition Waste Strategy for Malta is intended as an opportunity for efficient and an effective C&D waste management system. The success of such a Strategy however relies on monitoring of the implementation of the measures and its periodical review.

A key measure presented in the Construction and Demolition Waste Strategy for Malta 2021–2030 refers to the development and implementation of National Standards which are also integrated in the regulatory framework. Standards for deconstruction, recycling oriented deconstruction and classification of waste and recycled aggregate, need to refer to the characteristics of buildings and have to be designed for the waste groups in a territory. Standards SM810 and SM 820, developed for the Maltese Islands, are complimentary Standards and considered to be an important and strategic milestone, intended specifically to support the implementation of Circularity in the Construction Industry in Malta.

References

1. ERA (2021) Construction and demolition waste strategy for Malta 2021–2030: managing construction & demolition resources–2021, p 34. Environment and Resources Authority, Malta
2. Akhtar A, Sarmah AK (2018) Construction and demolition waste generation and properties of recycled aggregate concrete: a global perspective. J Clean Prod 186:262–281. https://doi.org/10.1016/j.jclepro.2018.03.085
3. López Ruiz LA, Roca Ramón X, Gassó Domingo S (2020) The circular economy in the construction and demolition waste sector–a review and an integrative model approach. J Clean Prod 248. https://doi.org/10.1016/j.jclepro.2019.119238
4. Zhang L, Liu B, Du J, Liu C, Wang S (2019) CO_2 emission linkage analysis in global construction sectors: alarming trends from 1995 to 2009 and possible repercussions. J Clean Prod 221:863–877. https://doi.org/10.1016/j.jclepro.2019.02.231
5. Scrivener KL, John VM, Gartner EM (2018) Eco-efficient cements: potential economically viable solutions for a low-CO_2 cement-based materials industry. Cement Concrete Res 114:2–26. https://doi.org/10.1016/j.cemconres.2018.03.015

6. Villoria Sáez P, Osmani M (2019) A diagnosis of construction and demolition waste generation and recovery practice in the European union. J Clean Prod 241. https://doi.org/10.1016/j.jclepro.2019.118400
7. Waste Framework Directive, Directive 2008/98/EC. https://environment.ec.europa.eu/topics/waste-and-recycling/waste-framework-directi;ve_en. Last Accessed 20 Feb 2023
8. Borg RP, Hajek P, Fernandez Ordonez D (2018) Sustainable concrete: materials and structures. Proceedings of Fib conference: sustainable concrete: materials and structures, materials science and engineering. IOP Conf Ser 442. https://doi.org/10.1088/1757-899X/442/1/011001
9. Borg RP, Cuenca E, Garofalo R, Schillani F, Nasner ML, Ferrara L (2021) Performance assessment of ultra-high durability concrete produced from recycled ultra-high durability concrete. Front Built Environ 7:648220. https://doi.org/10.3389/fbuil.2021.648220
10. Tax Credit (Construction Waste Recycling) Rules, Subsidiary legislation SL 123.186; 28th June, 2018, legal notice 209 of 2018. As amended by legal notice 47 of 2019
11. Borg RP (2018) Excavation, construction and demolition waste strategic framework-Malta (Unpublished report). Building Industry Consultative Council, Malta
12. Martin CAR, Guenter G, Willing E, Hans-Joerg Z (2008) Recycling of construction & demolition waste in Malta-strategy for short-term implementation. Twinning Project MT05-IB-EN-01, Umweltbundesamt, GmbH, Austia & Ministry for Rural affairs and the Environment, Malta
13. Borg RP (2021a) Excavation, construction and demolition waste standards for Malta: strategic framework (Unpublished report). University of Malta, BICC
14. Borg RP (2021b) Excavation, C&D waste standards for Malta: recycling oriented deconstruction & classification of waste (Report). University of Malta, BICC
15. Borg RP (2021c) Excavation, C&D waste standards for Malta: classification of recycled aggregate (Unpublished standard report). University of Malta, BICC
16. SM810 (2022) Recycling-oriented deconstruction, controlled excavation works and classification of waste-requirements for planning and execution. Malta Competitions and Consumer Affairs Authority (MCCAA)
17. Construction Management Site Regulations, Subsidiary legislation S.L. 623.08. Legal notice 340 of 2022, Government of Malta. (Published 16th December 2022)
18. DSM820 (2022) Classification of recycled aggregate. MCCAA, Malta
19. BS6187 (2011) Code of practice for full and partial demolition. British Standards Institution

SPRINGER NATURE

GPSR Compliance

The European Union's (EU) General Product Safety Regulation (GPSR) is a set of rules that requires consumer products to be safe and our obligations to ensure this.

If you have any concerns about our products, you can contact us on ProductSafety@springernature.com

In case Publisher is established outside the EU, the EU authorized representative is:

Springer Nature Customer Service Center GmbH
Europaplatz 3
69115 Heidelberg, Germany

The manufacturer's authorised representative in the EU is Springer Nature Customer Service Centre GmbH, Europaplatz 3, 69115 Heidelberg, Germany. If you have any concerns regarding our products, please contact ProductSafety@springernature.com

Printed and bound by CPI Group (UK) Ltd, Croydon, CR0 4YY
28/09/2026
02224056-0009